Cambridge Studies in Biological and Evolutionary Anthropology 68

The Foragers of Point Hope: The Biology and Archaeology of Humans on the Edge of the Alaskan Arctic

On the edge of the Arctic Ocean, above the Arctic Circle, the prehistoric settlements at Point Hope, Alaska, represent a truly remarkable accomplishment in human biological and cultural adaptations. Presenting a set of anthropological analyses on the human skeletal remains and cultural material from the Ipiutak and Tigara archaeological sites, *The Foragers of Point Hope* sheds new light on the excavations from 1939 to 1941, which provided one of the largest sets of combined biological and cultural materials of northern latitude peoples in the world.

A range of material items indicated successful human foraging strategies in this harsh Arctic environment. They also yielded enigmatic artifacts indicative of complex human cultural life filled with dense ritual and artistic expression. These remnants of past human activity contribute to a crucial understanding of past foraging lifeways and offer important insights into the human condition at the extreme edges of the globe.

CHARLES E. HILTON is an Assistant Professor in the Department of Anthropology at Grinnell College, Iowa. As a biological anthropologist with a background in human skeletal biology, functional morphology, human evolutionary ecology, and epidemiology, his research focuses on how small-scale human groups, particularly foragers, develop and evolve both short- and long-term biological and cultural responses within environmental settings offering limited resources.

BENJAMIN M. AUERBACH is an Associate Professor in the Department of Anthropology at The University of Tennessee, Knoxville. A functional anatomist, skeletal biologist, and evolutionary biologist, he has spent fifteen years collecting osteometric and anthropometric data to document morphological variation among modern humans within the context of evolutionary forces.

LIBBY W. COWGILL is an Assistant Professor in the Anthropology Department at the University of Missouri, Columbia. Her research interests include human growth, development, and functional morphology as well as Late Pleistocene human evolution. Her current research program explores the relationship between childhood behavior and selection pressure and the formation of adult skeletal morphology.

Cambridge Studies in Biological and Evolutionary Anthropology

Also available in the series

53 *Technique and Application in Dental Anthropology* Joel D. Irish & Greg C. Nelson (eds.) 978 0 521 870 610
54 *Western Diseases: An Evolutionary Perspective* Tessa M. Pollard 978 0 521 61737 6
55 *Spider Monkeys: The Biology, Behavior and Ecology of the genus Ateles* Christina J. Campbell 978 0 521 86750 4
56 *Between Biology and Culture* Holger Schutkowski (ed.) 978 0 521 85936 3
57 *Primate Parasite Ecology: The Dynamics and Study of Host-Parasite Relationships* Michael A. Huffman & Colin A. Chapman (eds.) 978 0 521 87246 1
58 *The Evolutionary Biology of Human Body Fatness: Thrift and Control* Jonathan C. K. Wells 978 0 521 88420 4
59 *Reproduction and Adaptation: Topics in Human Reproductive Ecology* C. G. Nicholas Mascie-Taylor & Lyliane Rosetta (eds.) 978 0 521 50963 3
60 *Monkeys on the Edge: Ecology and Management of Long-Tailed Macaques and their Interface with Humans* Michael D. Gumert, Agustín Fuentes & Lisa Jones-Engel (eds.) 978 0 521 76433 9
61 *The Monkeys of Stormy Mountain: 60 Years of Primatological Research on the Japanese Macaques of Arashiyama* Jean-Baptiste Leca, Michael A. Huffman & Paul L. Vasey (eds.) 978 0 521 76185 7
62 *African Genesis: Perspectives on Hominin Evolution* Sally C. Reynolds & Andrew Gallagher (eds.) 978 1 107 01995 9
63 *Consanguinity in Context* Alan H. Bittles 978 0 521 78186 2
64 *Evolving Human Nutrition: Implications for Public Health* Stanley Ulijaszek, Neil Mann & Sarah Elton (eds.) 978 0 521 86916 4
65 *Evolutionary Biology and Conservation of Titis, Sakis and Uacaris* Liza M. Veiga, Adrian A. Barnett, Stephen F. Ferrari & Marilyn A. Norconk (eds.) 978 0 521 88158 6
66 *Anthropological Perspectives on Tooth Morphology: Genetics, Evolution, Variation* G. Richard Scott & Joel D. Irish (eds.) 978 1 107 01145 8
67 *Bioarchaeological and Forensic Perspectives on Violence: How Violent Death is Interpreted from Skeletal Remains* Debra L. Martin & Cheryl P. Anderson (eds.) 978 1 107 04544 6

"This volume represents a true anthropological reconstruction of life among the prehistoric foragers from Point Hope, Alaska. It includes important perspectives regarding the ecological realities of adaptation in this harsh environment that are integrated into the perception of this landscape by the Ipiutak and Tigara people themselves. The work is a must-read for all who find interest in hunter-gatherer populations and scholars who value integrated anthropological research."

Daniel H. Temple, University of North Carolina, Wilmington

"Point Hope, a narrow spit of land on the Arctic Ocean, is something of an enigma in Alaskan prehistory. Since the pre-Second World War excavations of its varied habitation and cemetery sites the archaeology has been well chronicled, though not with unqualified acceptance. In addition, with exceptions, the ancient inhabitants themselves received little attention – as in the lack of research on some 500 recovered Ipiutak and Tigara skeletons. Finally, after more than 70 years, this superbly edited volume addresses that neglect. Between up-to-date accounts of the archaeological context and thoughtful comment by highly respected circumpolar researchers, a series of comprehensive yet highly readable chapters by biological anthropologists and bioarchaeologists give insight into the origins, affinities, and everyday lives of people who once called Point Hope home. Though long overdue, this much-needed biocultural insight was worth the wait."

Joel D. Irish, Liverpool John Moores University

"This volume provides the reader with almost everything one would want to know about the archaeology and skeletal biology of the prehistoric Ipiutak and Tigara samples from a tiny, but important strip of land in Point Hope, Alaska. With a wide array of well-written chapters on topics as diverse as the Ipiutak "spirit-scape" to dental microwear to paleopathology, from growth and development to the samples' genetic affinities inferred from cranial morphology, this book provides much-needed contextual knowledge on this fascinating skeletal sample, and will be a go-to resource for those interested in the bioarchaeology of circumpolar peoples."

Trenton W. Holliday, Tulane University

The Foragers of Point Hope

The Biology and Archaeology of Humans on the Edge of the Alaskan Arctic

Edited by

Charles E. Hilton
Grinnell College, Iowa, USA

Benjamin M. Auerbach
University of Tennessee, Knoxville, USA

Libby W. Cowgill
University of Missouri, Columbia, USA

CAMBRIDGE
UNIVERSITY PRESS

University Printing House, Cambridge CB2 8BS, United Kingdom

One Liberty Plaza, 20th Floor, New York, NY 10006, USA

477 Williamstown Road, Port Melbourne, VIC 3207, Australia

314-321, 3rd Floor, Plot 3, Splendor Forum, Jasola District Centre, New Delhi - 110025, India

79 Anson Road, #06-04/06, Singapore 079906

Cambridge University Press is part of the University of Cambridge.

It furthers the University's mission by disseminating knowledge in the pursuit of education, learning and research at the highest international levels of excellence.

www.cambridge.org
Information on this title: www.cambridge.org/9781108829410

First published 2014
First paperback edition 2020

A catalogue record for this publication is available from the British Library

Library of Congress Cataloging in Publication data
The foragers of Point Hope : the biology and archaeology of humans on the edge of the Alaskan arctic / edited by Charles E. Hilton, Benjamin M. Auerbach, Libby W. Cowgill.
pages cm. – (Cambridge studies in biological and evolutionary anthropology ; 68)
Includes bibliographical references and index.
ISBN 978-1-107-02250-8 (hardback)
1. Ipiutak Site (Point Hope, Alaska) 2. Tigara Site (Point Hope, Alaska) 3. Eskimos – Material culture – Alaska – Point Hope. 4. Human remains (Archaeology) – Alaska – Point Hope. 5. Excavations (Archaeology) – Alaska – Point Hope. 6. Point Hope (Alaska) – Antiquities. I. Hilton, Charles E., editor of compilation. II. Auerbach, Benjamin M. (Benjamin Miller) editor of compilation. III. Cowgill, Libby W., editor of compilation.
E99.E7F6218 2014
979.8′7 – dc23 2014007610

ISBN 978-1-107-02250-8 Hardback
ISBN 978-1-108-82941-0 Paperback

Additional resources for this publication at www.cambridge.org/9781108829410

Contents

Contributors

Benjamin M. Auerbach, Ph.D.
Department of Anthropology, The University of Tennessee, Knoxville, TN, USA

Libby W. Cowgill, Ph.D.
Department of Anthropology, University of Missouri, Columbia, MS, USA

Don E. Dumond, Ph.D.
Department of Anthropology, University of Oregon, Eugene, OR, USA

Sireen El Zaatari, Ph.D.
Wiener Laboratory, American School of Classical Studies, Athens, Greece

William W. Fitzhugh, Ph.D.
Department of Anthropology, National Museum of Natural History, Smithsonian Institution, Washington, D.C., USA

Sarah Gossett, M.S.N., F.N.P.-C.
Department of Endocrinology, Mayo Clinic Health System-Franciscan Healthcare, La Crosse, WI, USA

Charles E. Hilton, Ph.D., M.P.H.
Department of Anthropology, Grinnell College, IA, USA

Anne M. Jensen, Ph.D.
UIC Science LLC., Barrow, AK, USA

Kristin L. Krueger, Ph.D.
Department of Anthropology, Loyola University Chicago, Chicago, IL, USA

Megan Latchaw Czarniecki, M.S., M.A.
Maternal Fetal Medicine, University of Minnesota Medical Center, Minneapolis, MN, USA

Blaine Maley, Ph.D.
College of Osteopathic Medicine, Marian University, Indianapolis, IN, USA

Owen K. Mason, Ph.D.
Institute of Arctic and Alpine Research, University of Colorado, Boulder, CO, USA

Marsha D. Ogilvie, Ph.D.
Texas Archaeological Research Laboratory, The University of Texas at Austin, Austin, TX, USA

Laura L. Shackelford, Ph.D.
Department of Anthropology, University of Illinois at Urbana-Champaign, Urbana, IL, USA

Ian Tattersall, Ph.D.
Division of Anthropology, American Museum of Natural History, New York, NY, USA

David Hurst Thomas, Ph.D.
Division of Anthropology, American Museum of Natural History, New York, NY, USA

Foreword

Originating as the local name for a sandbar separating two lagoons on the north shore of Point Hope, at the western extremity of Alaska's northerly Lisburne Peninsula on the Chukchi Sea, the term Ipiutak is synonymous today with a world-famous art form and an ancient Arctic culture. We know about the lives, cultures, and biology of the extraordinary people who lived at Point Hope between about 1,600 and 500 years BP largely thanks to excavations carried out by the archaeologists Helge Larsen, of the Danish National Museum, and Froelich Rainey, of the University of Pennsylvania. These excavations ran from 1939 to 1941, until interrupted by World War II, and cultural results were rapidly published (Rainey, 1941, 1947; Larsen and Rainey, 1948). In three short but intensive field seasons, Larsen and Rainey began by excavating the ruined village of Old Tigara, adjacent to the modern settlement of Tigara (now Tikiġaq). They also investigated neighboring middens and graveyards, both near Tigara and in the vicinity of Jabbertown, about a kilometer to the east. These sites yielded mainly artifacts comparable to those found in "historic Eskimo" sites. Later, the researchers concentrated their attention on dunes a few hundred meters to the north and east of Tigara, which proved to cover the substantial habitation remnants (of around 600 houses) and graveyards that yielded the older classic "Ipiutak" materials.

For the 1941 season, Larsen and Rainey were joined by Harry Shapiro, a physical anthropologist at the American Museum of Natural History, whose participation assured that all of the skeletal remains recovered would go to New York City, while the artifacts were eventually dispersed among several institutions (and some were later lost in transit when the collection was divided after study at the American Museum). Also involved, from the beginning, was the dendrochronologist James Louis Giddings of the University of Arizona, who provided a valuable independent account of the work at Point Hope in his memoir (Giddings, 1968) published over two decades later.

The excavations proved to be extraordinarily productive. In all, some 10,000 artifacts and 500 skeletons were recovered, spanning multiple periods of sustained occupation of the Point Hope sand spit. The very sparse oldest remains, identified by Larsen and Rainey (1948) as "Near-Ipiutak," are currently of uncertain age, but may date as far back as 2,200 years BP (Mason, 2006a). A small series of conventional radiocarbon dates places the overlying Ipiutak culture proper at between 1,600 and 1,300 years BP (Gerlach and Mason, 1992; Mason, 2006b). Subsequent to this, there is some evidence of sporadic occupation of Point Hope through Birnirk and early Thule times, until settlement flourished again with the Tigara occupation that began at some point between about 600 and 500 years BP (Gerlach and Mason, 1992; Mason, 2006b). The break between the Ipiutak and Tigara occupation periods is marked by substantial cultural differences (Larsen and Rainey, 1948; Mason, 1998). The older Ipiutak material culture is highly characteristic, and is set apart particularly by distinct mortuary practices, exquisite stone working, and striking figurative depictions of mainly animal subjects in ivory. Although marine prey were extensively hunted in Ipiutak times, caribou also figured significantly in the subsistence economy. In contrast, the Tigara fit firmly within the whale-hunting Thule tradition, with clear affinities to later Inuit populations.

In their classic work *Ipiutak and the Arctic Whale Hunting Culture*, Larsen and Rainey (1948) explored the origin and development of Ipiutak culture, tracing the shifting environmental conditions and speculating about the attendant subsistence technologies that made life possible on the extreme northwestern tip of North America. The momentous discovery of the rich Ipiutak culture on Alaska's Arctic coast introduced a new piece to the cultural puzzle: an almost 2,000-year-old expression that, while subtle and complex, lacked such typical western Eskimo features as lamps, rubbed slate tools, sleds, bow drills, pottery, and harpoon floats and other evidence of whale hunting. Contemporary archaeologists continue to debate the nature of Ipiutak subsistence and adaptation (Mason, 2006a).

Despite their primary focus on the survival skills that are so evident in the material culture from Point Hope, Larsen and Rainey could not resist sharing their feelings about the extraordinary artworks they unearthed at Point Hope. There was just something special about those ancient Ipiutak carvers. Not only were they extraordinarily skilled at working ivory, but their originality and ingenuity find few counterparts, then or now. Virtually each example of this delicate work is a unique construction, reflecting its own beauty of form, and showing a deft and concise simplicity in decorative touch.

While the cultural assemblage from Point Hope benefited from the immediate monographic attention of Larsen and Rainey, the skeletal remains, intended for the attention of Harry Shapiro, languished effectively unstudied for decades. Shapiro's sole contribution to the publication of the American Museum of Natural History's large Point Hope collection was as nominal co-author, with the retired surgeon Charles Lester, of a short contribution on defects in the vertebral arch (Lester and Shapiro, 1968). As a consequence of this neglect the Point Hope skeletal collection was never monographed as it should have been. With the sole exception of a note in 1959 by the Russian anthropologist G. F. Debetz, short publications only began to appear in 1980, when Raymond L. Costa began to publish the results of his thesis research (Costa, 1977) on dental pathologies (Costa, 1980a, b, 1982). Soon thereafter, Point Hope additionally figured in Charles Utermohle's (1984) comparative study of ancient Alaskan populations. Subsequent publications on the Point Hope skeletons have included contributions on possible tooth drilling (Schwartz *et al.*, 1995), on dental microwear (El-Zataari, 2008; also an unpublished thesis by Krueger, 2006), body proportions (Holliday and Hilton, 2010), and thesis-based (Dabbs, 2009a) examinations of evidence in the samples for tuberculosis and general health status (Dabbs, 2009b, 2011). Guatelli-Steinberg *et al.* (2004) used the series as an important reference in a comparative study of Neanderthal adaptation, and an abstract was published on erosive arthropathy by Latchaw and Hilton (2004). An unpublished thesis on dental evidence for division of labor was additionally completed by Madimenos (2005). Genomic investigation began with the work of Blaine Maley (Maley *et al.*, 2006; Maley, 2007, 2011); preliminary work revealed that all members of an initial sample belonged to mtDNA haplogroup A, found in frequencies of up to 97% in modern Inuit populations (Maley *et al.*, 2006).

In 2006, a high point was reached when a well-attended symposium was devoted to the Point Hope skeletal series at the Annual Meeting of the American Association of Physical Anthropologists in Anchorage, Alaska (Hilton *et al.*, 2006). This book is in part an outgrowth of that symposium, and in equal part an attempt to address the dearth of available documentation of the outstandingly important Point Hope collection. The editors have gathered a broad-based team of researchers both to overview what is known, and to present the results of new research. Anne Jensen provides an invaluable review of the place of Point Hope in Alaskan archaeology, while Owen Mason delivers an ethnographic appraisal of the cult of shamans at which Larsen and Rainey darkly hinted. Blaine Maley uses cranial data to look at the important question of biological continuity between the Ipiutak and Tigara populations, and differences in diet are addressed by both Sireen El-Zataari and Kristin Krueger through the lens of dental microwear. Issues of growth and development are considered by Libby Cowgill, and the putative skeletal effects of physical activity by Laura Shackelford. General health and morbidity in the samples are examined by Hilton and colleagues, while global views of the samples are presented from varying perspectives by Ben Auerbach, Don Dumond, and Bill Fitzhugh. All in all, thanks to the efforts of the editors, we finally have before us the general appraisal that the hugely significant Point Hope skeletal collections have for so long, and so richly, deserved.

Ian Tattersall, Curator Emeritus, Biological Anthropology
David Hurst Thomas, Curator, North American Archaeology
Division of Anthropology
American Museum of Natural History

References

Costa, R. L. (1977). *Dental Pathology and Related Factors in Archaeological Eskimo Samples from Point Hope and Kodiak Island, Alaska.* Ph.D. University of Pennsylvania.

Costa, R. L. (1980a). Incidence of caries and abscesses in archaeological Eskimo skeletal samples from Point Hope and Kodiak Island, Alaska. *American Journal of Physical Anthropology*, 52, 501–14.

Costa, R. L. (1980b). Age, sex and antemortem loss of teeth in prehistoric Eskimo samples from Point Hope and Kodiak Island, Alaska. *American Journal of Physical Anthropology*, 53, 579–87.

Costa, R. L. (1982). Periodontal disease in the prehistoric Ipiutak and Tigara skeletal remains from Point Hope, Alaska. *American Journal of Physical Anthropology*, 59, 97–110.

Dabbs, G. (2009a). *Health and Nutrition at Prehistoric Point Hope, Alaska: Application and Critique of the Western Hemisphere Health Index.* Ph.D. University of Arkansas.

Dabbs, G. (2009b). Resuscitating the epidemiological model of differential diagnosis: Tuberculosis at prehistoric Point Hope, Alaska. *Paleopathology Newsletter*, 148, 11–24.

Dabbs, G. R. (2011). Health status among prehistoric Eskimos from Point Hope, Alaska. *American Journal of Physical Anthropology*, 146, 94–103.

Debetz, G. F. (1959). The skeletal remains of the Ipiutak cemetery. *Actas del XXXIII Congreso Internacional de Americanistas*, San José, pp. 57–64.

El Zaatari, S. (2008). Occlusal molar microwear and the diets of the Ipiutak and Tigara populations (Point Hope) with comparisons to the Aleut and Arikara. *Journal of Archaeological Science*, 35, 2517–22.

Gerlach, S. C. and Mason, O. K. (1992). Calibrated radiocarbon dates and cultural interaction in the Western Arctic. *Arctic Anthropology*, 29, 54–81.

Giddings, J. L. (1968). *Ancient Men of the Arctic.* London: Secker & Warburg.

Guatelli-Steinberg, D., Larsen, C. S. and Hutchinson, D. L. (2004). Prevalence and the duration of linear enamel hypoplasia: a comparative study of Neandertals and Inuit foragers. *Journal of Human Evolution*, 47, 65–84.

Hilton, C. E., Guatelli-Steinberg, D. and Mowbray, K. (2006). Symposium: Pre-Contact Forager Adaptations to Northwest Coastal Alaska: The Bioarchaeology of Point Hope. *American Journal of Physical Anthropology*, S42, 25.

Holliday, T. W. and Hilton, C. E. (2010). Body proportions of circumpolar peoples as evidenced from skeletal data: Ipiutak and Tigara (Point Hope) versus Kodiak Island Inuit. *American Journal of Physical Anthropology*, 142, 287–302.
Krueger, K. L. (2006). *Incisal Dental Microwear of the Prehistoric Point Hope Communities: A Dietary and Cultural Synthesis.* M.A. Western Michigan University.
Larsen, H. E. and Rainey, F. (1948). *Ipiutak and the Arctic Whale Hunting Culture.* Anthropological Papers of the American Museum of Natural History 42. New York, NY: American Museum of Natural History.
Latchaw, M. R. and Hilton, C. E. (2004). Bilateral erosive arthropathy in the upper limbs: An Inuit case from Pt. Hope, Alaska. *American Journal of Physical Anthropology*, S30, 132.
Lester, C. W. and Shapiro, H. L. (1968). Vertebral arch defects in the lumbar vertebrae of pre-historic American Eskimos. *American Journal of Physical Anthropology*, 28, 43–8.
Madimenos, F. (2005). *Dental Evidence for Division of Labor Among the Prehistoric Ipiutak and Tigara of Point Hope, Alaska.* M.A. Louisiana State University.
Maley, B. (2007). Using nested clade analysis to explore temporal change in ancient population structure with aDNA sequence data. *American Journal of Physical Anthropology*, S44, 163.
Maley, B. (2011). *Population Structure and Demographic History of Human Arctic Populations Using Quantitative Cranial Traits.* Ph.D. Washington University in St. Louis.
Maley, B., Doubleday, A., Kaestle, F. and Mowbray, K. (2006). Sorting out population structure and demographic history of the Tigara and Ipiutak cultures using ancient DNA analysis. *American Journal of Physical Anthropology*, S42, 124.
Mason, O. K. (1998). The contest between Ipiutak, Old Bering Sea and Birnirk polities and the origin of whaling during the First Millennium A.D. along Bering Strait. *Journal of Anthropological Archaeology*, 17, 240–325.
Mason, O. K. (2006a). Contradictions in the archaeological construction of the Ipiutak culture: Sedentary, stratified walrus hunters and/or nomadic caribou hunters. *American Journal of Physical Anthropology*, S42, 127.
Mason, O. K. (2006b). Ipiutak remains mysterious: A focal place still out of focus. In *Dynamics of Northern Societies*. Proceedings of a Symposium. Copenhagen: Danish National Museum and Danish Polar Center, pp. 106–20.
Rainey, F. (1941). The Ipiutak culture at Point Hope, Alaska. *American Anthropologist*, 43, 364–75.
Rainey, F. (1947). *The Whale Hunters of Tigara.* Anthropological Papers of the American Museum of Natural History 41(2). New York, NY: American Museum of Natural History, pp. 227–84.
Schwartz, J. H., Brauer, J. and Gordon-Larsen, P. (1995). Brief communication: Tigara (Point Hope, Alaska) tooth drilling. *American Journal of Physical Anthropology*, 97, 77–82.
Utermohle, C. (1984). *From Barrow Eastward: Cranial Variation of the Eastern Eskimo.* Ph.D. Arizona State University.

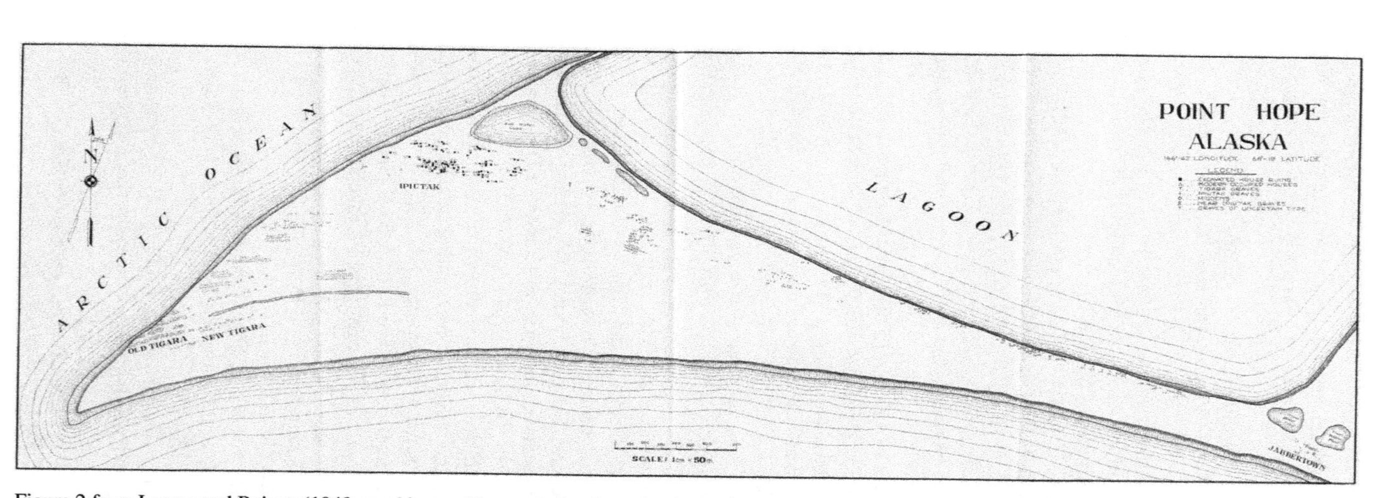

Figure 2 from Larsen and Rainey (1948, used here with permission from the American Museum of Natural History) highlighting the archaeological sites and the original archaeological excavations during the 1939–1941 field seasons. Subsequent erosion of the western shoreline forced the Point Hopers to relocate the entire town to a spot approximately 2.5 km eastwards. A single airplane runway (3992 × 75 ft) has been built near the western shoreline of the peninsula and has an approximate north-south orientation that parallels the western shoreline. The runway is perpendicular to the east-west oriented beach ridges and cuts across the locations of former Old and New Tigara houses. (The original caption reads: "Fig. 2. Map of Point Hope, Alaska, showing position of Ipiutak village and graves, the Old and New Tigara graves, and the Jabbertown site.")

Preface

In 2006, one of us (C. Hilton) co-organized a symposium focused on the Point Hope skeletal remains at the meetings for the American Association of Physical Anthropology. Given the location of the meetings in Anchorage, Alaska, that year, this represented the ideal time and place to bring together a group of scholars interested in the bioarchaeology of this important, yet relatively understudied, site. At the same time, it allowed for living inhabitants of the Point Hope community to interact with the scholars interested in their history. In many ways, this symposium was long overdue. Although Dr. Harry Shapiro initiated formal analyses of the Point Hope skeletal material in the 1940s, this research was never completed. In the late 1990s, as a result of encouragement by Dr. Ian Tattersall of the American Museum of Natural History, skeletal research on this collection resumed. Therefore, the 2006 symposium represented the first focused group of scholars specifically interested in the biology of Point Hope in over fifty years.

The primary result of the symposium was a renewed understanding of the wide body of questions that could be addressed using the Point Hope material, and how little progress had been made in exploring these potential research avenues. The symposium served to highlight the myriad ways that Point Hope had been underutilized. While a large body of literature had previously been written about the archaeology of Point Hope, the archaeological analyses had not been revisited in many years. In addition, very little research had been undertaken on the skeletal remains themselves. While early hypotheses about cultural change at Point Hope had been suggested based on archaeological data, the evidence from biological anthropology had not been evaluated in this context, and it remained unclear if the subsistence transitions detected in the archaeological material from the site could be detected in behavioral analyses of the skeletal remains. Lastly, and perhaps most significantly, the site of Point Hope represented a truly unique opportunity to gain insight into lifeways in an Arctic foraging system. The recovery of the skeletal remains of almost 500 Arctic foragers makes the Point Hope skeletal collection one of the largest samples of northern latitude skeletal remains in the world. Because of the large size of the collection, its completeness and antiquity, as well as its unique geographic location, the human skeletal remains from this site represent an important source of knowledge for studies of hunting and gathering populations in general, and analyses of Arctic subsistence and biological adaptation in particular.

As editors of this volume, we sincerely hope that we have been able to rectify some of the more notable gaps in our knowledge about both Arctic foragers in general and Point Hope in particular. Given that, the goals of this volume are twofold. First, we hope to expand our current understanding of subsistence, adaptation, and stress at Point Hope, and also to broaden our knowledge of precontact life in Arctic Alaska as a whole. The biological anthropology contributions to this volume focus on relatively narrow parts of the picture of life at Point Hope, but aim to situate their research within the wider context of foraging adaptations and the biosocial demands of Arctic subsistence. Second, we hope this research more fully integrates both archaeological and biological perspectives on the Point Hope material into a cohesive whole. By including both lines of evidence in this volume, we may be able to resolve prior questions on the cultural transition that occurred at the site, and provide an additional viewpoint on culture change in foraging populations that can be applied to other groups in the future. In brief, we aim to create an example of research that is more comprehensively bioarchaeological, not only through the techniques of skeletal analysis, but also through a more complete integration of two diverse lines of evidence.

Chuck, Benjamin and Libby

Acknowledgements

First and foremost, we thank Drs. Ian Tattersall and David Hurst Thomas of the American Museum of Natural History (AMNH) for their consistently positive encouragement and support throughout all phases of the work associated with this volume. They have graciously not only provided us with access to the Point Hope collections and images but have given generously of their time through hours of discussion on issues related to the original Point Hope archaeological excavations and the artifacts curated by the AMNH. Drs. Tattersall and Thomas made every research trip to the AMNH a true pleasure. We also express our thanks to Dr. Ken Mowbray, Gary Sawyer, and Giselle Garcia-Pack. These integral members of the AMNH's Physical Anthropology Section have always provided additional support and conversations that facilitated the completion of this volume. Others have also contributed in significant ways to completing this volume, including Martin Gomberg, Drs. John (Jack) Martin Campbell (deceased), Charles Merbs, Clark S. Larsen, and Debra Guatelli-Steinberg. We would like to thank our outside reviewers who helped to improve the quality of this valume. Additional thanks are extended to the wonderful people at Cambridge University Press, including Martin Griffiths, Ilaria Tassistro, and Rachel Cox, who have provided us with superior advice, direction, help, and patience as the various parts of this volume came together as a unit.

1 *Introduction: Humans on the edge of the Alaskan Arctic*

CHARLES E. HILTON, BENJAMIN M. AUERBACH, AND LIBBY W. COWGILL

The ability to make a living in the Arctic represents one of humanity's truly exceptional achievements. The fact that it was accomplished by people who depended exclusively on wild (that is, uncultivated) resources attests to the levels of human imagination and ingenuity in creating cultural systems that effectively buffered against the physical elements of the Arctic. For these reasons, studies of North American Arctic foraging peoples have long played an important role in the early foundations and development of American anthropology (Collins, 1984; Dumond, 1987; Burch, 1988). North American Arctic foragers represent one end of the range of variation of the human foraging spectrum (Bettinger, 1991; Kelly, 1995; Binford, 2001), but, as they exist at one end of that range, they especially present insight into human adaptability.

Arctic peoples clearly push the boundaries of human adaptation and resiliency by living in areas marked by long periods of cold temperatures, limited availability of plant resources for food, and marked seasonal variation in daylight hours. North American Arctic foragers, both today and in the past, present a broad range of sophisticated technology, dietary flexibility, social organization, and residential mobility as they expanded across large areas of the North American Arctic landscapes. Given this remarkable resiliency in the face of adverse conditions, researchers have examined numerous aspects of cultural and biological characteristics of North American Arctic foragers in order to understand the complex problems related to survival in these circumpolar environments, and the many solutions taken by humans to overcome them.

As separate cultural entities, Arctic forager groups often possess distinctive organizational strategies and technological skills that provide an impressive array of innovative solutions for survival in their circumpolar landscapes

---

The Foragers of Point Hope: The Biology and Archaeology of Humans on the Edge of the Alaskan Arctic, eds. C. E. Hilton, B. M. Auerbach, L. W. Cowgill. Published by Cambridge University Press.

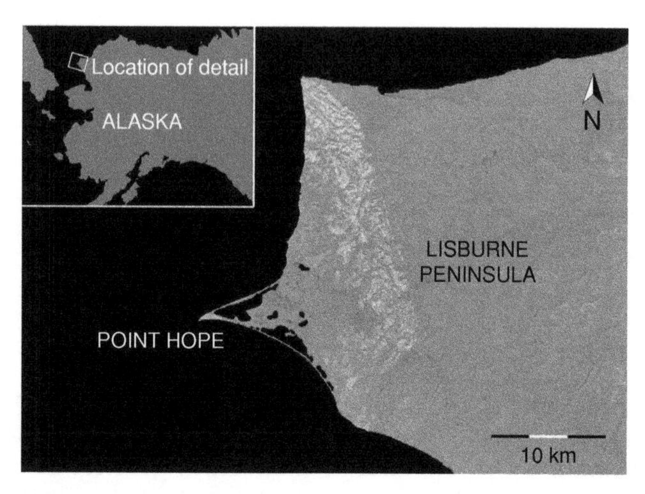

Figure 1.1. Location of Point Hope on the Lisburne Peninsula of Alaska (inset). Image is modified from Google Maps.

(Murdoch, 1892; Nelson, 1899; Stefansson, 1914; Rainey, 1941; Spencer, 1959, 1977, 1984; Nelson, 1969; Binford, 1978; Sheehan, 1985; Fagan, 2005). Different groups of North American Arctic foragers relied on large terrestrial fauna – namely the organizational strategies developed by some hunters that took advantage of caribou behavior during their annual migrations – as well as other, smaller terrestrial animals (Binford, 1978; Hall, 1984). Other Arctic foragers learned to rely heavily on marine resources, exploiting a variety of species and utilizing multiple technologies. In certain cases, cultural groups developed hunting methods that utilized kayaks and umiaks in order to engage in ice-lead hunting of marine mammals, whereas others utilized similar watercraft for open-water hunting of sea mammals (Spencer, 1977; Dumond, 1987; Burch, 1988; Fagan, 2005). The successful nature of all of these strategies allowed for the rapid expansion of these human groups from Bering Strait to the western shores of Greenland (Spencer, 1977; Dumond, 1987; Burch, 1988; Fagan, 2005).

Nowhere is the success of these different modes of cultural organization and faunal exploitation of the Arctic more evident and juxtaposed than the focal region of this volume – approximately 300 kilometers north of the Arctic Circle, at the far western end of Alaska's Lisburne Peninsula, on a spit of land jutting into the Chukchi Sea called Tikiġaq ("finger" in Iñupiat; see Figure 1.1). Currently known as Point Hope, this location on the northwest coast of Alaska has served as a settlement location for Arctic hunters of different cultures. With evidence of repeated prehistoric occupations spanning nearly two millennia, Point Hope has been a critical place where people gained access to key resources from the surrounding environments. Nevertheless, only a small body of analyses, predominantly archaeological, has been offered on Point Hope to date. Thus, this volume integrating evidence from both archaeological and skeletal contexts aims to bring a more comprehensive understanding of the biology and culture of humans practicing foraging systems at high northern latitudes.

As a place on the landscape, the numerous prehistoric houses and settlement structures found distributed along its beach ridges confirm Point Hope's importance to past groups, a prominence maintained today by the Tikiġaġmiut who live adjacent to (and atop) the archaeological sites. The prehistoric houses, which represent many generations of occupants and construction episodes, indicate a concentration of people who were able to take advantage of predictable resources found nearby. Moreover, Point Hope appears to have been occupied for longer time periods than other prehistoric human communities of northwest coastal Alaska.

The atypical aggregation, within the broader region, of prehistoric people at the Point Hope locality has compelled anthropological researchers to investigate the role of this settlement and its inhabitants, mostly in relationship to the development of the modern Iñupiat and their foraging systems. The archaeological sites at Point Hope have yielded a large set of human skeletal remains and a rich archaeological assemblage dating to before European contact (Larsen and Rainey, 1948; Rainey, 1971). Excavated by the Rainey–Larsen Point Hope Expedition between 1939 and 1941, Point Hope is one of the oldest "continuously" occupied sites in North America, although the occupations likely do not represent any linear ancestor–descendant temporal sequences (Collins, 1984; see contributions by Jensen, Mason, and Maley in this volume). Approximately 10,000 artifacts and the remnants of over 500 permanent dwellings were uncovered at Point Hope (Larsen and Rainey, 1948; Rainey, 1971). In addition, the excavation of the skeletal remains of almost 500 individuals makes the Point Hope human skeletal remains one of the largest northern latitude samples in the world. These individuals provide an important resource for the study of hunting and gathering populations in general, and analyses of Arctic subsistence, the morphology of high latitude North American groups, and the population history of the western North American Arctic specifically.

As noted, to date, the vast majority of the anthropological information available about precontact Point Hope is derived from purely archaeological investigations. The main cultural periods at Point Hope consist of the Ipiutak (*c.* 1,600 to 1,100 years BP) and Tigara (*c.* 800 to 300 years BP). Even though these groups lived in the same location, they practiced distinct forms of subsistence and had richly distinct cultures (Larsen and Rainey, 1948; Rainey, 1971; Schwartz *et al.*, 1995). Larsen and Rainey (1948) proposed that Ipiutak individuals relied seasonally on caribou found in more inland areas of Alaska. Their conclusions were based on the fact that Ipiutak deposits held what they considered to be high frequencies of caribou skeletal remains, caribou derived artifacts such as antler flint flakers, and the presence of bows and arrows, implements that were traditionally used to hunt caribou (Murdoch, 1892; Nelson, 1899; Larsen and Rainey, 1948; Rainey, 1971). In contrast, Tigara archaeological artifacts point to a heavy dependence on maritime resources including walruses, seals, and, especially, whales. The Tigara component displays a high frequency of whale-hunting implements, such as large harpoons and floatation devices, and whale bones were frequently utilized for house construction (Larsen and Rainey, 1948; Spencer, 1959, 1977; Rainey, 1971).

Despite previous research (much is summarized by Tattersall and Thomas in the Foreword), the nature of prehistoric life at Point Hope and the interaction between its inhabitants and their surrounding environments remain largely unknown. While analyses of the artifacts have yielded insight into the development and spread of the Ipiutak and Thule cultures in the region (see Jensen, Chapter 2, and Mason, Chapter 3, this volume), as well as an appreciation for shamanistic art and practice (see Fitzhugh, Chapter 11, this volume), limited study has utilized the human skeletons. To this point, research regarding Point Hope prehistory has generally centered around three ethnohistorical questions, and a fourth comparative one: (1) what were the origins of the people who inhabited Point Hope; (2) what was the scope of subsistence activities and their impact on the people who utilized those activities; (3) how did social networks related to kin, marriage, and forager mobility operate; and (4) how did prehistoric Point Hope communities fit into the broader scheme of foraging lifeways in Alaska and beyond? While these questions have provided a useful framework for initial work on Point Hope, the body of work done to date has not sufficiently positioned the data with respect to issues of broad anthropological and cross-disciplinary relevance. Thus, this volume builds upon and importantly expands previous work to add analytic insights pertaining to forager evolutionary history and climatic adaptations of circumpolar foragers. More specifically, these essays contextualize the morphology, diet, and disease of these communities within a diverse mosaic of populations. In doing so, the volume contributes to timely questions about the ways that humans transform even the harshest environments into habitable landscapes.

The volume's biological and archaeological approach is particularly salutary. These contributions arrive at an opportune time; new theoretical approaches in the last twenty-five years in the areas of human behavioral ecology and ethnoarchaeology have generated reappraisals of hunter-gatherer lifeways, giving us a more expansive understanding of behavioral variation in association with modern human foragers. A key outcome of these reassessments indicates that modern human foragers exhibit a greater range of behavioral variability than previously discussed in the anthropological literature. Human foraging groups are now recognized as possessing greater responsiveness to ecological changes and cultural innovations. This behavioral flexibility also means that foraging populations are not monolithic but, rather, are technologically dynamic and culturally diverse. These perspectives are particularly relevant as we consider how such behavioral flexibility in a cultural system manifests itself in skeletal variation. This lived experience-on bone approach permits holistic assessments of the histories and livelihoods of past populations.

The foragers of Point Hope: Addressing the unresolved and the unexplored

This volume, then, is focused on addressing questions and topics that have not been thoroughly considered, or have been overlooked. While other researchers have discussed the morphological distinctions between the Ipiutak population and Tigara population that occupied Point Hope, their affinities with populations across the western Alaskan Arctic region are unclear. Indeed, an enduring mystery that Point Hope is especially well positioned to address is the origin and place of the Ipiutak culture within the western North American Arctic, as well as, potentially, the fate of that group. As noted above, researchers have established that the Ipiutak and Tigara peoples practiced dissimilar modes of subsistence, based on archaeofaunal analyses and hunting artifacts, but the dental and skeletal evidence for the dietary and activity differences incurred with this distinction have not been as explicitly examined. And while a majority of biological anthropology research on the skeletons buried at Point Hope has considered disease, trauma, and their effects, the consequences of these pathological lesions for understanding the lifeways and cultural differences between the groups bear further consideration in light of the many sources of evidence provided in this volume. Finally, even though researchers have used the morphology of the past groups living at Point Hope as a "representative" sample for human adaptations to high latitudes, their body shape, size, proportions, and limb mechanical properties have generally not been considered within the context of population history in the Arctic, and the specific lifeways of these groups.

Thus, this volume seeks to bring these topics into conversation, and organizes its contributions into three thematic parts to aid their synthesis. The first part provides archaeological and historical context for population and cultural history in the western North American Arctic. The second part derives evidence from the skeletal remains at Point Hope to illuminate similarities and differences between the individuals associated with the Ipiutak and Tigara cultures in light of regional (or global) biological variation. Finally, the third part provides three syntheses of this evidence.

The first part of the volume provides readers with the context necessary to establish reasons for the biological explorations that follow in the second part. Readers will find Anne Jensen's chapter (Chapter 2) useful for establishing the chronology of traditions in western and northern Alaska; her comprehensive review of archaeological sites and cultures connects and elucidates a complex history (both of peoples and of archaeologists) that provides the background necessary to understand the importance of evidence gleaned from Point Hope. Owen Mason builds on this general background in Chapter 3 by focusing on the Ipiutak culture, contextualizing the place of Point Hope within that archaeological tradition, and exploring the setting of the Ipiutak within the broader regional history. Then, in Chapter 4, Blaine Maley examines the population relationships evidenced by the archaeological record using cranial data to assess biological affinities. Together, these chapters provide the archaeological, population, and cultural foundations for biological analyses of the peoples of Point Hope.

In the second part, the contributors collectively demonstrate that, while clear biological distinctions existed between the Ipiutak and the Tigara populations, they were not wholly dissimilar. Kristen Krueger, in Chapter 5, provides information on non-dental tooth use and diet via her assessment of levels and patterns of anterior dental microwear between the Point Hope Ipiutak and Tigara individuals. The analyses by Sireen El Zaatari in Chapter 6 complement Kruger's study by investigating the evidence for dietary differences between the two cultural periods on occlusal molar microwear patterns. Together, Krueger and El Zaatari distinguish the subsistence and diets of various Arctic foragers. Charles Hilton and colleagues add further evidence to this division, as they examine morbidity differences between the Ipiutak and the Tigara peoples through the lens of postcranial skeletal lesions in Chapter 7. They argue that physical demands exerted on the postcranial skeleton increased with the transition to whaling subsistence, a finding that is further assessed mechanically in the next chapter. Laura Shackelford's analysis of limb strength properties in Chapter 8 reflects on the mechanical repercussions of high marine and terrestrial mobility on the postcranial remains of adults at Point Hope, and adds another source of evidence for understanding the lifeways of these peoples. Lastly, Libby Cowgill explores these activity patterns and variation in ecogeographic body proportion in the large sample of juvenile skeletons from Point Hope, establishing distinctions and similarities between the Point Hope groups within a global context. In total, these chapters provide the first collection of holistic skeletal analyses of the past occupants of Point Hope.

Finally, in the third part, Auerbach, Fitzhugh, and Dumond provide thoughtful analyses and commentaries integrating the evidence presented in the previous two parts. Benjamin Auerbach's contribution (Chapter 10) synthesizes the biological analyses of the second part, and brings them into conversation with previously published findings about Point Hope, as well as emphasizing the need to place this biological understanding within a broader, continental context. William Fitzhugh echoes the importance of maintaining a broader perspective when reaching conclusions about archaeological sites, as he provides a unique circumpolar context for understanding the place of the Point Hope Ipiutak culture among many archaeological and living Arctic cultures in Chapter 11. In Chapter 12, Don Dumond brings these elements together, as he unites the archaeological, biological, and ethnohistorical evidence to yield broad conclusions about the lifeways, history, and culture of the foragers who lived on the edge of the Alaskan Arctic.

Together, these analyses from archaeological and biological perspectives provide an approach for evaluating and reconstructing the origins, variation, and demands on Arctic foragers. Overall, they provide a testament to and insight on the adaptive success of humans within one of the world's inhospitable environments. Incontrovertibly, one conclusion drawn from these many perspectives is that humans arrived at a variety solutions to subsistence in the Arctic – even at the same location – reflected in their teeth, bones, and artifacts, and the rituals and lifeways those represent. Given the success of human occupation of northwest Alaska, this volume provides a unique contribution from an archaeological context regarding the adaptive variability that encompasses the modern human condition.

References

Bettinger, R. L. (1991). *Hunter-Gatherers: Archaeological and Evolutionary Theory*. New York, NY: Plenum Press.

Binford, L. R. (1978). *Nunamuit Ethnoarchaeology*. New York, NY: Academic Press.

Binford, L. R. (2001). *Constructing Frames of Reference: An Analytical Method for Archaeological Use of Hunter-Gatherer and Environmental Data Sets*. Berkeley, CA: University of California Press.

Burch, E. S. (1981). *Traditional Eskimo Hunters of Point Hope, Alaska: 1800–1875*. North Slope Borough.

Burch, E. S. (1988). *The Eskimos*. Norman, OK: University of Oklahoma Press.

Collins, H. B. (1984). History of research before 1945. In *Handbook of North American Indians, Volume 5: Arctic*. Washington, D.C.: Smithsonian Institution Press, pp. 8–16.

Dumond, D. E. (1987). *The Eskimos and Aleuts*. Second Edition. London: Thames and Hudson.

Fagan, B. M. (2005). *Ancient North America.* Fourth Edition. New York, NY: Thames and Hudson.
Hall, E. S. (1984). Interior north Alaska Eskimo. In *Handbook of North American Indians, Volume 5: Arctic.* Washington, D.C.: Smithsonian Institution Press, pp. 338–46.
Kelly, R. L. (1995). *The Foraging Spectrum: Diversity In Hunter-Gatherer Lifeways.* Washington, D.C.: Smithsonian Institution Press.
Larsen, H. and Rainey, F. (1948). *Ipiutak and the Arctic Whale Hunting Culture.* Anthropological Papers of the American Museum of Natural History 42. New York, NY: American Museum of Natural History.
Murdoch, J. (1892). Ethnological results of the Point Barrow expedition. *Annual Report of the Bureau of American Ethnology*, 9, 19–441.
Nelson, R. K. (1969). *Hunters of the Northern Ice.* Chicago, IL: The University of Chicago Press.
Nelson, W. E. (1899). The Eskimo about the Bering Strait. *Annual Report of the Bureau of American Ethnology*, 18, 3–518.
Rainey, F. (1941). The Ipiutak culture at Point Hope, Alaska. *American Anthropologist* 43 (3), 364–75.
Rainey, F. (1971). *The Ipiutak Culture: Excavations at Point Hope, Alaska.* Reading, MA: Addison-Wesley.
Schwartz, J. H., Brauer, J. and Gordon-Larsen, P. (1995). Brief communication: Tigaran (Point Hope, Alaska) tooth drilling. *American Journal of Physical Anthropology,* 97, 77–82.
Sheehan, G. W. (1985). Whaling as an organizing focus in Northwestern Alaskan Eskimo societies. In *Prehistoric Hunter-Gatherers: The Emergence of Cultural Complexity.* New York, NY: Academic Press, pp. 123–53.
Spencer, R. F. (1959). *The North Alaskan Eskimo: A Study in Ecology and Society.* Bureau of American Ethnology, Bulletin 171. Washington, D.C.: Smithsonian Institution Press.
Spencer, R. F. (1977). Arctic and Sub-Arctic. In *The Native Americans*, Second Edition. New York, NY: Harper & Row, pp. 56–113.
Spencer, R. F. (1984). North Alaska Coast Eskimo. In *Handbook of North American Indians, Volume 5: Arctic.* Washington, D.C.: Smithsonian Institution Press, pp. 320–37.
Stefansson, V. (1914). *The Stefansson-Anderson Arctic Expedition of the American Museum: Preliminary Ethnological Report.* Anthropological Papers of the American Museum of Natural History 14(1). New York, NY: American Museum of Natural History.

Part I

Regional archaeological and biological context

2 *The archaeology of north Alaska: Point Hope in context*

ANNE M. JENSEN

Point Hope has been significant in the history of Arctic archaeology, first and foremost as the location of the spectacular type-site of the Ipiutak culture. The modern residents of the Alaskan village of Tikiġaq (Point Hope) often refer to their community as the longest continuously inhabited village in North America. This statement is difficult to evaluate definitively, especially given the sea level rise and coastal erosion the region has seen, but it is clear that Point Hope has been continuously inhabited for many centuries.

Previous research and expeditions

The Arctic and its residents have long been a source of fascination to anthropologists, and the literature on archaeology and ethnography in the North American Arctic is voluminous. In the interests of space, I concentrate here on those sources that are most important to the definition and understanding of the Ipiutak and/or Thule cultures, and to setting Point Hope sites in a broader context. Mason provides a more detailed discussion of Point Hope in the next chapter. All locations and sites discussed in this chapter are located on the map presented in Figure 2.1.

Early exploration and whaling in north Alaska

The first recorded visit of non-natives to the North Slope of Alaska took place in 1826, in the form of an expedition led by Captain Frederick Beechey of the Royal Navy, on HMS *Blossom* (Beechey, 1831). Although *Blossom* did not make it far past Icy Cape, nearly 300 kilometers northeast of Point Hope, due to ice, *Blossom*'s barge reached Point Barrow. Exploring and mapping expeditions followed, with minimal contact with North Slope Iñupiat continuing through

The Foragers of Point Hope: The Biology and Archaeology of Humans on the Edge of the Alaskan Arctic, eds. C. E. Hilton, B. M. Auerbach, L. W. Cowgill. Published by Cambridge University Press.

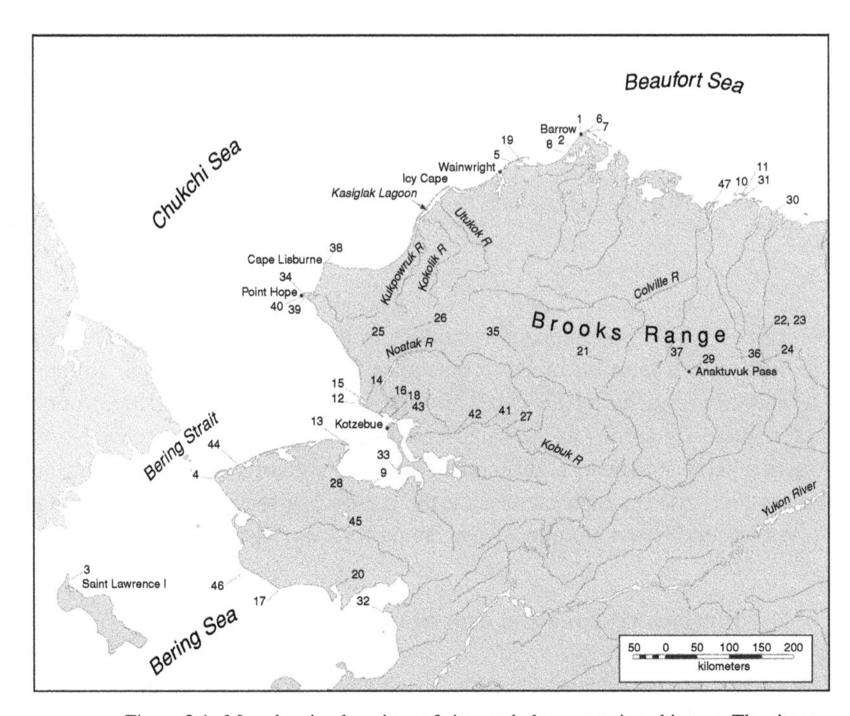

Figure 2.1. Map showing locations of sites and places mentioned in text. The sites are numbered as follows: (1) Utqiaġvik; (2) Kugusugaruk; (3) St. Lawrence Island; (4) Wales; (5) Nunagiak; (6) Nuvuk; (7) Birnirk; (8) Walakpa; (9) Deering; (10) Thetis Island; (11) Pingok Island; (12) Cape Krusenstern; (13) Cape Espenberg; (14) Palisades; (15) Battle Rock; (16) Sisualik; (17) Cape Nome (Bockstoce sites); (18) Tulaagiaq; (19) Piŋusugruk; (20) Qitchauvik; (21) Mesa; (22) Putu; (23) Bedwell; (24) Hilltop; (25) Raven's Bluff; (26) Tuluaq Hill; (27) Onion Portage; (28) Trail Creek; (29) Tuktu; (30) Putuligayuk River Delta Overlook; (31) Kuparuk Pingo; (32) Iyatayet; (33) Choris; (34) Ipiutak; (35) Feniak Lake; (36) Bateman; (37) Kayuk; (38) Uiivaq; (39) Jabbertown; (40) Tigara; (41) Ahteut; (42) Ekseavik; (43) Old Kotzebue; (44) TEL-104; (45) West Village Kuzitrin Lake; (46) Ayak (sledge Island); (47) Niġliq.

the 1850s. Trips by Simpson and Dease in 1837 (Simpson, 1843) and a Russian, Kashevarov, in 1838 (Van Stone, 1977) were the first to follow *Blossom.* The early 1850s saw trips by Moore in 1849 and 1850 (Moore, 1851; Hooper, 1853), Sheddon in 1849 (Murdoch, 1892), McClure in 1850, and Collinson in 1851 and 1854.

These expeditions provided some useful information in their accounts about settlement locations and population sizes of people living on the North Slope. For example, HMS *Plover* overwintered twice in the Elson Lagoon in the early 1850s. The captain's log, rediscovered and edited by John Bockstoce, is a valuable account of an indigenous coastal whaling village at a time when many effects of contact with the Europeans had not yet been felt among groups living in the Arctic (Maguire, 1988).

North Alaskan ethnography and archaeology

The earliest primarily scientific expedition to north Alaska was the U.S. Army Signal Corps expedition under Lt. Patrick Henry Ray, which was conducted as part of the first International Polar Year (Ray, 1885). Among other accomplishments, the members compiled a basic Iñupiat dictionary, conducted a census of the community of Utqiaġvik (modern Barrow), took anthropometric measurements, wrote an ethnographic sketch, and, most significantly, collected hundreds of items of material culture, along with notes on their manufacture and use (Ray, 1885; Murdoch, 1892). This material, particularly valuable due to the extensive documentation that accompanies it, is now at the Smithsonian Institution's National Museum of Natural History. The work by the Army Signal Corps further resulted in one of the most important references for north Alaskan archaeology, John Murdoch's *Ethnological Results of the Point Barrow Expedition* (1892).

Not all late nineteenth century accounts were written as part of formal, scientific expeditions. Charles D. Brower, a whaler and trader who spent over half a century living on the North Slope, notably recorded observations about indigenous life during the period marked by increasingly regular interactions with Western cultures. When he settled in northern Alaska in 1884, he joined a native Point Hope whaling crew, and later moved to Barrow, where he managed a shore-based whaling station and trading company. Brower learned to speak Iñupiaq fluently, which greatly aided his endeavors in Alaska, and his first wife, Toctoo, was an Iñupiaq woman from Point Hope. Except for several trips to the continental United States, Brower remained in northern Alaska for the rest of his life. His autobiography (1942) contains extensive ethnographic information on North Slope Iñupiat people, starting a few years after Yankee whalers began coming to the Alaskan Arctic and ending shortly before World War II. This book was edited from a much longer manuscript (Brower, n.d.).

Other sources documented incidental ethnographic information about the coastal inhabitants of northern Alaska, though they were less extensive than Murdoch's or Brower's publications. Among these are U.S. Revenue Marine (predecessor service to the U.S. Coast Guard) reports. Some of their published cruise reports include information on the various coastal Iñupiat villages (e.g., Healy, 1887); see also the logs of the U.S. Revenue cutter *Bear* (U.S. Treasury Department, 1899).

Vilhjálmur Stefánsson (1913, 1914), a Canadian ethnographer, was the first professional anthropologist to work on the Alaskan North Slope. His editor published his ethnographic observations in 1914 from miscellaneous notes gleaned from field books while Stefánsson was on another expedition. These have since been revisited and reedited (Pálsson, 2001). Stefánsson also collected some artifacts and cultural materials, mostly by purchase, beginning in 1912. These collections were reported in several works (Stefánsson, 1914; Wissler, 1916; Ford, 1959).

In addition to the ethnographic research notes and descriptions published by Stefánsson, Diamond Jenness, who served as part of Stefánsson's Canadian Arctic Expedition, provided important cultural observations about people living in northern Alaska. Jenness overwintered with Iñupiat families on the Beaufort Sea coast in 1913–14. Publications based on this fieldwork (Jenness, 1918, 1957) were especially important in their documentation of the considerable population movement that occurred during the latter part of the 1800s and the early 1900s. Ethnographic fieldwork and cultural studies have continued to document Native cultures in the North American Arctic in the century since these first studies, showing how the Inuit cultures have changed and adapted, but a full review of these studies is beyond the scope of this chapter.

Though Stefánsson provided some documentation of cultural objects in addition to Murdoch's 1892 volume, neither of their expeditions performed archaeological investigations. William Van Valin (1941, n.d.), who moved to Barrow with his family to serve as a schoolteacher, was among the first to excavate any locations in northern Alaska, albeit without training. He is noted for digging at Kugusugaruk, south of Barrow, where he unearthed a large number of human remains. Unfortunately, his excavation techniques and documentation were poor, even for the time; this has left researchers without the context necessary for interpreting the material, which is currently located at the University Museum at the University of Pennsylvania. Ford (1959) concluded that the excavated materials were early Birnirk, a conclusion supported by radiocarbon dating (Ralph and Ackerman, 1961).

Therkel Mathiassen (1927) carried out the first systematic archaeological excavations anywhere in the Arctic (Collins, 1984). While people such as Van Valin conducted "digs" at Arctic sites without formal training, Mathiassen was the first professional archaeologist to practice in the region. His excavations, conducted during the course of the Fifth Thule Expedition, were consequential in defining the origins of Thule culture. Mathiassen saw Thule as a prehistoric culture originating in Alaska, ancestral to many (if not all) of the living Inuit groups in the North American Arctic. Among his contributions was the first systematic harpoon head typology (Mathiassen, 1927), which at the time and subsequently was used to define Arctic cultures, forming the basis for subsequent classifications of these artifacts and cultures from north Alaska eastward. Ford (1959) later based his own typology for the Point Barrow region in large part on that of Mathiassen, supplemented by Collins' (1937) typology from the Bering Strait area.

In addition to Mathiassen's excavations, Knud Rasmussen, the leader of the expedition and noted Danish anthropologist, bought several collections of archaeological specimens from the North Slope as he traveled through the area on the Fifth Thule Expedition, like Stefánsson had done in smaller quantities a decade before. These collections were described and illustrated by Therkel Mathiassen (1930). Yet, because none of the material was stratigraphically excavated, it addressed the geographic distribution of certain types of artifacts, but did little to illuminate chronology.

In addition to the Canadian and northern Alaskan Arctic, the Bering Strait region also saw increasing archaeological activity during the 1920s and 1930s. Jenness (1928), Collins (1935, 1937), and Geist (Geist and Rainey, 1936) worked at stratified sites on St. Lawrence Island, located in the Bering Sea far north of the Aleutian Islands, identifying the antecedent cultures to Thule, such as Okvik, Old Bering Sea, and Punuk. They, moreover, found evidence of the Thule occupation leading up to the present-day inhabitants of the island. These individuals also conducted excavations on the Alaskan coast of Bering Strait, namely in Wales, Alaska, near to Nome; Jenness excavated there in 1928, and Collins in 1936 (Dumond and Collins 2000).

Farther north, James A. Ford worked on the North Slope between 1931 and 1936, excavating at several sites, including Nunagiak (Point Belcher), Nuvuk, Birnirk, and Walakpa. In 1936, Ford traveled up the coast from Kotzebue (see Figure 2.1) by small boat, surveying sites along the way. Ford's report on his Chukchi coast excavations defined harpoon head typology for this region, and thus temporal sequences for cultures, as well as providing extensive and well-illustrated descriptions of artifacts from Birnirk through contact period sites (Ford, 1959). These descriptions would later prove crucial for interpreting the historical sequences of peoples living in the Alaskan Arctic.

This was the archaeological context and work that existed at the time of the excavations at Point Hope, which are reviewed in detail by Mason (Chapter 3). Like Ford, Helge Larsen and colleagues surveyed the coast of Alaska, traveling from Point Hope to a point between Wainwright and Point Franklin. These individuals tested along the Utukok River and at Icy Cape (Larsen and Rainey, 1948). While the survey results were archaeologically promising, and included several inland settlements (Larsen, 1981, personal communication), they were overshadowed by the work that Larsen and Froelich Rainey did at Point Hope from 1939 to 1941 (Larsen and Rainey, 1948). After his work at Point Hope, Larsen (2001) conducted excavations at

Deering, including the excavation of the largest Ipiutak men's house known to date.

During the post-war years, many archaeological surveys were carried out on Alaska's North Slope, and most were associated with exploration within the Naval Petroleum Reserve-4 (PET-4). None of these returned to the Ipiutak and Tigara sites at Point Hope for many years, however, and discussion here is limited to research that dealt in some way with coastal peoples. Raymond Thompson surveyed the Utukok River, uncovering a fluted point from the upper Utukok, and finding several villages on the middle Utukok, as well as one village at the mouth of the Utukok on Kasiglak Lagoon (Thompson, 1948). He thought the villages were recent, although one reportedly contained pottery, which suggests a precontact or early contact date. Following a similar archaeological survey pattern, Ralph Solecki (1950) accompanied a geological party in 1949, which also traveled along rivers. Among the sites on the Kukpowruk and Kokolik Rivers (both of which also drain into Kasiglak Lagoon, near Point Lay) was a village near the mouth of the Kukpowruk, as well as a village about 35 miles inland, which Solecki believed to have been Nunamiut (Solecki, 1950). Two years later, in 1951, William Irving began several seasons of fieldwork on the North Slope. The first season was interior work, but in 1952 he surveyed the lower Colville River, the Colville Delta and eastward along the Beaufort Sea coast, excavating Thule and later sites on Thetis and Pingok Islands (Irving, n.d.).

Also in 1951, Wilbert Carter began extensive excavations at Piġniq (Birnirk) and Nuvuk. Like Point Hope, these were coastal sites, and represented groups living up to 500 kilometers north of the latitude of Point Hope. The work resulted in sizable collections, which contain Birnirk and Thule material, but his excavations remain unpublished. Carter produced a series of interim technical reports (Carter, 1953a, 1953b, 1962) and a final report (Carter, 1966). Unfortunately, some of his reports are missing (as cited in his 1962 report), and the remaining reports contain internal contradictions concerning the numbers and types of artifacts recovered, as well as omissions (e.g., burials from Nuvuk are not mentioned, although they are drawn in the field notes). However, James Ford worked with the Carter party in 1953, making very detailed (1′ contour) maps of the Birnirk and Nunagiak sites. These maps were published together with his description of his 1930s work (Ford, 1959); this remains the best published information on the Birnirk site. (See Maley, Chapter 4 of this volume, for more discussion about the Birnirk site.)

In 1956, J. L. Giddings began work in the Cape Krusenstern vicinity. Cape Krusenstern, like Point Hope, consists of a beach ridge complex. These beach ridges contain archaeological remains from a long series of cultures, ranging at Cape Krusenstern from recent Iñupiat back to the Denbigh Flint Complex, generally arranged in chronological order corresponding to the development of the beach ridges. Giddings and Anderson excavated twenty-one Ipiutak house ruins, located in tightly clustered settlements. They also identified several areas as campsites, and excavated one of them (Giddings and Anderson, 1986), which consisted primarily of concentrations of fire-reddened pebbles or oil-soaked gravel. Giddings and Anderson suggested that these might be the remains of seal hunters' campsites. Cape Krusenstern also contained a number of Western Thule features, which Giddings and Anderson were able to divide into early and late Western Thule on the basis of house styles and artifact typology. Giddings and his crew also surveyed or tested at other sites in the region (Giddings and Anderson, 1986). These included Battle Rock, where both Western Thule and Ipiutak burials were encountered; Cape Espenberg, which included an Ipiutak locale; and Sisualik, where two Western Thule houses were located.

Fieldwork on the Northern Slope declined during the 1960s and 1970s, but a few excavations are important to note. In 1968 and 1969, Dennis Stanford excavated at Walakpa (Stanford, 1976), clarifying the chronological relationship between the Birnirk and Thule archaeological cultures. These cultures were fairly well defined, but no prior work had recognized them in stratigraphic relationship. Stanford's work involved the excavation of seventeen superimposed levels running from Early Birnirk through Late Thule, indicating that, at least at Walakpa, Neoeskimo culture developed *in situ* in a gradual transition from Early Birnirk through Late Thule (Stanford, 1976). John Bockstoce conducted excavations at Cape Nome (Bockstoce, 1979) that revealed sites ranging in age from Denbigh through modern Iñupiat. While most of these excavations focused on artifacts, some additional human skeletons were uncovered during this period. For example, Douglas Anderson's (1978) work at Tulaagiaq, a transitional Near Ipiutak–Ipiutak archaeological site from Kotzebue Sound, included the excavation of burials.

Archaeological excavations increased from the 1980s into the 1990s, and have continued to improve the understanding of past human cultures in the Alaskan Arctic. Some of these focused on the region around Point Barrow. In 1981 and 1982, major excavations were carried out at the Utqiaġvik site in Barrow (Hall and Fullerton, 1990). The majority of the material excavated was late precontact and post-contact, although some earlier lithics were recovered. The project resulted in good descriptions of a great deal of material culture (e.g., Reynolds, 1990). The results of the excavations from Mound 34, which included a precontact *qargi* as well as thirteen houses, were published as a monograph (Sheehan, 1997). A block of tundra from the Utqiaġvik site slumped onto the beach and was excavated in 1990. Given the limitations of the original context, the report (Mason *et al.*, 1991) provides excellent artifact descriptions and illustrations and analysis of faunal remains. The investigators also obtained a number of ^{14}C dates on diagnostic artifacts. In 1986, Sheehan, Reinhardt and Jensen (1991) carried out survey and limited excavation at Piŋusugruk, located west of Barrow on the northern coast. That season demonstrated the presence of a fairly deep stratified midden, and led to further excavations of the midden, several Late Western Thule houses and a pair of post-contact houses in 1994–6. In 1994, erosion of bluffs in Barrow led to the exposure of the frozen body of a girl. Excavation focused on the recovery of the individual, although a small collection of artifacts resulted (Zimmerman *et al.*, 2000), ranging in age from post-contact glass beads to two harpoon heads and a throwing board that were clearly Birnirk. The Birnirk artifacts lay on the floor of a house, undisturbed by the later digging of the meat storage pit used for the girl's burial through the house floor.

Since 1997, work has been renewed at the Nuvuk site, at Point Barrow (Jensen, 2007b, 2009a, b). An extensive Thule graveyard has been partially excavated, dating between 1,000 and 300 years BP (calendar years before present), and these human remains are part of ongoing isotopic and aDNA studies. In addition, an Ipiutak occupation was encountered, extending the known Ipiutak range almost 500 kilometers north of Point Hope.

Excavations of Ipiutak period sites have also extended to the south of Point Hope in the last two decades. Additional work took place in Deering in the late 1990s, in connection with the installation of a water and sewer system in the village. Several spectacular Ipiutak graves were found. In addition, an Ipiutak house was located and partially excavated (Reanier *et al.*, 1998). The project also resulted in the excavation of two Western Thule houses (Reanier *et al.*, 1998). Recently, excavations at Qitchauvik, in Golovin Bay, revealed an Ipiutak men's house (Mason, 2000; Mason *et al.*, 2007). This site is currently the southernmost manifestation of Ipiutak, located over 400 kilometers south of Point Hope.

To the south of Point Hope, Wales, Cape Krusenstern, and Cape Espenberg have all also seen renewed research interest (e.g., Dumond and Collins, 2000; Darwent, 2006; Darwent *et al.*, 2013), leading to major projects in the last few years. The projects have mapped large numbers of additional structures, carried out additional testing and dating, and conducted excavations of some houses (Harritt, 2004, 2010b; Hoffecker and Mason, 2010a, b, 2011; Darwent *et al.*, 2013).

North Alaskan cultural history

As implied in the discussion of research history, archaeology in the Arctic has always been difficult and expensive, and has lagged behind archaeology in more accessible parts of the world. For that reason, even cultural chronologies are not completely resolved. Despite some uncertainties, a cultural chronology of north/northwest Alaska has been established by the analysis of evidence gathered over the decades of excavations summarized in the preceding pages. This chronology is reviewed here to provide a context for the more detailed discussions of Point Hope in the rest of this volume. Dates are provided as calendar years before present (BP) when possible, or as radiocarbon years before present (RCYBP) when calendar year equivalences are not available.

Paleoindian tradition

Investigations at the Mesa site in the northern foothills of the Brooks Range (Kunz and Reanier, 1994, 1995), and at the Bedwell, Putu, and Hilltop sites in the Sagavanirktok and Atigun River region of the Brooks Range (Alexander, 1974, 1987; Reanier, 1995) document the presence of a Late Pleistocene/Early Holocene occupation of northern Alaska by peoples using tools remarkably similar to those of Paleoindian cultures identified from central North America, especially Agate Basin. Similarly early sites are also located near the Brooks Range divide, notably Raven's Bluff, which has the oldest archaeofauna from the Alaskan Arctic, dominated by caribou (Rasic, 2012, personal communication). These sites are considered part of a Sluiceway complex; the known sites are all located on hills or knolls, and evidently served as hunting lookouts and weapon repair stations. Site locations and lithics tool kits suggest a terrestrial hunting orientation. There are no documented coastal sites, but this is not surprising, as sea level was far lower at that time, and any sites preserving evidence of coastal occupations would now be underwater. The Northern Paleoindian tradition dates to 12,600–11,900 years BP, with a possible earlier component between 13,700 and 13,000 years BP (Bever, 2006).

American Paleoarctic tradition

Until recent decades, the earliest cultural period recognized from north/northwest Alaska was the American Paleoarctic tradition (Anderson, 1984), defined from the Akmak and Kobuk complexes at Onion Portage and Trail Creek (Anderson 1968, 1970, 1988; Larsen, 1968). This tradition consists of large core bifaces and cores used to yield blade preforms, which were used to produce end scrapers, gouges, burins, knives, and shaft smoothers (Anderson 1984, 1988). Wedge-shaped cores yielded microblades, which were apparently used as insets for antler arrowheads or spear points. The flint-knapping methods used to produce these tools appear to be derived from techniques present, if not developed, in Siberia. In general, American Paleoarctic sites are thought to date between 9,700 and 8,000 RCYBP. Again, there are no known coastal sites, either due to sea level rise or an interior orientation.

Northern Archaic tradition

The Northern Archaic tradition (6,600–3,000 RCYBP) also was defined from assemblages found at Onion Portage (Anderson, 1968) and the Palisades site (Giddings, 1962, 1967). This tradition is characterized by side- and corner-notched and stemmed projectile points, bifacial knives, unifacial flakeknives, side scrapers, end scrapers, and notched pebbles thought to be net sinkers (Anderson, 1968, 1988). Researchers previously thought that this tradition was confined to the boreal forest, but Northern Archaic assemblages are now known from tundra environments (Davis *et al.*, 1981; Lobdell, 1986; Esdale, 2008; Rasic and Slobodina, 2008). One of the oldest sites of the Northern Archaic tradition is the Tuktu site, located near Anaktuvuk Pass in the Brooks Range. Farther north, evidence for the tradition comes from the Putuligayuk River Delta Overlook (Lobdell, 1985, 1995) and Kuparuk Pingo (Lobdell, 1986, 1995) sites. There is only one coastal site that has been recorded in north Alaska (Palisades), suggesting that Northern Archaic had a strong interior orientation, although coastal manifestations may be underrepresented in the archaeological record, again due to sea level rise.

Arctic Small Tool tradition

In north Alaska, the primary culture representative of the Arctic Small Tool tradition (ASTt) is the Denbigh Flint Complex. The Denbigh Flint Complex was defined by Giddings (1951, 1964) based on excavations at the Iyatayet site at Cape Denbigh on Norton Sound. Since then, many more Denbigh sites, both coastal and interior, have been located. Small, delicately flaked end and side blades for insetting into antler arrowheads are diagnostic for the Denbigh Flint Complex (4,200–3,600 RCYBP). Microblade cores and abundant microblades are also prominent features of the complex, as are the flaked mitten-shaped Denbigh burins. Other artifact types include end scrapers, bifacial knives, ground slate adze blades and burins, and notched pebbles, thought to represent percussion tools, rather than fishing gear. ASTt cultures that appear to be related to the Denbigh Flint Complex are found across the North American

Arctic. They include the Pre-Dorset in Canada, and Independence I and Saqqaq in Greenland (Maxwell, 1984, 1985).

The Denbigh Flint Complex is the first culture in north/northwest Alaska for which clear evidence exists of more than casual use of the maritime environment (Ackerman, 1998). In addition to the number of sites that are located along the coast, the Denbigh tool kit includes items Giddings (1964) interpreted to be harpoon end blades, and the very limited faunal remains include a few fragments of seal bone. Other ASTt sites to the east also show evidence of maritime subsistence adaptations.

The Norton/Paleoeskimo tradition

Some researchers, particularly those who work primarily in Alaska, refer to the following cultures as part of the Arctic Small Tool tradition. Others consider them part of a successor tradition known variously as the Paleoeskimo tradition or the Norton tradition. In either case, these cultures appear to follow Denbigh chronologically and have related technology.

The Choris culture (3,600–2,500 RCYBP) has inset end and side blades for arrowheads similar to those found in Denbigh, but some Denbigh forms, including microblades, Denbigh chipped and ground burins, and Denbigh style flakeknives disappear during Choris. Moreover, the Choris culture includes important new tools and technologies not known to be present in Denbigh, such as pottery, stone lamps, chipped adze blades, burin spalls from irregular bifaces, and new stemmed and shouldered projectile point forms (Anderson, 1984, 1988; Giddings and Anderson, 1986). An occupation that Stanford (1976) termed transitional between Denbigh ASTt and Choris was present at Walakpa, although he is now inclined to consider the level Choris (Stanford, 2013, personal communication). Importantly, Choris people clearly exploited the marine environment. Archaeofaunas at coastal sites contain large amounts of seal bone (e.g., Giddings and Anderson, 1986), although the significant amount of caribou bone and the existence of inland sites suggest a mixed subsistence economy.

The Old Whaling culture (2,900–2,700 RCYBP), defined from a small cluster of houses at Cape Krusenstern (Giddings, 1961), is not well distinguished or understood. Ackerman (1984) noted similarities between harpoon heads at Cape Krusenstern and the Chertov Ovrag site on Wrangell Island. Furthermore, recent work (Darwent and Darwent, 2003) at the site found additional structures, which date between 1,300 and 1,050 years BP. Careful analysis of faunal material by the authors suggested that the site may have been a winter and spring seal hunting camp for groups that were otherwise interior focused.

Giddings (1964) first defined Norton culture (2,500–2,000 RCYBP) at Iyatayet. Norton culture, which is far less ambiguous than the Old Whaling culture, continued many of the elements of Choris, including end and side blade insets, notched pebbles, ground burins and adzes, pottery, and clay and stone lamps. The Norton culture marked the debut of ground slate as major raw material for tools in the region. Large numbers of notched pebbles have been recovered, suggesting a subsistence economy with considerable emphasis on fishing in some areas. There have been a few large end blades found at Norton sites, which suggest to some that Norton people may have been casual whalers, in addition to hunting smaller marine mammals. A few of these occur at Point Hope, as part of a culture Larsen and Rainey (1948) had defined as "Near Ipiutak" because of its similarities to Ipiutak. Near Ipiutak appears to be essentially the same as the culture later defined by Giddings as Norton. Several Norton sites, as well as sites of Norton age that did not have any diagnostic materials in them, have also been found in the Barrow area (Stanford, 1976; Jensen, 2007a).

Ipiutak culture (1,600–1,050 years BP), as noted previously, was first identified at Point Hope. Noted for its elaborately carved ivory objects (Larsen and Rainey, 1948), other artifacts associated with Ipiutak culture include side blade insets and flakeknives similar to those from the Denbigh Flint Complex, and discoidal scrapers, large bifacial knives, and ground burins or chisels like those of Norton culture. Surprisingly, Ipiutak appears to lack the pottery, ground slate, lamps, and whale-hunting equipment common to both the earlier Norton and the later Eskimo cultures. Coastal Ipiutak sites show a strong reliance on hunting of marine mammals, particularly walrus. The abundance of ivory found in many Ipiutak sites is an indication of how successful they were at this pursuit.

The Ipiutak culture was defined on the basis of the excavations conducted by Larsen and Rainey in the vicinity of Point Hope. They encountered houses and burials that they identified as belonging to a previously unknown culture which they named Ipiutak, after the immediately adjacent small lagoon and spit. Rainey reported their discovery first in the professional literature in a 1941 article, which was followed the next year by a more popular article in *National Geographic* (Rainey, 1942), and by two popular accounts in *Time* magazine (1941, 1948). Subsequently, Larsen and Rainey (1948) published their extremely detailed monograph on their work at Point Hope, focusing on the Ipiutak finds.

Despite a fairly extensive history of survey and excavation on the North Slope, summarized in part in the previous section, Ipiutak culture has only recently been found on the Arctic coastal plain (cf. Mason, 2006). Diagnostic Ipiutak artifacts have been found in three locations within the North Slope Borough. The first, of course, is the type-site itself, Ipiutak (Larsen and Rainey,

1948; Jensen, 1997). The site is located on five beach ridges adjacent to Marryat Inlet and Ipiutak Lagoon, near the present-day community of Point Hope. The second is a caribou kill or processing site at Anaktuvuk Pass (Mills *et al.*, 1999), an example of the inland manifestation of Ipiutak. The third is Nuvuk, located at the tip of Point Barrow (Jensen, 2009a).

In contrast with the few North Slope coastal Ipiutak sites, there are many more inland sites and coastal sites located south of Point Hope. Interior Ipiutak sites include Feniak Lake (Hall, 1974) and Bateman (Reanier, 1992), as well as Ipiutak components at Trail Creek and Onion Portage. There is also clear artifactual evidence of an Ipiutak occupation at the Kayuk site in Anaktuvuk Pass, although the site appears to be culturally mixed (Campbell, 1959). The other coastal Ipiutak occupations are known from Deering (Reanier *et al.*, 1998; Larsen, 2001), Tulaagiaq (Anderson, 1978), Cape Krusenstern (Giddings and Anderson, 1986), Cape Espenberg (Hoffecker and Mason, 2010a, 2011) and Cape Nome (Bockstoce, 1979).

Neoeskimo tradition

The Neoeskimo tradition (1,500 years BP to Present) has also been referred to as the Northern Maritime tradition. In north Alaska, Birnirk culture (1,500–1,000 years BP) is the earliest of the Neoeskimo cultures, and is exclusively coastal in its known distribution. Named for its type site near Barrow (Ford, 1959; Carter, 1966; Stanford, 1976), Birnirk material culture includes ground slate weapon points and knives, chipped chert knives, discoidal scrapers, and projectile points, as well as bifurcated and trifurcated spur harpoon heads with some self-barbed and some containing chipped stone side blades. Pottery and lamps persist in Birnirk and subsequent Neoeskimo cultures. Birnirk sites are known from the north coast of the Chukotsk Peninsula in the Russian Far East (northeastern Siberia) – as well as from St. Lawrence Island and a few sites on the north shore of Norton Sound – and extend north along the north Alaska coast at least as far as Point Barrow. Birnirk people subsisted on a wide variety of marine mammals, as well as caribou and other land mammals. Reflecting on their tool kit and faunal evidence, while Birnirk people could successfully hunt whales, they do not appear to have been sufficiently successful as whalers to re-orient their entire socioeconomic system toward whaling (Stanford, 1976). Some evidence of Birnirk occupation has been recovered from Point Hope, but the area where it was loacted had eroded away by the time Larsen and Rainey excavated the site.

The relationship between peoples with Ipiutak and Birnirk material cultures is not resolved. Recent refinements in radiocarbon dating suggest there is an overlap in time between people with Ipiutak and Birnirk material cultures (Gerlach and Mason, 1992). Whether these are two entirely different groups of people, or whether they are simply related people who used different toolkits is not clear. Birnirk sites have only been found very close to the coast, while later Ipiutak sites appear to be primarily located inland. Mason (1998) has suggested that they may represent two separate political groupings.

Elsewhere in Alaska, especially on St. Lawrence Island along the Bering Sea coast, several other cultures with a maritime focus are known to have existed during this period. These include Old Bering Sea, Okvik, and Punuk cultures, and all are characterized by elaborate decorative arts styles and by a focus on hunting of marine mammals. Old Bering Sea art, in particular, can be quite similar to Ipiutak art. Old Bering Sea materials are also found in Siberia, as are Punuk materials. Punuk materials are also known from Uiivaq, just north of Cape Lisburne (Mason, 2003), and Nunagiak, a coastal site between Wainwright and Barrow (Ford, 1959). Several types of technology important for successful large-scale whaling, including drag float parts, are known from these cultures. Since these items are also useful in hunting other large marine mammals, such as walruses, it cannot be assumed that they were deliberately employed in whaling immediately upon their introduction. However, archaeological evidence has been used to argue that the Punuk people were doing some whaling (Mason, 2009). Contact between Punuk and Birnirk peoples may have led to the refinement of Birnirk whaling techniques.

The Thule culture (1,100–500 years BP) followed Birnirk; its tools represent an elaboration and diversification of the Birnirk examples. Knife blades and weapon blades were primarily made of ground slate, harpoon head styles became less elaborate, and fishing and bird hunting equipment became more diversified. Self-pointed and barbed arrowheads replaced the end-slotted Birnirk forms. Larsen and Rainey defined the Western Thule phase of what they referred to as the Arctic Whale Hunting culture (1948: 37–9), based on their work at Point Hope. This definition was based on two Western Thule burials (1948: 61, 170) and the excavation of Jabbertown House 2, a large multi-room house (1948: 170–3) at the Jabbertown site at the base of the Point Hope spit. Western Thule, as they defined it, was clearly antecedent to Thule in the eastern Arctic, which they suggested be referred to as Eastern Thule (1948: 39, 170–4), and considered to be another phase of the Arctic Whale Hunting culture. These peoples were highly successful whalers, who tended to live in larger communities oriented toward whale hunting, although they also hunted the same marine mammals and caribou that their Birnirk forebears had pursued. Subsequent to their initial appearances in the archaeological record, Thule people developed more specialized regional cultures from Alaska to Greenland. In Alaska, this regional variant is the Late Western Thule (LWT) culture, which led directly to the diverse regional late prehistoric Eskimo cultures of northern Alaska (Sheehan, 1985, 1990, 1997). Larsen and Rainey argued that Western Thule was succeeded at Point Hope by a culture they termed Tigara, another phase of the Arctic Whale Hunting culture, which they distinguished from the preceding Western Thule on the basis of artifactual differences (1948: 175). Tigara and the subsequent Modern phase, representing the Tikerarmiut (contact and later residents of Point Hope), were clearly a continuum; the distinction between the two was essentially arbitrary (1948: 175).

Securely dated Thule sites in north and northwest Alaska with occupations dating prior to 650 years BP are located along the coast, with a few exceptions. Nuvuk, Birnirk, Utqiaġvik, Walakpa, Jabbertown (Qimiarzuk), and Deering all were inhabited between 750 and 650 years BP. The large coastal sites of Cape Espenberg and Cape Krusenstern were almost certainly occupied during this period as well. Ahteut, located on the Kobuk River, is the only securely dated site located away from the coast, with at least five dendrochronological dates from a single structure falling in the first half of the thirteenth century. The nearby caribou crossing/hunting site of Onion Portage has several dates on what is described as Ahteut material (Anderson, 1988: 134), one of which coincides with this time span. As evidenced by its citing in affiliation with other cultures, Onion Portage was first used as a caribou hunting location *c.* 9,000 years ago, and various groups continued to use the site until the present day (Anderson, 1988). Artifact typology (Irving, n.d.; Mathiassen, 1930) also suggests that there was an occupation of the Beaufort Sea coastline contemporaneously with the coastal sites listed above, but there are no reliable dates.

Coastal occupations continued at the major sites throughout the LWT period. In the earlier part of the LWT period, fewer sites have established dates, but this is probably attributable in part to researchers' tendency to date initial occupations only when measuring a limited series of dates. It may also reflect reduced occupations at places such as Cape Krusenstern and Cape Espenberg, as these locations proved less suitable for whaling than locations such as Point Hope and Point Barrow (cf. Savelle and McCartney, 2003). There are indications that the overall LWT population was growing; people were beginning to diversify the focus of their subsistence economies and establish settlements in locations not previously used for permanent settlements by Thule people, although many sites have indications of use by earlier cultures or transient/seasonal use. Notably, Thule peoples occupied Ekseavik, located on the middle Kobuk River, by about 550 years BP (Giddings, 1952), and the Old Kotzebue site seems to have been occupied around the same time. Inhabitants at these sites appear to have shifted subsistence focus from whaling to a combination of fishing and sealing, with beluga drive hunts playing a role in communities around Kotzebue Sound. There are also indications that coastal locations away from the prime whaling locations were being occupied. For example, TEL-104, in the Ikpek area of the Seward Peninsula coast, has one date that calibrates to 1355 ± 80 CE (Harritt, 1994), and Walakpa (located near to Barrow) continued to be used (Stanford, 1976).

The period between 450 and 350 years BP was marked by population growth and a change in technological emphasis. In the Kotzebue Sound/Kobuk River area, fishing appears to have moved to greater reliance on use of nets, based on the artifacts found at sites such as Intermediate Kotzebue (Giddings, 1952), and may reflect a need to feed growing populations. At the same time, the amount of slat armor found in coastal sites such as Kotzebue (Van Stone, 1955), Tigara (Larsen and Rainey, 1948) and Utqiaġvik (Sheehan, 1990, 1997) increased greatly from the few pieces scattered in earlier sites, suggesting heightened social tensions if not full-scale warfare. Coincidentally, sites in less accessible locations, including interior settlements away from major rivers, seem to develop at this time. For example, the West Village at Kuzitrin Lake, in the middle of the Seward peninsula, has a ^{14}C date of 415 ± 80 years BP. Another site, Ayak, located on Sledge Island just off the coast of Cape Nome, was first occupied around this period (Harritt, 2010a). In addition to these settlement expansions, infilling occurred along the coast, with occupations at Pingusugruk on Point Franklin (Sheehan *et al.*, 1991) and at additional sites in the Ikpek area (Harritt, 1994) dating from this time period.

By the time of contact with Europeans, there was an extensive traditional trading network operating throughout the North. Western goods such as bronze, copper, and glass beads (Murdoch, 1892; Kunz and Reanier, 2005) reached north and northwest Alaska through this network long before the arrival of Euro-Americans, most likely through middlemen from St. Lawrence Island. Iron from Asia is known from Neoeskimo sites on St. Lawrence Island (Collins, 1937; Bandi and Blumer, 2004), as well as in Ipiutak sites (Larsen and Rainey, 1948: 83), and was regularly available in small quantities for those who had the means to trade for it.

In addition to these exotic goods, there was a thriving trade in Alaskan products. These included lithic raw materials such as jade (sourced from Jade Mountain and perhaps other locations), obsidian, and perhaps cherts; large logs that could serve for entryway boards (traded from the mouth of the Mackenzie River in Canada to people in north Alaska, where wood was scarce); and material for tools, such as ivory and beaver teeth (e.g., Jensen, 2007b, 2009a). Significant amounts of marine mammal oil and other marine mammal products were traded from the coast to inland groups, and were exchanged for skins from caribou. It has been argued that this trade was necessary to support sustained inland occupation (Sheehan, 1997), which in turn was helpful in procuring the quantities of interior products necessary to sustain major whaling settlements.

Some of these items, such as beaver teeth, probably were traded across ethnic boundaries, since beavers seem to have been almost nonexistent in Western Thule/Iñupiaq territory.

Moreover, by contact, established trade fairs occurred at Sisualik in the Kotzebue area (Burch, 2005), at Point Spencer on the southern Seward Peninsula (Ray, 1975), at Niġliq near the mouth of the Colville River (Simpson, 1855; Rasmussen, 1952; Spencer, 1959; Maguire, 1988; Campbell, 1998), on Barter Island (Franklin, 1828; Simpson, 1855), and probably also in the Icy Cape area (Burch, 2005) and at Piġniq or Birnirk (Burch, 2005). It is likely that trade fairs took place in some form during LWT times, and probably earlier. Point Hope participated in this network primarily through winter trade with Barrow-area residents, and at Sisualik.

References

Ackerman, R. E. (1984). Prehistory: The Asian Eskimo Zone. In *Handbook of North American Indians, Volume 5: Arctic*. Washington, D.C.: Smithsonian Institution Press, pp. 106–18.

Ackerman, R. E. (1998). Early maritime traditions in the Bering, Chukchi, and East Siberian Seas. *Arctic Anthropology*, 35 (1), 247–62.

Alexander, H. L. (1974). The association of Aurignacoid elements with fluted point complexes in North America. In *International Conference on the Prehistory in Paleoanthropology of Western North American Arctic and Subarctic*. Calgary: University of Calgary Press, pp. 21–31.

Alexander, H. L. (1987). *Putu: A Fluted Point Site in Alaska*. Publication No. 17, Department of Archaeology, Burnaby, BC: Simon Fraser University.

Anderson, D. D. (1968). A stone age campsite at the gateway to America. *Scientific American*, 218 (6), 24–33.

Anderson, D. D. (1970). Microblade traditions in northwestern Alaska. *Arctic Anthropology*, 7 (2), 2–16.

Anderson, D. D. (1978). Tulaagiaq: a transitional Near-Ipiutak-Ipiutak period archaeological site from Kotzebue Sound. *Anthropological Papers of the University of Alaska*, 19 (1), 45–57.

Anderson, D. D. (1984). Prehistory of north Alaska. In *Handbook of North American Indians, Volume 5: Arctic*. Washington, D.C.: Smithsonian Institution Press, pp. 80–93.

Anderson, D. D. (1988). Onion Portage: the archaeology of a stratified site from the Kobuk River, northwest Alaska. *Anthropological Papers of the University of Alaska,* 22 (1–2), i–163.

Bandi, H.-G. and Blumer, R. (2004). Matériel funéraire exceptionnel du Vieus Béringien découvert récemment près de Gambell. In *Physische Anthropologie und Archäo-Chronologie der S. Lorenz Insel-Urbevölkerung, Alaska*. Bern: Haupt Verlag, pp. 161–6.

Beechey, F. W. (Captain). (1831). *Narrative of a Voyage to the Pacific and Beering's Strait, Volume 1*. London: Henry Colburn and Richard Bentley.

Bever, M. (2006). Too little, too late? The radiocarbon chronology of Alaska and the peopling of the New World. *American Antiquity*, 71 (4), 595–620.

Bockstoce, J. R. (1979). *The Archaeology of Cape Nome, Alaska*. Philadelphia, PA: The University Museum, University of Pennsylvania.

Brower, C. D. (1942). *Fifty Years Below Zero: A Lifetime of Adventure in the Far North*. New York, NY: Dodd Mead and Company.

Brower, C. D. (n.d.). Charles D. Brower manuscripts. Hanover, NH: Dartmouth College Library.

Burch, E. S., Jr. (2005). *Alliance and Conflict: The World System of the Iñupiaq Eskimos*. Lincoln, NE: University of Nebraska Press.

Campbell, J. M. (1959). The Kayuk Complex of Arctic Alaska. *American Antiquity*, 25 (1), 94–105.

Campbell, J. M. (1998). *North Alaska Chronicle: Notes from the End of Time: The Simon Paneak Drawings*. Santa Fe, NM: University of New Mexico Press.

Carter, W. (1953a). *Archaeological Survey of Eskimo, or Earlier, Material in the Vicinity of Point Barrow, Alaska. Final Report: 30 April, 1953*. Report to Office of Naval Research and Harvard University.

Carter, W. (1953b). *Archaeological Survey of Eskimo, or Earlier, Material in the Vicinity of Point Barrow, Alaska. Status Report: 1953 Field Activities*. Report to Office of Naval Research and Arctic Institute of North America.

Carter, W. (1962). *Archaeological Survey of Eskimo, or Earlier, Material in the Vicinity of Point Barrow, Alaska. Status Report: 1 January 1962*. Report to Office of Naval Research and Arctic Institute of North America.

Carter, W. (1966). *Archaeological Survey of Eskimo, or Earlier, Material in the Vicinity of Point Barrow, Alaska. Final Report: 31 January 1966*. Report to Office of Naval Research and Arctic Institute of North America.

Collins, H. B. (1935). Archaeology of the Bering Sea region. In *Annual Report Back to the Smithsonian Institution for 1933*. Washington, D.C.: Smithsonian Institution Press, pp. 453–68.

Collins, H. B. (1937). *Archaeology of St. Lawrence Island, Alaska*. Smithsonian Miscellaneous Collections 96 (1). Washington, D.C.: Smithsonian Institution Press.

Collins, H. B. (1984). History of research before 1945. In *Handbook of North American Indians, Volume 5: Arctic*. Washington, D.C.: Smithsonian Institution Press, pp. 8–16.

Darwent, C. M. (2006). Reassessing the Old Whaling locality at Cape Krusenstern, Alaska. In *Dynamics of Northern Societies: Proceedings of the SILA/NABO Conference on Arctic and North Atlantic Archaeology*. Copenhagen: National Museum of Denmark, pp. 95–102.

Darwent, J. and Darwent, C. (2005). Occupational history of the Old Whaling Site at Cape Krusenstern, Alaska. *Alaska Journal of Anthropology*, 3 (2), 135–54.

Darwent, J., Mason, O. K., Hoffecker, J. F. and Darwent, C. M. (2013). 1,000 years of house change at Cape Espenberg, Alaska: a case study in horizontal stratigraphy. *American Antiquity*, 78 (3), 433–55.

Davis, C. W., Linck, D. C., Schoenberg, K. M. and Shields H. M. (1981). *Slogging, Humping and Mucking through the NPR-A: An Archaeological Interlude.* Five volumes. Occasional Paper 25, Anthropology and Historic Preservation Cooperative Park Studies Unit. Fairbanks, AK: University of Alaska.
Dumond, D. E. and Collins, H. B. (2000). *Henry B. Collins at Wales, Alaska 1936.* University of Oregon Anthropological Papers No. 56. Eugene, OR: Department of Anthropology and the Museum of Natural History, University of Oregon.
Esdale, J. A. (2008). Early maritime traditions in the Bering, Chukchi, and East Siberian Seas. *Arctic Anthropology*, 35 (1), 247–62.
Ford, J. A. (1959). *Eskimo Prehistory in the Vicinity of Point Barrow, Alaska.* Anthropological Papers of the American Museum of Natural History 47:1. New York, NY: American Museum of Natural History.
Franklin, J. (1828). *Narrative of a Second Expedition to the Shores of the Polar Sea in the Years 1825, 1826, and 1827, with Map.* London and Philadelphia, PA: John Murray.
Geist, O. W. and Rainey F. G. (1936). *Archaeological Excavation at Kukulik, St. Lawrence Island, Alaska: Preliminary Report.* University of Alaska Miscellaneous Publication 2. Washington, D.C.: U.S. Government Printing Office.
Gerlach, S. C. and Mason, O. K. (1992). Calibrated radiocarbon dates and cultural interaction in the western Arctic. *Arctic Anthropology*, 29 (1), 54–81.
Giddings, J. L. (1951). The Denbigh Flint Complex. *American Antiquity*, 16 (3), 193–203.
Giddings, J. L. (1952). *The Arctic Woodland Culture of the Kobuk River.* University Monographs. Philadelphia, PA: University Museum.
Giddings, J. L. (1961). Cultural continuities of Eskimos. *American Antiquity*, 27 (2), 155–73.
Giddings, J. L. (1962). Onion Portage and other flint sites of the Kobuk River. *Arctic Anthropology*, 1 (1), 6–27.
Giddings, J. L. (1964). *The Archeology of Cape Denbigh.* Providence, RI: Brown University Press.
Giddings, J. L. (1967). *Ancient Men of the Arctic.* New York, NY: Knopf.
Giddings, J. L. and Anderson, D. D. (1986). *Beach Ridge Archaeology of Cape Krusenstern.* National Park Service Publications in Archaeology 20. Washington, D.C.: National Park Service.
Hall, E. S., Jr. (1974). Archaeological investigations in the Noatak River valley, Summer 1973. *Contributions from the Center for Northern Studies*, No. 1, 460–523.
Hall, E. S., Jr., and Fullerton, L. eds. (1990). *The Utqiaġvik Excavations.* Barrow, AK: The North Slope Borough Commission on Iñupiat History, Language and Culture.
Harritt, R. K. (1994). *Eskimo Prehistory on the Seward Peninsula, Alaska.* Resources Report NPS/ARORCH/CRR-93-21. Anchorage, AK: National Park Service, Alaska Region.
Harritt, R. K. (2004). A preliminary reevaluation of the Punuk-Thule interface at Wales, Alaska. *Arctic Anthropology*, 41 (2), 163–76.

Harritt, R. K. (2010a). *An Ethnohistorical Study of Sledge Island, Alaska: Historical Records Research and Field Report.* Anchorage, AK: U.S. Fish and Wildlife Service, Alaska Region.

Harritt, R. K. (2010b). Variations of late prehistoric houses in coastal Northwest Alaska: a view from Wales. *Arctic Anthropology*, 47 (1), 57–70.

Healy, M. A. (1889). *Report of the Cruise of the Revenue Marine Steamer CORWIN in the Arctic Ocean 1884.* Washington, D.C.: Government Printing Office.

Hoffecker, J. F. and Mason, O. K. (2010a). *Human Response to Climate Change at Cape Espenberg: A.D. 800–1400. Field Investigations at Cape Espenberg, 2010.* Annual Report submitted to the National Park Service, U.S. Department of the Interior.

Hoffecker, J. F. and Mason, O. K. (2010b). Research notes: Cape Espenberg Thule Origins Project. *Alaska Journal of Anthropology*, 8 (2), 143–4.

Hoffecker, J. F. and Mason, O. K. (2011). *Human Response to Climate Change at Cape Espenberg: A.D. 800–1400. Field Investigations at Cape Espenberg, 2011.* Annual Report submitted to the National Park Service, U.S. Department of the Interior.

Hooper, W. H. (1853). *Ten Months Among the Tents of the Tuski, with Incidents of an Arctic Boat Expedition in Search of Sir John Franklin as Far as the Mackenzie River and Cape Bathhurst.* London: John Murray.

Irving, W. (n.d.). *An Archaeological Reconnaissance of the Lower Colville River and Delta Regions.* Report submitted to the Chief of Naval Research.

Jenness, D. (1918). The Eskimos of Northern Alaska: A study in the effect of civilization. *The Geographical Review*, 5 (2) (February), 89–110.

Jenness, D. (1928). Archaeological Investigations in Bering Strait, 1926. Annual Report of the National Museum of Canada for the Fiscal Year 1926. *National Museum of Canada Bulletin*, 50, 71–80.

Jenness, D. (1957). *Dawn in Arctic Alaska.* Minneapolis, MN: University of Minnesota Press.

Jensen, A. M. (1997). *An Archaeological Field Survey in Connection with Proposed Construction at the Ipiutaq Site and Old Town Site, Point Hope, Alaska.* UIC Science Division Technical Report #1. Barrow, AK: UIC Cultural Resources, prepared for LCMF, Inc.

Jensen, A. M. (2007a). *Archaeological and Cultural Resources Reconnaissance for the Proposed Intrepid Prospect, North Slope, Alaska.* Barrow, AK: UIC Science LLC, prepared for ConocoPhillips Alaska, Inc. Anchorage, Alaska.

Jensen, A. M. (2007b). Nuvuk burial 1: an Early Thule hunter of high status. *Alaska Journal of Anthropology*, 5 (1), 119–22.

Jensen, A. M. (2009a). *Nuvuk: Point Barrow, Alaska: The Thule Cemetery and Ipiutak Occupation.* Ph.D. Bryn Mawr College.

Jensen, A. M. (2009b). Radiocarbon dates from recent excavations at Point Barrow, Alaska and their implications for Neoeskimo prehistory. In *On the Track of the Thule Culture From Bering Strait to East Greenland.* Publications from the National Museum, Studies in History and Archaeology 15. Copenhagen: National Museum of Denmark, SILA, pp. 45–62.

Kunz, M. L. and Reanier, R. E. (1994). Paleoindians in Beringia: evidence from Arctic Alaska. *Science*, 263, 660–2.
Kunz, M. L. and Reanier, R. E. (1995). The Mesa Site: a Paleoindian hunting lookout in Arctic Alaska. *Arctic Anthropology*, 32 (1), 5–30.
Kunz, M. L. and Reanier, R. E. (2005). The Denbigh Flint Complex at Punyik Point, Etivlik Lake, Alaska. *Alaska Journal of Anthropology*, 3 (2), 101–15.
Larsen, H. (1968). Trail Creek: Final report on the excavation of two caves on Seward Peninsula, Alaska. *Acta Arctica*, 15, 7–79.
Larsen, H. (2001). *Deering: A Men's House from Seward Peninsula, Alaska.* Copenhagen: National Museum of Denmark.
Larsen, H. and Rainey, F. (1948). *Ipiutak and the Arctic Whale Hunting Culture.* Anthropological Papers of the American Museum of Natural History 42. New York, NY: American Museum of Natural History.
Lobdell, J. E. (1985). *The Putuligayuk River Delta Overlook Site: Fragile Traces of Ancient Man at Prudhoe Bay, Beaufort Sea, Alaska.* Second Edition. Alaska: ARCO Alaska Inc.
Lobdell, J. E. (1986). Kuparuk Pingo Site: A Northern Archaic hunting camp of the Arctic Coastal Plain, North Alaska. *Arctic*, 39 (1), 47–51.
Lobdell, J. E. (1995). North Alaska Pingos: ephemeral refugia in prehistory. *Arctic Anthropology*, 32 (1), 62–81.
Maguire, R. (Captain). (1988). *The Journal of Rochfort Maguire, 1852–1854: Two Years at Point Barrow, Alaska, Aboard HMS Plover in the search for Sir John Franklin.* Works issued by The Hakluyt Society, Second Series No. 169. London: The Hakluyt Society.
Mason, O. K. (1991). Chronological inferences from harpoon and arrowheads from Mound 44 Slump. In *Coastal Erosion and Salvage Archaeology at Utqiagvik, Alaska: The 1990 Excavation of the Mound 44 Slump Block.* Occasional Papers of the Alaska Quaternary Center No. 4. Fairbanks, AK: Alaska Quaternary Center, pp. 64–80.
Mason, O. K. (1998). The contest between Ipiutak, Old Bering Sea and Birnirk polities and the origin of whaling during the First Millennium A.D. along Bering Strait. *Journal of Anthropological Archaeology*, 17, 240–325.
Mason, O. K. (2000). Archaeological Rorschach in delineating Ipiutak, Punuk and Birnirk in NW Alaska: masters, slaves, or partners in trade? In *Identities and Cultural Contacts in the Arctic.* Danish Polar Center Publication No. 8. Copenhagen: Danish Polar Center, pp. 229–51.
Mason, O. K. (2003). *Uivvaq Heritage Project: Field Season 2002.* Final report to Aglaq/CONAM, J.V. No.2 in fulfillment of Contract #2103–004, Cape Lisburne Clean Sweep Remedial Action. Anchorage, AK: GeoArch Alaska.
Mason, O. K. (2006). Ipiutak remains mysterious: a focal place still out of focus. In *Dynamics of Northern Societies: Proceedings of the SILA/NABO Conference on Arctic and North Atlantic Archaeology.* Copenhagen: National Museum of Denmark, pp. 103–19.
Mason, O. K. (2009). Flight from Bering Strait: did Siberian Punuk/Thule military cadres conquer Alaska? In *The Northern World A.D. 900–1400.* Salt Lake City, UT: University of Utah Press, pp. 76–127.

Mason, O. K., Gerlach, S. C. and Ludwig, S. L. (eds.) (1991). *Coastal Erosion and Salvage Archaeology at Utqiagvik, Alaska: The 1990 Excavation of the Mound 44 Slump Block*. Occasional Papers of the Alaska Quaternary Center No. 4. Fairbanks, AK: Alaska Quaternary Center.

Mason, O. K., Ganley, M. L., Sweeney, M. A., Alix, C. and Barber, V. (2007). *An Ipiutak Outlier: A 1,500-Year Old Qargi at Qitchauvik on the Golovin Lagoon, the Golovin Heritage Field School, 1998–2000*. NPS Technical Report Number: NPS/AR/CRR/2007–67. Shared Beringian Heritage Program. Anchorage, AK: National Park Service, Alaska Region.

Mathiassen, T. (1927). *Archaeology of the Central Eskimos*. Report of the 5th Thule Expedition, 1921–1924, Volume 4 (1). Copenhagen: Reitzels.

Mathiassen, T. (1930). *Archaeological Collections from the Western Eskimos*. Report of the 5th Thule Expedition, 1921–1924, Volume 4 (1). Copenhagen: Reitzels.

Maxwell, M. S. (1984). Pre-Dorset and Dorset Prehistory of Canada. In *Handbook of North American Indians, Volume 5: Arctic*. Washington, D.C.: Smithsonian Institution Press, pp. 359–68.

Maxwell, M. S. (1985). *Prehistory of the Eastern Arctic*. New York, NY: Academic Press.

Mills, R., Gerlach, S. G., Bowers, P. M. and MacIntosh, S. J. (1999). *Final Report to the Cultural Resources Mitigation of the 1998 Anaktuvuk Pass Runway Realignment Project*. NLUR: report prepared for LCMF, Inc.

Moore, T. E. L. (1851). General Proceedings of Cmdr. T. E. L. Moore, HMS *Plover*, Sept. 1949–Sept. 1950. *British Blue Books*, 33, 28–40.

Murdoch, J. (1892). Ethnological results of the Point Barrow Expedition. In *Ninth Annual Report of the Bureau of Ethnology to the Secretary of the Smithsonian Institution 1887–88*. Washington, D.C.: U.S. Government Printing Office.

Pálsson, G. (2001). *Writing on Ice: The Ethnographic Notebooks of Vilhjálmur Steffansson*. Hartford: University Press of New England.

Rainey, F. G. (1941). The Ipiutak culture at Point Hope, Alaska. *American Anthropologist*, 43 (3), 364–75.

Rainey, F. G. (1942). Discovering Alaska's oldest arctic town. *The National Geographic Magazine*, 82 (3), 318–26.

Ralph, E. K. and Ackerman, R. E. (1961). University of Pennsylvania Radiocarbon Dates IV. *Radiocarbon*, 3, 4–14.

Rasic, J. T. and Slobodina, N. S. (2008). Weapon systems and assemblage variability during the Northern Archaic Period in Northern Alaska. *Arctic Anthropology*, 45 (1), 71–88.

Rasmussen, K. (1952). The Alaskan Eskimos: as described in the posthumous notes of Dr. Knud Rasmussen. In *Report of the 5th Thule Expedition, 1921–1924 X(3)*. Copenhagen: Gyldendalske Boghandel, Nordisk Forlag.

Ray, D. J. (1975). *The Eskimos of Bering Strait, 1650–1898*. Seattle, WA: University of Washington Press.

Ray, P. H. (Lieutenant). (1885). *International Polar Expedition to Point Barrow, Alaska*. Senate Committee on Military Affairs. Washington, D.C.: U.S. Government Printing Office.

Reanier, R. E. (1992). *Refinements to K-means Clustering: Spatial Analysis of the Bateman Site, Arctic Alaska.* Ph.D. University of Washington.

Reanier, R. E. (1995). The antiquity of Paleoindian materials in Northern Alaska. *Arctic Anthropology*, 32 (1), 31–50.

Reanier, R. E., Sheehan, G. W. and Jensen, A. M. (1998). *Report of 1997 Field Discoveries: City of Deering Village Safe Water Cultural Resources Project.* UIC Science Division Technical Report #3.

Reynolds, G. L. (1990). Harpoon heads in the Utqiaġvik Collection. In *The Utqiaġvik Excavations. 1.* Barrow, AK: The North Slope Borough Commission on Iñupiat History, Language and Culture, pp. 37–359.

Savelle, J. M. and McCartney, A. P. (2003). Prehistoric bowhead whaling in the Bering Strait and Chukchi Sea regions of Alaska: a zooarchaeological assessment. In *Indigenous Ways to the Present: Native Whaling in the Western Arctic.* Edmonton: Canadian Circumpolar Institute Press, pp. 167–84.

Sheehan, G. W. (1985). Whaling as an organizing focus in Northwestern Alaskan Eskimo societies. In *Prehistoric Hunter-Gatherers: The Emergence of Cultural Complexity*. New York, NY: Academic Press, pp. 123–54.

Sheehan, G. W. (1990). Excavations at Mound 34. *The Utqiaġvik Excavations. 2.* Barrow, AK: The North Slope Borough Commission on Iñupiat History, Language and Culture, pp. 181–353.

Sheehan, G. W. (1997). *In the Belly of the Whale: Trade and War in Eskimo Society. Aurora, Alaska.* Anthropological Association Monograph Series VI. Anchorage, AK: Alaska Anthropological Association.

Sheehan, G. W., Reinhardt, G. A. and Jensen, A. M. (1991). *Pingasagruk: A Prehistoric Whaling Village on Point Franklin, Alaska.* Report prepared for Alaska Maritime National Wildlife Refuge, Alaska. Barrow, AK: SJS Archaeological Services. Report is available from UIC Science Division, Barrow, AK.

Simpson, J. (1855). Observations on the Western Esquimaux and the country they inhabit; from notes taken during two years at Point Barrow, by Mr. John Simpson, Surgeon, R. N., Her Majesty's Discovery Ship "Plover". Great Britain, Parliament, House of Commons. *Sessional Papers, Accounts and Papers 1854–55*, 35 (1898), pp. 917–42.

Simpson, T. (1843). *Narrative of the Discoveries on the North Coast of America, Effected by the Officers of the Hudson's Bay Company During the Years 1836–1839.* London: Richard Bentley.

Solecki, R. S. (1950). A preliminary report of an archaeological reconnaissance of the Kukpowruk and Kokolik Rivers in Northwest Alaska. *American Antiquity*, 16 (1), 66–9.

Spencer, R. F. (1959). *The North Alaskan Eskimo: A Study in Ecology and Society.* Bureau of American Ethnology Bulletin 171. Washington, D.C.: U.S. Government Printing Office.

Stanford, D. J. (1976). *The Walakpa Site, Alaska: Its Place in the Birnirk and Thule Cultures.* Smithsonian Contributions to Anthropology 20. Washington, D.C.: Smithsonian Institution Press.

Stefánsson, V. (1913). *My Life with the Eskimo*. New York, NY: Macmillan.
Stefánsson, V. (1914). *The Stefansson-Anderson Arctic Expedition of the American Museum: Preliminary Ethnological Report*. Anthropological Papers of the American Museum of Natural History 14 (1). New York, NY: American Museum of Natural History.
Thompson, R. M. (1948). Notes on the archaeology of the Utukuk River, Northwestern Alaska. *American Antiquity*, 14 (1), 62–5.
Time magazine (1941). Arctic metropolis. *Time*, XXXVII, 11, 58–9 [issue of March 17].
Time magazine (1948). Diggers. *Time*, LII, 16, 70–2 [issue of October 18].
United States Treasury Department (1899). *Report of the Cruise of the U.S. Revenue Cutter Bear and the Overland Expedition for the Relief of the Whalers in the Arctic Ocean from November 27, 1897, to September 13, 1898*. Washington, D.C.: U.S. Government Printing Office.
Van Stone, J. W. (1955). Archaeological excavations at Koztebue, Alaska. *Anthropological Papers of the University of Alaska*, 3 (2), 75–155.
Van Stone, J. W. (1977). *A.F. Kashevarov's Coastal Explorations in Northwest Alaska, 1838*. Fieldiana Anthropology 69. Chicago, IL: Field Museum of Natural History.
Van Valin, W. B. (1941) *Eskimoland Speaks*. Caldwell, ID: The Caxton Printers.
Van Valin, W. B. (n.d.) Manuscript on file. Philadelphia, PA: University Museum of the University of Pennsylvania.
Wissler, C. (1916). *Harpoons and Darts in the Stefansson Collection*. Anthropological Papers of the American Museum of Natural History 14 (2). New York, NY: American Museum of Natural History, pp. 397–443.
Zimmerman, M., Jensen, A. M. and Sheehan, G. W. (2000). Aġnaiyaaq: the autopsy of a frozen Thule mummy. *Arctic Anthropology*, 37 (2), 52–9.

3 *The Ipiutak cult of shamans and its warrior protectors: An archaeological context*

OWEN K. MASON

The specter of skulls with bulging ivory eyes (Figure 3.1) and prophylactic mouth shields (Rainey, 1941, 1942: 318; Giddings, 1967: 119)[1] has hovered over the archaeology of Point Hope for over seventy years. From this disturbing image, the Ipiutak mortuary cult at Point Hope retains its allure and mystique, yielding readily to hyperbole: Its discovery recalled, to J. Louis Giddings (1967: 102), the opening of the fabled 3,500-year-old tomb of Tutankhamun. The facts from the ground have suffered tremendous neglect in the face of the Ipiutak narrative constructed by one of its discoverers (Rainey, 1941). Windswept and austere, Point Hope, 300 kilometers north of the Arctic Circle, is a low, treeless gravel spit thrust into the perennially ice-covered Chukchi Sea. The discovery of hundreds of abandoned houses lined up on shore-parallel beach ridges suggested to Rainey (1941, 1942) the contours of a single deliberate community, an "arctic metropolis" (*Time*, 1941), the homes of thousands of Ipiutak people. Did several thousand people live at Point Hope over 2,000 years ago? And how could such a considerable populace obtain sufficient wood for fuel, building, subsistence foods and even water? Although the "city" model was disavowed in the final report (Larsen and Rainey, 1948: 47; Larsen, 1952: 23), Rainey (1971: 7) remained wedded to the possibility thirty years later and many in the local community still hold to it. The narrative produced in the flush of discovery serves as a cautionary tale for archaeologists, admitting the theoretical and methodological advances unavailable in the 1940s. The discovery of graves with ivory eyes resonated within the oral history of local residents (Giddings, 1967: 118) and the site was christened *Ipiutak*, anglicizing its innocuous, descriptive place name *Ipiutaq*: "strip of land" (Burch, 1981: 71).

The Foragers of Point Hope: The Biology and Archaeology of Humans on the Edge of the Alaskan Arctic, eds. C. E. Hilton, B. M. Auerbach, L. W. Cowgill. Published by Cambridge University Press.

Figure 3.1. Close up photograph of Ipiutak Burial 51, showing the inlaid ivory eyes within the orbits of the cranium. The image is reproduced from Larsen and Rainey (1948) from their Plate 98 with permission of the American Museum of Natural History.

Glowing accounts of the immense scale of the late prehistoric (1800 CE) Tikiġaq settlement had originally induced the Danish–American expeditionary effort of 1939 (Rainey, 1941: 364; Larsen and Rainey, 1948: 5). The Danish explorer Knud Rasmussen (Ostermann, 1952: 47) had pronounced Tikiġaq "... incontestably the largest ruined Eskimo town in existence. The old settlement, now deserted, consists of 122 very large houses. But as the sea is constantly eating into the spit... this figure is no indication of how large the settlement used to be." In theory, 600 people lived at Tikiġaq *c.* 1800 CE as estimated by Burch (1981: 44), occupying seventy-five structures.[2] The number of human remains at Point Hope was far greater by orders of magnitude: the Episcopal Reverend Thomas claimed to have re-interred over 4,000 crania in just two years, many of whom were "chiefs... interred on top of the upright jaws of the whales they once had killed" (Burch, 1981). Nearly 1,000 whale mandibles that presently surround the Point Hope cemetery were re-assembled in the 1920s by the Reverend Hoare (Lowenstein, 2008: 309ff). Certainly, Larsen and Rainey were justified in their great expectations about Point Hope.

Despite the magnitude of their expectations, the first archaeology of Ipiutak at Point Hope was adventitious; although interested in the *longue durée* of Inuit origins (Larsen, 1952: 27ff),[3] the Larsen and Rainey team was small, three or four archaeologists, unprepared for the magnitude of the discovery. Not until 1941 did the expedition secure assistance from the Civilian Conservation Corps and field a crew of up to thirty "able bodied" men (Larsen and Rainey, 1948: 17). The entirety of the 575 houses were mapped in a mere six days using a plane table – a feat that still limits its cartographic reliability (Larsen and Rainey, 1948: 16); no one has yet superimposed, using GIS, the 1940 map (reproduced at the beginning of this edited volume on page xvi) atop aerial photographs subsequently available. Visiting the site in 1966, Ed Hosley (1967: 13) counted thirteen additional houses, mostly on the oldest ridge despite twenty-five years of erosion at the site's margin.[4] The discovery of the Ipiutak burial grounds was similarly problematic; local residents were employed as a shovel-wielding posse, placing thousands of shovel probes across the spit, lured by "prize money" (Larsen and Rainey, 1948: 17) – the "stimulus" of three U.S. dollars per grave (Rainey, 1942: 320) and a bonus of up to fifteen U.S. dollars for ivory eye burials (Kowanna in Hall, 1990: 38).[5] Discovering graves required, on occasion, digging up to one-and-a-half meters below the surface (Kowanna in Hall, 1990: 38). Fortunately, the note-keeping was, for the period, of good quality (Mason, 2010; notes on Larsen field books archived in the Danish National Museum). In all, over seventy houses and 176 burials were uncovered by Larsen and Rainey (1948: 15–17) – with 60% of the burials classified as Ipiutak, despite the circumstance that 20% lacked human bones.[6] Although one might expect that more graves lie undiscovered, a program of extensive shovel testing (Hall, 1990), conducted by local assistants in the late 1980s, failed to disclose any graves near Point Hope, and the extensive gravel pad underlying the relocated village (cf. Hosley, 1967) shelters over two square kilometers from further scrutiny.

Ipiutak mobiliary art presents a Rorshach-blot canvas that has offered prehistorians myriad speculations: the carved walrus bones are pregnant with the visages of polar bears, of loons, of seals and the larvae of botflies. A shamanistic use is likely (Larsen, 1952: 26) for the numerous ($n = 200$) complex and abstract open work carvings that resemble metal work (Larsen and Rainey, 1948: 119, 131ff), termed "impractical objects" by Newton (2002: 46). Upon its discovery, the Ipiutak phenomenon was deconstructed, unraveled in two opposite directions, one far to the west toward the horse-riding plains of Central Asia, to trace its intellectual shamanistic origins in Scytho-Siberian animal art (Bunker *et al.*, 1970), the other to the east into the boreal forest to trace its material culture, particularly its use of birch bark and a predilection for burning wood instead of seal oil (Larsen, 2001). Ipiutak is one of the rare archaeological phenomena to be defined by its absences, most prominently pottery and the oil lamp (Anderson, 1988: 123ff, Mason, n.d. a), although many of the typical Paleoeskimo practices were lacking (Collins, 1964: 103).

The distribution of Ipiutak sites: Eighty years of discovery

As noted in Chapter 2 by Jensen, following Ipiutak's discovery at Point Hope in 1940, over a dozen significant Ipiutak localities were identified from the shores of the Chukchi Sea across the Brooks Range, a distribution that only partially reflects the advantages of helicopter survey in the 1960s and 1970s. Despite aerial surveys, the coast remains the focal area for Ipiutak. The summer of 1948 produced tantalizing evidence of Ipiutak settlement across Seward Peninsula, uncovering a birch-bark-covered *qargi* at Deering (Larsen, 2001), Trail Creek Caves (Larsen, 1968b), and Point Spencer (Larsen, 1979/80). The second densest occupation was uncovered in the early 1960s at Cape Krusenstern by Giddings and Anderson (1986). A relationship with migrating caribou was established in the early 1960s, primarily at Feniak and Desparation Lakes (Irving, 1962), at Onion Portage (Anderson, 1988) and at Anaktuvuk Pass by Campbell (1962: 47ff, Plate 5) and by Mills *et al.* (2005). Cultural resource surveys in the 1970s and 1980s produced only a handful of localities across the Brooks Range, typically situated on small lakes in the northern foothills, Croxton (Gerlach, 1989) and Itkillik (Reanier, 1992). Only a single site, Hahanudan Lake, discovered in 1971, occurs wholly in the boreal forest (Clark, 1977), although Onion Portage, discovered in 1940, lies at the tree line along the Kobuk River.

Discoveries across northwest Alaska in the last fifteen years have considerably refined the universe of Ipiutak archaeology; these events were further catalyzed by the publication of Larsen's (2001) monograph on the *qargi* at Deering. The major discovery ("world class archaeology," according to Glenn Sheehan) occurred at Deering in 1997, with the uncovering of a small cemetery precinct in association with cultural resource mitigation during a water and sewer project (Steinacher, 1998). In iconography, the ornately decorated grave goods from Deering rival those from Point Hope (Bowers *et al.*, 2009; Mason, 2009c). Since 1985, four coastal localities have produced evidence of small settlements: Cape Espenberg (found in 1988), Qitchauvik, near Golovin (found in 1998), at Kivalina (found in 2005), and at Nuvuk (found in 2006). At Cape Espenberg, several houses occur for which lithic traces were first encountered in 1960 by Giddings and Anderson (1986: 24), were relocated and tested in 1988 by Harritt (1994: 113ff), and further recorded by Hoffecker *et al.* (2010). Two locations present range extensions (see also Jensen, Chapter 2): at the south at Qitchauvik within Golovnin Bay on Norton Sound (Mason *et al.*, 2007) and to the north, at Nuvuk, past Point Barrow on the Beaufort Sea. The Nuvuk locality lies several hundred kilometers beyond Point Hope (Jensen, 2009). The Kivalina discovery is among the dozen graves discovered in the last twenty years, and contains part of a structure, excavated to mitigate construction impact (Stern *et al.*, 2010).

The place of Ipiutak in Alaska prehistory: Origins and fate

Many archaeologists have mused that without its unique mortuary assemblage, or its bone technology, Ipiutak would offer far less insight into prehistory. To trace its origins, its predecessors only offer stone tools, so in its deep ancestry Ipiutak is viewed from its distinctive flint-knapping, which "is strongly reminiscent of the [microlithic] Arctic Small Tool tradition" (Dumond, 1977: 103; see discussion in Giddings, 1964: 226ff), "in large part a continuation from the [5,000] year old Denbigh Flint Complex [DFC]" (Collins, 1964: 89). Ignoring its ideological and aceramic idiosyncracies, Ipiutak is unanimously placed within a post DFC technological continuum from Choris to Norton (Willey, 1966: 430ff). The Choris–Norton–Ipiutak societies (Anderson, 1983: 88) were the progenitors of the Yup'ik societies of the Yukon Delta – a position argued by Collins (1964: 91), but which is obscured by Dumond's (1977) inclusive use of the "Thule tradition" to describe the cultural entirety of the last millennium.

Between the DFC and Choris, a 500-year "break" occurs "everywhere" in Alaska (Dumond, 1978: 64), so that the development of Choris and Norton is uncertain – did it occur in southern or northern Alaska? If Ipiutak indeed has its origin in the widespread Norton culture (Giddings, 1964; Larsen, 1982; Dumond, 1987, 2000), the process of inventing Ipiutak occurred either on the way to Point Hope – or indeed at Point Hope – or one of the other Ipiutak sites. The artistic and intellectual culture of Ipiutak bears close resemblance to the Old Bering Sea (OBS) mentality (Collins, 1937) that prevailed along Bering Strait – which implies the effects of conversion, political conquest or alliance (Mason, 1998). Very likely, Ipiutak arose *in situ* at Point Hope from the Norton "Near Ipiutak" community (Larsen, 1982), attracted to a prime walrus location, like many other OBS polities across Bering Strait huddled near walrus haul-outs (Hill, 2011; Mason, n.d. b). The relationship of Ipiutak with younger cultures is one of profound cultural and, likely, genetic disjuncture, especially with those of the Northern Maritime tradition, particularly Birnirk and Thule (Collins, 1964: 98ff). The fate and relationship of Ipiutak to younger Alaska cultures is discussed more fully later; suffice it to say that Mason (1998, 2000; Mason and Bowers, 2009) has argued for population replacement, with Birnirk and Thule peoples entering an empty niche and a barren landscape.

The prehistory of Point Hope: An **in situ** *origin?*

From Near Ipiutak to Ipiutak, 2,200 years BP to 1,550 years BP

The earliest archaeological evidence at Point Hope is thin and ambiguous, consisting of only the single undated, H24 house presumed to be early (Larsen and Rainey, 1948: 165, Plate 83) and two dated small camps of the ceramic-using Norton culture (Dumond, 1982), confusingly termed "Near Ipiutak" (Larsen, 1982), that was possibly as early as 2,200 years BP to 1,750 years BP (Larsen, 1968a: 82–3). Despite the sparse domestic evidence from this occupation, the community was probably much larger, considering the survival of thirty middens and twenty-three graves across 500 meters on the oldest beach ridge at Point Hope (Larsen and Rainey, 1948: 225ff). Near Ipiutak, and Ipiutak as well, shares a considerable part of its inventory with the Norton culture (Larsen, 1982), widespread from the Yukon Delta to the southern Alaska Peninsula (Dumond, 1987, 2000; Maschner, 2008). Especially noteworthy are its lithic technology, its use of labrets, the atlatl, the brow band, its house construction, and its organizational imperative in the men's house (*qargi*). The "many animal bones" in House 24 indicated to Larsen and Rainey (1948: 165) that seal, walrus, and caribou served as the basis for subsistence, with salmon spears reflecting fishing. More controversial are two whaling harpoon heads from two burials (B83, B85) that led Larsen and Rainey (1948: 163, Plate 79:1,2) to affirm "beyond a doubt" that Near Ipiutak people were whalers. The two graves lack age estimates, however, and conceivably could be Ipiutak. However, so few whale bones were recovered that the degree of whaling, if any, was only a "casual interest" in Harritt's (1995: 36) view. The implications of Near Ipiutak or Ipiutak whaling remain imperfectly documented – or dated – but such a surfeit of food had to have social consequences, as well as implications for the expertise and the organization embedded in the culture. The loss of the whaling imperative distinguishes Near Ipiutak from Ipiutak, and begs for a fuller consideration by archaeologists.

Grave goods from the Near Ipiutak multiple and single burials may be argued to for an *in situ* development of Norton into Ipiutak – a proposition that remains only hypothetical. For one, the taphonomy of the burials argues against primary context: objects range vertically more than one meter, and the sparsity of bone indicates scavenging or secondary interment (Larsen and Rainey, 1948: 251). Two disturbed graves, B98 and JB7, have early dates, between 2,200 years BP and 1,860 years BP (Newton, 2002: 99), based on ^{14}C ages on antler open work carvings with "Ipiutak" motifs (*sensu* Larsen and Rainey, 1948: 245, 251; Plates 59, 66–68). If the graves were associated with the Near Ipiutak camps, then conceivably Norton shamans or artisans prefigured or even developed the Ipiutak cult. Two graves (B102 and B134) of "uncertain type" (Larsen and Rainey, 1948: 102, 250) may also reflect a pre-Ipiutak presence on the Tikiġaq spit in the first centuries CE (Newton, 2002: 99). Nevertheless, the size and scale of the Norton occupation at Tikiġaq remains uncertain; the possible presence of shamans, dedicated cemetery precincts, and casual whaling are powerful arguments for sedentism at Tikiġaq prior to the florescence of Ipiutak elsewhere. Nonetheless, the presence of outsiders remains a major possibility in view of the seemingly major technological shift away from oil lamps and the use of ceramics that delineate Ipiutak from Near Ipiutak Norton. The occurrence of walrus with the earliest occupations at Point Hope is part of a region-wide pattern (Hill, 2011; Mason, n.d. a).

The ultimate origins of Ipiutak mortuary practices may be far to the south within Cook Inlet or on Kodiak Island, considering the idiosyncratic practice of artificial eye-coverings in Kachemak III burials (de Laguna, 1934: 42ff; Workman, 1992: 23), "exotic burial ceremonialism" which occasioned Workman (1982) to speculate on southern affinities. Workman's (1982: 110) comments remain relevant: ". . . some of the complex late Kachemak tradition [2,350 years BP to 1,750 years BP] and Ipiutak ceremonialism are linked. Since direct contact is scarcely likely, a link through as yet poorly documented Norton practices seems [more] likely." In the last thirty years, still no definitively Norton graveyard has surfaced in southern Alaska, due, possibly, to a lack of archaeological effort. Distinctive stone points offer another line of evidence for southern affinities; as re-iterated by Maschner (2008: 174), affirming an original insight linking Ipiutak points with nearly identical "fishtail" points common in lower Cook Inlet at Chirikof Island. The widespread appearance of the weapons system employed for warfare occurs between 2,350 years BP and 1,850 years BP (Maschner, 2008: 175), linked to the development of the Kachemak burial complex (Maschner and Reedy-Maschner, 1998: 36). The southern origin model is supported by temporal priority, if not one of immediate geographic proximity. However, the precise cultural linkages across western Alaska remain unarticulated due to a dearth of archaeological discoveries in the Yukon Delta and Norton Sound. The "ubiquity" of warfare across the entire North Pacific basin should not be underestimated as a pan-regional evolutionary driver (Maschner and Reedy-Maschner, 1998).

Spread of classic Ipiutak, 1,550 years BP to 1,100 years BP

In the first centuries CE, nearly contemporaneous with the OBS culture on St. Lawrence Island (Collins, 1937) and along the Chukotka coast (Mason, 1998, n.d. c; Aruitunov and Sergeev, 2006a, b), the Ipiutak cult arose on the Alaskan mainland, possibly as an ally or rival of OBS for temporal and sacral power in Bering Strait, seeking access to highly valued trade items such as iron and obsidian (Mason, 1998). The Ipiutak cult had several epicenters, representing shamans and communities that interacted within a common *zeitgeist* (Mason, 1998, 2006, 2009b, c, d). While coastal settlements were large and dominated by houses, the Ipiutak phenomenon extended across the Brooks Range, focusing on single large structures interpreted as *qargi*, men's or community structures (e.g., Feniak Lake [Hall, 1974]). Isolated tent rings with a small amount of lithic debitage are known from major caribou crossings (Anaktuvuk Pass [Mills *et al.*, 2005]) or along lake margins (Croxton [Speiss, 1979; Mason and Gerlach, 1995]; Bateman [Reanier, 1992]). The middle Kobuk treeline site of Onion Portage was exceptional in that a notable caribou crossing co-occurred with at least two winter houses and fifty-eight hearths related to temporary fall camps (Anderson, 1988: 113ff).

The large Ipiutak settlement at Point Hope lies toward, if not at, its northern extreme, its size biases our conception of Ipiutak paleoeconomy. In a critical sense, the Ipiutak center of gravity, in demographic terms, was equally within Kotzebue Sound and to the south and into the mountains. The Ipiutak settlement at Kivalina is the closest to Point Hope, less than 120 kilometers southeast, but is known only from a single grave and part of a structure (Stern *et al.*, 2010). Farther south, Ipiutak occupation was notably dense within Kotzebue Sound and across Seward Peninsula, reported at Cape Krusenstern (Giddings and Anderson, 1986), Cape Espenberg (Harritt, 1994; Mason *et al.*, 2007), Kotzebue (Shinabarger, 2009, personal communication), Deering (Larsen, 2001; Bowers, 2009), Point Spencer (Larsen, 1979/80), and within the Trail Creek Caves (Larsen, 1968b). Sites at its farthest extremes, north and south, show a wide contrast: To the south, the Qitchauvik *qargi* along Norton Sound has a sizable inventory (Mason *et al.*, 2007) while the far north limit, a storm-collapsed driftwood structure at Nuvuk (Point Barrow) consists only of a handful of objects with Ipiutak characteristics, e.g., linear motifs on bone arrow points and in sled runner construction (Jensen, 2009).

Most Ipiutak sites were probably little more than homesteads or brief campsites; only at two places, Point Hope and Cape Krusenstern, does Ipiutak reveal the depth and breadth that establishes it as an equal to the OBS culture (Collins, 1937; Arutiunov and Sergeev, 2006a, b; Mason, n.d. a) which straddled Bering Strait. The several villages at Cape Krusenstern and at Point Hope had such considerable intellectual and material significance for the Ipiutak world, considering the profound similitude between motifs at far-distant sites. A most expeditious conclusion is that distant communities participated within a wider cultural system bound together by trade, war, and cosmology (Mason, 1998).

Architecture and domestic activities

Establishing Ipiutak chronology is closely associated with documenting the development of its architecture and in reconstructing its population history. The architecture of Ipiutak, by Alaska standards, is exceedingly well documented, with abundant collections (albeit imprecisely excavated in 1940–1) from nearly a hundred excavated houses and other structures. The sedentary signature of Ipiutak architecture at Point Hope is fairly consistent: Single room houses were built above ground, with many containing short entryways, likely serving as winter vestibules. Ipiutak houses at Point Hope employ stacked driftwood logs to form a single room that varies around a mean of sixteen square meters, with a quarter (25%) that are larger, between twenty-five and thirty square meters, but with just as many that are smaller than sixteen square meters (Mason, 1998: 27). Many houses contain precious few artifacts, as at Cape Espenberg (Harritt, 1994: 111ff), Hahanudan Lake (Clark, 1977), as within the house excavated by Jensen (1997) at Point Hope, and at Deering (Bowers, 2009).

Variability in Ipiutak house sizes presumably reflects household composition and/or comparative social status. The inventories within the excavated Ipiutak houses at Point Hope were so variable that residential craft specialization in ivory working and clothing manufacture was inferred by Mason (2006, 2009a), based on skewed representation of needles, ivory debitage, and stone points. High numbers of arrow points, stockpiled within houses, also reflect the critical need for defense from intruders (Mason, 2006), not the caribou hunting inferred by Larsen and Rainey (1948: 66–7). At Cape Krusenstern the excavation of twenty square wood-heated central-hearth driftwood houses produced lithics (scrapers, discoids), a few decorated pieces, and eighty-five incised pebbles with enigmatic designs, most in the House 40 *qargi* (Giddings and Anderson, 1986: 140–1). One, possibly typical, short-term occupation is the Ipiutak house at Deering excavated in 1999, which covered nine to twelve square meters, was outlined by narrow timbers, and had a short arctic entry (Bowers, 2009). The Deering structure contained only a small inventory of objects, some with definitively Ipiutak motifs and lithics that included discoidal scrapers and finely wrought end blades (Reuther, 2009b). The most prominent artifacts included a seal-decorated iron-inset engraver and a basally bi-spurred sealing dart (Bowers, 2009). At Cape Espenberg, a bark-floored winter house with a central hearth contained only rather sparse lithic assemblages

($n = 53$), with several structural wood posts but no organic artifacts (Harritt, 1994: 111ff).

The culture also offers a previously unparalleled range of other features, in caches, summer and winter houses, and community structures (i.e., *qargi*). While cache pits were not documented at Point Hope by Larsen and Rainey (1948), the pits were relatively common at Cape Krusenstern, and may co-occur with summer camp structures of ephemeral construction (Giddings and Anderson, 1986: 146). The contents of a small pit, presumed a cache pit, at Hahanudan Lake contained but few bones and little else (Clark, 1977). Architectural variability is best documented from Cape Krusenstern (Giddings and Anderson, 1986) and from two structures meticulously excavated at Deering (Larsen, 2001; Bowers, 2009). Of the three types of Ipiutak house at Cape Krusenstern (Giddings and Anderson, 1986: 119), the most substantial was excavated nearly one meter subsurface, and was sided with horizontal logs and covered with sod. Many, if not most, have short entries presumably indicative of winter use. The storm-destroyed fifth-century CE structure at Nuvuk seems to be a temporary shelter – a lean-to – in a brief camp (Jensen, 2009).

Chronology and settlement pattern

Unfortunately, the chronology of Ipiutak house construction or mortuary practices is not well authenticated; very few Ipiutak houses or graves are even dated, let alone reliably dated (e.g., using absolute dating methods). Further, the coastal and inland manifestations of Ipiutak differ in the quantity and reliability of data (cf. Gal, 1982; Giddings and Anderson, 1986; Gerlach and Mason, 1992; Harritt, 1994; Larsen, 2001; Newton, 2002; Mason, 2006; Mason *et al.*, 2007). On the coast, the approximately fifty ^{14}C assays are dominated by the dozen from Deering (Larsen, 2001; Reuther and Bowers, 2009). The two largest settlements, Point Hope and Cape Krusenstern, have chronologies that are very thin (less than ten dates each) and rely on ^{14}C assays run over fifty years ago, except for eight assays more recently obtained from Point Hope on archived Danish National Museum samples from the 1940s excavations (Newton, 2002). The original five Point Hope radiocarbon ages are from only five houses; these range from 1,520 to 1,260 years BP from antler within House 50 on the oldest ridge (Ridge E), to 1,342 to 1,057 years BP from wood in House 3 on Ridge C (Mason 2006: 105), overlapping within the seventh century CE.

Point Hope remains the largest known Ipiutak locality: about 600 closely spaced single-room houses cluster tightly on the four earliest beach ridges (for its geomorphology, see Hosley, 1967; Shepard and Wanless, 1971). The amalgamation of houses is more properly considered a *series* of Ipiutak sites at a critical resource nexus, most prominently, a walrus haul-out. None of the houses apparently overlie other houses, a pattern that led to the extreme model of a single large town, popularized by Rainey (1941, 1942, 1971).[7] The spatial array of houses precludes a single settlement since the packing of houses was so tight and some houses had entryways that conflicted with adjacent structures (Larsen and Rainey, 1948: 44). This density suggested to Larsen (1952: 23) that Point Hope served only as a temporary "spring and summer" encampment – with inland winter residence; however, such a comparably sized interior winter occupation cannot be established. An even more minimalist, and arguably fanciful, approach was that of McGhee (1976), who proposed that *all* of the houses and graves resulted from repeated generations of only two families, who were also highly motivated to produce art.

Outside of Point Hope, Ipiutak may have commenced earlier, between 1,700 years BP and 1,350 years BP, based on the recently obtained ^{14}C ages from Nuvuk (Point Barrow [Jensen, 2009]), Kivalina (Stern *et al.*, 2010), Qitchauvik (Mason *et al.*, 2007), and those from Cape Krusenstern (Giddings and Anderson, 1986: 30). The collapsed structure at Nuvuk is not definitively a winter house, and at Kivalina only a single grave, only slightly older (see below) is firmly authenticated. The domestic arrangements at Cape Krusenstern offer more support for sedentism. The Deering Ipiutak cache also produced evidence of an earlier use, possibly *c.* 1,700 years BP (Reuther and Bowers, 2009). Equivocal evidence of earlier Ipiutak settlement, prior to 1,750 years BP, occurs in the western Brooks Range at Feniak and Desperation Lakes (Gal, 1982); the former consists of a purported *qargi* hastily excavated in the early 1970s by Hall (1974). The earliness of the lakeside occupations may be a function of definition ("What is Ipiutak?"), its progenitors, and of the inherent imprecision of ^{14}C ages. The definition of Ipiutak is related to the elaborate mortuary complex that only developed late in the culture. Nonetheless, most Ipiutak sites were occupied between 1,350 years BP and 1,050 years BP (Gerlach and Mason, 1992; Mason, 2000, 2006, 2009d), with no ages following that date, a circumstance that suggests a sudden and dire end to the culture.

The ruins of other Ipiutak communities are orders of magnitude smaller than that at Point Hope; only that of Cape Krusenstern recorded sizable settlements where seventy houses occur across five ridges within eight discrete groupings that suggest villages (Giddings and Anderson, 1986: 118). The site was dated by only seven assays from five houses.[8] The Krusenstern Ipiutak occupation may have been earlier than that at Point Hope, possibly as early as the second century CE (1,820 years BP to 1,520 years BP, at House 60). However, for the most part, Ipiutak thrived at Krusenstern between 1,450 years BP and 1,250 years BP, as at Houses 11 and 17, while House 18 was occupied after 1,250 years BP (Giddings and Anderson, 1986: 30). The largest Ipiutak structure at Cape Krusenstern, House 17, covered forty-two square meters and had a central hearth. Twice as large as other houses, House 17 possibly served as a *qargi* for whaling crews, if one accepts the association of a harpoon head (Giddings and Anderson, 1986: 123, 150ff). Each discrete Krusenstern village had between five and fourteen houses and the entire beach ridge complex was possibly "more heavily populated than at any other time" (Giddings and Anderson, 1986: 116).[9] This contention is supported by S. Anderson's (2011: 162) survey-derived data that record a major peak between 2,000 and 1,500 RCYBP. Beach Ridge 35, an early and high ridge, supported one large structure, seven-by-six meters, possibly a *qargi* (House 30). In addition to the villages at Cape Krusenstern, at least thirty-five *isolated* houses were also located, as well as myriad temporary tent camps. The summer structures may represent a temporary aggregation for trade or feasting, as around the year 1880 at Sisualik "trade fairs" (Nelson, 1899: 260–2). Nonetheless, the degree of Ipiutak commitment to Cape Krusenstern throughout the year remains uncertain, given its dearth ($n = 19$) of burials, six of which were within structures (Simmons, 1986: 360).

Ipiutak dominated nearly all of the Kotzebue Sound coast between 1,550 years BP and 1,050 years BP, with most areas inhabited in the later centuries. A single feature and a burial at Kivalina dated between 1,483 years BP and 1,295 years BP (1,470 $\pm$ 70 years BP, Beta-266435) (Stern *et al.*, 2010).[10] At Cape Espenberg, one winter house was occupied between 1,350 years BP and 1,200 years BP (Harritt, 1994: 111ff). The critically important Deering settlement contained several structures and ten graves, and dated between 1,350 years BP and 1,050 years BP. Deering is notable for its large birch-bark-roofed log *qargi*, *c.* 100 square meters (Larsen, 2001) in size, and in a small sparsely supplied driftwood structure (Reuther, 2009a). At the Deering burial precinct, of the ten interments (cf. Bowers, 2009) only the maskoid Burial 4 is dated, at between 1,370 years BP and 1,100 years BP (Reuther and Bowers, 2009). Otherwise, Ipiutak peoples were present across the Seward Peninsula, albeit in only scattered areas (Mason, 2009d), e.g., within the Trail Creek Caves (Larsen, 1968b) and at the western margin, based on a midden from Point Spencer, possibly associated with Ipiutak, dated between 1,350 years BP and 1,050 years BP (Ganley, 2011, personal communication). Ipiutak was notably absent from the crucial Wales region, the long-standing pivot of the Seward Peninsula, although a single Oopik sealing harpoon head, at Cape Prince of Wales, is attributed to Ipiutak by Stanford (1976: 103). At its southern extreme, two large *qargi* structures at Qitchauvik within Golovin Bay and at Unalakleet contain diagnostic objects (especially at Qitchauvik, in lithics, open work carvings and designs) and may reflect the amalgamation of Ipiutak with the neighboring Norton culture: dated between 1,550 years BP and 1,350 years BP at Qitchauvik

(Mason *et al.*, 2007) and prior to 1,750 years BP at Unalakleet – although this occupation is considered Near Ipiutak (Lutz, 1972).

The Ipiutak penetration up the Noatak River valley was apparently negligible (Anderson, 1972: 100), with evidence limited to its mouth, from a purported (but undated) Near Ipiutak occupation at Tulaagiaq, a richly appointed burial of two children (Anderson, 1978), and a dense concentration of well-flaked end blades from a small riverside occupation in the lower Noatak canyon at NOA-287 (Hoff *et al.*, 1991), dated as 1,694–1,646 years BP and 1,637–1,366 years BP (Beta-41832) and 1,262–1,196 years BP and 1,191–921 years BP (Beta-41831).

Ipiutak occupation farther up in the mountains is broadly contemporaneous with that on the coasts, although a possibility exists that antecedent occupation occurred along the tundra lakes, for example at Feniak Lake. However, the Feniak ^{14}C ages that precede 1,750 years BP derive from rough, slap-dash shovel excavations into complex stratigraphy (Hall, 1974). Most other interior locales occur within the sixth to eighth centuries CE (Gerlach and Mason, 1992): e.g., at Onion Portage (Anderson, 1988: 48); the Croxton site (Mason and Gerlach, 1995); the Kayuk and Avingak sites, and tent rings dated between 1,705 years BP and 1,410 years BP (Mills *et al.*, 2005: 33, 44–45); near Anaktuvuk Pass (Campbell, 1962: 47; Mills *et al.*, 2005); and the eastern Brook Range (Reanier, 1992). In the boreal forest, near the Yukon River at its farthest southeast limit, Hahanudan Lake (Clark, 1977: 69) revealed several houses, at least two of Ipiutak affinity.[11] The central Brooks Range Avingak site, excavated by Campbell in 1961, consists of six house depressions along Tuluak Lake in Anaktuvuk Pass. A large winter house (21 square meters) was occupied between 1,538 years BP and 1,266 years BP (Gal, 1982: 170), based on one of the thousand caribou bones from at least twenty-seven carcasses.

Demography

At the large community at Point Hope, the spatial distribution of houses on the five ridges is weighted heavily toward the middle ridges (Larsen and Rainey, 1948: Figure 2; see the map reproduced at the front of this volume on page xvi), with the earliest and latest containing the fewest houses. This pattern suggests a demographic cycle with initial settlement, maturation, and decline (Mason, 2006); this consideration will need to be factored into demographic estimates. Individual houses are so close, typically less than five to ten meters apart, that it is unlikely many were contemporaries, although this assumption remains unsubstantiated due to insufficient radiometric dating. Of the seventy-three houses excavated in 1939–41, only five offer single ^{14}C assays (Mason, 2006:

105), with the oldest age on caribou antler from House 50 falling within the early fifth to seventh centuries CE on the most landward, twenty-eighth ridge. The remaining four ^{14}C ages overlap within a range between 1,370 years BP and 1,080 years BP.

The average winter population of the Ipiutak community can be estimated by employing two essential assumptions: (a) the length of the dated occupation is placed at approximately 300 years, and (b) total living space can be calculated from the 10% sample listed in the appendices in Larsen and Rainey (1948). These two referents enable us to reconstruct the average population of Ipiutak. First, I assume that either twenty or thirty years define each generation; so that between twelve and fifteen generations lived at Point Hope between 600 CE and 900 CE. Cross-cultural ethnographic data suggest that each person required 2.325 square meters (Narroll's [1962] rule). Total living space for the 575 houses is *c*. 6,000 square meters. Arithmetic calculation (total living space divided by generation number) yields an estimate of between 175 and 215 individuals (Mason, 1998: 274; 2006) on average, and possibly more if Ipiutak people preferred or tolerated tighter domestic arrangements. One last factor alters the average population estimate. Since the spatial distribution of houses across the four beach ridges suggests a normal distribution, a reasonable expectation is that the population was lower by possibly 10% in the earliest phases. However, the length of surface stability must be addressed by additional dating before this hypothesis can be accepted.

Subsistence

Ipiutak communities were predominantly coastal, reliant on sea-mammal hunting for food, but the inland areas served a vital function in providing caribou skin and flesh. The economic basis of the Ipiutak resembled most Holocene Arctic maritime hunters in northwest Alaska: a reliance on ringed and bearded seal, with some attention to caribou. Quantitative faunal analyses are a recent tool and remain few, outside of Point Hope and of recent interest (e.g., Cape Krusentern [Anderson, 1962], Deering [Moss and Bowers, 2007; Saleeby *et al.*, 2009], or the Croxton site [Gerlach, 1989]). Walrusing was important locally, as at Point Hope (Larsen and Rainey, 1948), as was whaling, possibly, at Cape Krusenstern (Giddings and Anderson, 1986: 153). Early researchers, following standard practice at the time, inferred dietary choices only from the percentage counts of animal bones (Larsen and Rainey, 1948: 68) indicating that the Ipiutak at Point Hope relied on ringed seal (53%), walrus (23%), bearded seal (12%), and only a small amount of caribou (10%). Recent quantitative archaeofaunal analyses show that small mammals and, especially, birds played a larger role in Ipiutak subsistence than formerly thought – at least, locally, at Deering (Moss and Bowers, 2005; Saleeby *et al.*, 2009). Fishing was a significant pursuit at many locales, considering the fish spears even at Point Hope (Larsen and Rainey, 1948: 78–79). However, despite the evidence of maritime hunting, early researchers emphasized the role of caribou hunting – even on the coast (Larsen and Rainey, 1948: 147ff), although much of this inference is based on the functional categorization of arrow points for hunting, not warring (Larsen and Rainey, 1948: 66; cf. Mason, 1998, 2006).

Interior Ipiutak hunters had a high mobility strategy and possibly some coastal groups or families ventured inland seasonally for caribou, but more likely coastal communities had trade relationships with interior ones. Camp sites at Onion Portage suggest sixty repeated visits by Ipiutak people (Anderson, 1988: 121) while Anaktuvuk Pass witnessed a similar pattern (Mills *et al.*, 2005). The charred caribou bone and artifact scatters at Onion Portage had such a complexity as to impress Anderson (1988: 121–2) with one house floor, the residua of a temporary pole-framed tent used in spring or winter–spring. Sophisticated archaeofaunal data and analyses of caribou hunting are restricted to Tukuto Lake, where drives into lakes may have occurred in the autumn (Speiss, 1979: 158), and at Anaktuvuk Pass, where a highly fragmented fauna records the "procurement of mature animals . . . [during] a summer or fall encampment, based on the location of tent structure[s]" (Mills *et al.*, 2005: 44–5). Other small settlements are scattered across the entire Brooks Range; at the farthest limit, a house was found at the Bateman site on Itkillik Lake more than 1,000 kilometers inland (Reanier, 1992). Ipiutak people did penetrate the boreal forest, where three households were settled within dunes on Hahanudan Lake east of the Koyukuk River (Clark, 1977: 69ff), less than fifty kilometers south of the Batza Tena obsidian source. Lakes were employed as caribou drive sites, although fishing cannot be ruled out. The thinness of interior Ipiutak occupations is observable from the undated Trail Creek Caves 2 and 9, two labyrinthine limestone cavities in interior Seward Peninsula, where Ipiutak hunters left seven broken decorated-antler barbed arrowheads (Larsen, 1968b: 35, Plate V: 4–10) as well as four end or side blades (Larsen, 1968b: 34, Plate IV: 14, Plates IV, V: 12) – evidence of several visits, with tools discarded during brief visits, sheltering from poor weather to re-tool their spears or arrow points; again, this is evidence of considerable mobility.

Mortuary patterns

The treatment of the dead offers innumerable clues to the social structure of the Ipiutak community, and reflects cultural practices that include warfare and shamanism. Questions of descent and chronology could be resolvable, however, if more graves were dated and if aDNA were analyzed; regrettably, both those research domains remain for future researchers (but see Maley, Chapter 4, this volume). As in establishing the origins of Ipiutak, it is often a matter of archaeological preconception and definition, rather than hard data, that defines the chronology of mortuary patterns. Significantly, not a single burial is reported from an interior settlement. The circumstance that mortality was most common within the large villages conflicts with the Larsen and Rainey (1948: 44) model of a transient trade-fair population agglomeration and supports a view that the coast was the locus of substantial winter/early spring settlements – not the interior.

Ipiutak burials are concentrated adjacent to the large coastal settlements, with approximately 120 from Point Hope (Larsen and Rainey, 1948); nineteen across several Cape Krusenstern ridges, mostly in or near houses (Simmons, 1986: 356ff); ten from Deering (Bowers, 2009); only two isolated likely graves at Cape Espenberg (Harritt, 1994); two subadult burials from Tulaagiaq (Anderson, 1978); and single graves from Kivalina (Stern *et al.*, 2010) and Kotzebue (Gannon, 1987). Unique among the mortuary remains is one presumptively early but undated battlefield assemblage, termed Battle Rock, which contains at least four headless "Ipiutak" graves of mixed character (Simmons, 1986: 359). The chronological placement of Battle Rock would be a major datum toward understanding the development of Ipiutak mortuary patterns; however, the human remains were repatriated without such analyses. One burial – Burial 4 – stands apart from the several dozen possibly war-battered bodies atop this knoll near Cape Krusenstern. Burial 4 is deep and stone lined, containing at least three individuals and 348 artifacts, mostly antler arrowheads (65%). Were these associated with heroic warrior(s) who mobilized their neighbors and relatives as a military effort? Two other burials are attributed, without clear justification, to Near Ipiutak: (a) a single individual at Kotzebue interred with ten arrowheads with end blade slots (Gannon, 1987: 2); and (b) at Tulaagiaq near the mouth of the Noatak, two children between five and thirteen years old were interred with a plentiful array of apparent grave goods, including more than 250 antler arrowheads (Anderson, 1978: 53). The amount of grave goods associated with children is remarkable.

Burials at Point Hope vary in construction and in the representation of associated grave goods. There were fifteen discrete groups or precincts, with three clusters accounting for over half the graves, as discussed by Newton (2002) and Mason (2006: 109). The grave precincts vary in the amounts of lithic armament (whether of hunting or military) and in the amount of open work carvings or mystical capital. The most expeditious conclusion, revising Newton's interpretation of chi-square analyses,[12] is that graves were placed

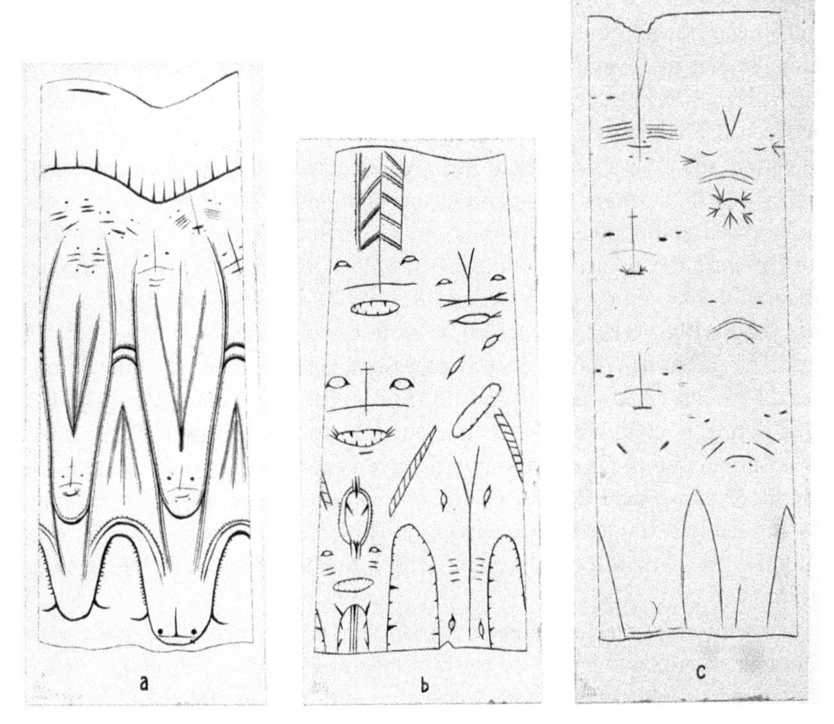

Figure 3.2. Line drawings of carvings made on caribou antler tubes associated with the child buried in Ipiutak Burial 21. These artifacts (60.1–7451, 60.1–7452, and 60.1–7454) depict various abstract human faces in repeating patterns. The images are reproduced from Larsen and Rainey's (1948) Figure 28 with permission of the American Museum of Natural History.

in a systematic fashion with degrees of power and influence affecting their placement (Mason, 2006: 109). A small fraction of the Point Hope graves were constructed of wood, with many of these bearing traces of substantial amounts of offerings and hence higher status individuals.

Multiple burials, often of an adult and a child, make up about one-quarter of the interments at Point Hope, with a few burials consisting of between four and six individuals. Several burials contained a substantial quantity of weaponry – more than twenty-five arrowheads – likely employed in warfare (Larsen and Rainey, 1948: 225, 227, 231, 240, 244, 247). For example, Burial 26 is an "unusually large" complex array of at least two individuals and fifty-one arrowheads (Larsen and Rainey, 1948: 231). Implements of shamanic curing were present in some burials as well, such as the antler tubes that occur above the child who was positioned between the legs of Burial 21 (see Figure 3.2).

Eight chronometric ages for Point Hope Ipiutak mortuary practices employed archived artifacts from the Danish National Museum (Newton, 2002: 99). The dates extend to several centuries earlier than that of the Ipiutak houses at the village site, but also overlap with them, with dates from graves being between 2,350 years BP and 1,550 years BP. Unfortunately, for ease of classification, the eight assays from five graves include two "Jabbertown" interments that had pottery. In addition, the relationship of grave goods to the burials was ambigiuous in Burials 74, 102 and JB7; these were inconclusive because the bodies were deposited within a midden possibly reworked or modified by animals (Larsen and Rainey, 1948: 245, 246, 251). In several cases (B98, B102 and B134), objects had inconclusive cultural affinities. The grave B74, however, had an assemblage most clearly Ipiutak, and dated between 1,680 years BP and 1,535 years BP (Larsen and Rainey, 1948). This dating is confirmed, based on four ^{14}C ages, as early as 2,225 years BP, continuing through the first two to three centuries CE, which precede the admittedly few dated houses. Of the five grave clusters closest to the Ipiutak houses, the single dated interment is Burial 74, which, as cited, is considerably older than the only house with a pre-1,500 years BP age (House 50).[13]

Grave offerings that hold numinous properties are rare among Ipiutak graves, although an association of these few offerings with children may prove crucial. Only one maskoid at Point Hope was placed atop the chest of a child buried at the knees of two adults, but with few other offerings, in a log "tomb" (Larsen and Rainey, 1948: 240–1). Deering Burial 4 also bore a maskoid along with the remains of seven children within a log coffin capped by stone slabs (Mason, 2010). Burial 51, depicted in Figure 3.1 with ivory eyes, faced east and was not associated with other grave goods or a wooden coffin (Larsen and Rainey, 1948: 236). Several objects, mainly ivory eyes or noseplugs, or loon skulls, were employed as prophylactic devices, especially the antler "death masks" in B107a and JB22, or the mouth covers in Burials 8, 41 and 61 (Larsen and Rainey, 1948: 122, Plate 49).

The radiocarbon dates from the Point Hope burials imply a far greater antiquity for its settlement, preceding by centuries the age of the houses from the beach ridge surface. Mason (2006: 108) speculated that houses were indeed stratified, contrary to conventional wisdom that no houses were superimposed. From that observation, one may conclude that older houses should be discoverable beneath the surface-revealed house depressions and outlines, especially noting the observation that graves were typically buried deeper than one meter into the gravel, some of which may be storm deposited. Quite possibly, some of the potentially older, underlying houses may belong to the presumed ceramic-using Near Ipiutak ancestors of the Ipiutak people.

Trade and exchange

Ipiutak people maintained numerous long-distance contacts, either direct or indirect, ranging from Kamchatka and Chukotka, across the Brooks Range, and amongst themselves, most in evidence between Point Hope and Deering. Metallurgical (not meteoric) iron and obsidian[13] likely associated with the Deering Ipiutak house were obtained from Kamchatka, several hundred kilometers distant (Reuther, 2009a: 209, 216; 2009b: E-79ff). Evidence that iron was possibly widely used in engraving is documented from three diverse occurrences: an iron burin in Point Hope House 51 (Larsen and Rainey, 1948: 83, Plate 8: 24); within Ipiutak House 1 at Deering (Reuther, 2009a: 209), and at Hahanudan Lake (Clark, 1977: 69). Brooks Range chert was in considerable demand across the Ipiutak world, as was coal jet, employed for decorative insets in mask-like carvings and figurines (Mason 2009c). Commodities that traded hands included walrus tusk, possibly traded to their Bering Strait neighbors,[14] especially the OBS, and probably also caribou skins from the interior (Mason, 1998). However, in comparison to other, later "Thule" peoples, the Ipiutak traded in far fewer commodities (i.e., lacking is evidence of an Ipiutak trade in jade, slate, or clay).

The identical motifs that occur on Deering, Point Hope, and Ekven objects, especially spear heads, are evidence either of the exchange of ideas across frontiers or possibly the curation of war trophies. Common intellectual legacies are apparent in the stylistic similitude of the OBS and Ipiutak design motifs, and the common use of atlatl stabilizers. Direct regional ties are evident in identical line and dot motifs, as exemplified by two widely separated cases: dual-rayed circle and dot motifs on the Deering Burial 4 maskoid and an identical design on a bone flaker from House 32 at Point Hope (Larsen and Rainey, 1948: 94). Both the Deering burial and the Point Hope house are dated by ^{14}C assays as between the seventh and ninth centuries CE (Mason, 2006, 2009b: 120; 2009c). Additional cross-regional ties can be established between Deering and Chukotka, observing two emblematic designs on double biface-inset spears, one from Deering Burial 1, that are nearly identical to those on a spear from Ekven. However, is this evidence of a common cultural language or is it an enemy spear, i.e., a war trophy, buried with its victim? (Mason 2009c: Figure 10–19a, b and c). Or is the Ekven piece a war trophy from Deering?

Shamanic vision in Ipiutak art

The Ipiutak cult has maintained its grip on the archaeological and popular imagination through the profundity of its iron-engraved walrus ivory mortuary objects (such as those depicted on the cover art of this volume), tied very likely to shamanic performances within the *qargi* (Rainey, 1941, 1942, 1947, 1971; Collins, 1973; Auger, 2005; Mason, 2006, 2009b, c). The techniques, psychic energy and cosmologies of shamanism are widespread across Eurasia and the New World (Eliade, 1964), and were observed in the late ninteenth century along Bering Strait (Nelson, 1899; Bogorz, 1904–9). Shamans possess healing knowledge and employ trance in order to contact other realms, considered to control sickness, famine, and unbalance (Lommel, 1967, Rouget, 1985: 125ff). The flight of the spirit is the essence of shamanic trance (Eliade, 1964: 5) that involves a journey to the "other world" (Pearson, 2002: 75). The ability to see one's skeleton, reducing one's self to the core, is a prerequisite of shamanic learning (Eliade, 1964: 62ff); this is a belief portrayed by Ipiutak iconography (Larsen and Rainey, 1948: 125ff).

The cosmic dimensions of Ipiutak iconography can be tracked through its participation in the pan-Arctic bear cult (Larsen, 1968/69, 1979/80; Sutherland, 2001b), with firm links to OBS, namely in the Okvik "Madonna" (Collins, 1976). The vision quest is well exemplified by the all-seeing, high-flying Arctic Loon whose sight replaces that of the blind human (Morrow and Volkman, 1975; Auger, 2005: 45). Second sight is a prime characteristic of the shaman, often associated with predicting the future and seeing into the souls of others (Eliade, 1964: 60–1). The three "masks" or maskoids – two from Point Hope (Burials 64 and 77; Larsen and Rainey, 1948: 123ff, Plates 54 and 55), and one from Deering (Mason, 2009c) – are the profound representation of cosmic transformation, replete with multiple animal faces and gaping mouths, although many Ipiutak objects reveal multiple animal and human identities (Auger, 2005: 64, cf. Mason, 2009c, d). Shamanic flight is allegorized by the transformative, nearly invisible botfly larvae (*Cephenemyia trompe*) issuing from the two most iconic masks (Mason, 2009b, c). Dozens of larvae gestate within a single caribou nasal passage and induce a lunacy in the animal (Whitney, 2004) that suggests shamanic possession (Mason, 2009b), recalling that, like shamans, botflies are so incredibly fast as to prevent human sight (*Time*, 1926).

A shamanic mindset is evident in a wide variety of conspicuous and copiously decorated Ipiutak objects, some likely in daily use (Auger, 2005: 40ff, 64). Sucking tubes for healing are known from Point Hope (Larsen and Rainey, 1948: Plates 26, 29). Nearly 100 open work carvings were likely used as shamanic accouterment (cf. Nelson, 1899: 359; Collins, 1973: 12) occurring primarily at Point Hope (Larsen and Rainey, 1948: 119, 127ff), but also at Kivalina (Stern *et al.*, 2010) and Qitchauvik (Mason *et al.*, 2007). The polar bear comb or "rake" from Point Spencer is among the most complex and accomplished objects of Ipiutak iconography (Larsen, 1979: 80; Collins, 1973:

24). This polar-bear open work carving has a fanged and gaping mouth, evidence of a pan-Arctic bear cult (Larsen, 1968/69, 1979/80); the piece has shaman-transformative aspects as well, in its multiple orientations, e.g., accessory seal figures (Collins, 1973: 25). A similar, roughly executed, undecorated "rake" occurs within Point Hope Burial 89 (Larsen and Rainey, 1948: 145).

Objects likely used in shamanic garb were often interred, although the precise clothing use and context are ambiguous in cases where human remains are lacking. One iconic, often illustrated, copiously engraved baby walrus, which was placed under the skull of an adult in Burial 42 (Larsen and Rainey, 1948: 125–6, Plate 53:1), was likely sewn to a shaman's garb (Anderson, 1984: 89). The three renowned maskoids vary in context; one lacked any human remains (B64), while the other two (B77 at Point Hope and Deering Burial 4) were, as already noted, associated with children's bones. Other mortuary offerings have parallels in folk tales, especially the story of the "blind man and the loon," suggested by the false (all-seeing!) eyes inset into a loon cranium placed within Burial 21 (Larsen and Rainey, 1948: Plate 49:7; Morrow and Volkman, 1975).

Browbands, apparently attached to wooden hunting hats (cf. Nelson, 1899: 166ff), offered a canvas for iconic representations of gaping mouthed creatures, of fearful awesomeness, similar to the "rakes" or the maskoids. Rear-fastening troughs on the ivory Deering maskoid suggest its attachment to skin clothing (Mason, 2009c), with a similar use for the openwork pieces. Drawings of human figures on sucking tubes and browbands suggest tattoos or labretifery, e.g., an antler tube from Burial 21 (Larsen and Rainey, 1948: 115; see Figure 3.2) and a carved figurine in Burial 108 (Larsen and Rainey, 1948: 116ff, Plate 52:3). Ipiutak tattooing likely resembled that of the OBS culture (Larsen and Rainey, 1948: 116ff; Smith and Zimmerman, 1975), reflecting a common ancestry. On the other hand, labrets were possibly standard accessories, present on several human figures and portraits (Larsen and Rainey, 1948: 114) but are rare ($n = 7$) in graves, serving as treasured heirlooms. The labret is a peculiarly southern trait, a very profound ethnic marker that branded Ipiutak people as firmly different than OBS people (cf. Dumond, 2009). Further, its successors, the Birnirk/Thule people, did not sport labrets until 1400 CE.

The qargi *institution: Social integration evident in architecture*

The significance of the oversized Ipiutak house aggregations remains mystifying: Mason (1998) argued for a fully sedentary character, but with only 150 to 250 winter residents, not thousands. The alternative is that of seasonal aggregation for trading (Larsen and Rainey, 1948) – though the season of occupation includes the presence of walrus. Can the two views be reconciled? One indication of such occasional aggregation is architectural: Very likely, several adjacent communities gathered seasonally to conduct ceremonies in the structures, *qargi*, that occur at evenly spaced intervals across northwest Alaska. The placement of *qargi* appears to have a systematic geographic logic, as if access involved a principle of regional affinity, with *qargi* constructed on the coast at Point Hope, Deering, Cape Krusenstern, Qitchuavik, and, possibly, Unalakleet. Across the western Brooks Range were *qargi* at Feniak Lake and nearby Desperation Lake. Another argument for the systemic character of the *qargi* is that many were contemporaneously occupied, accepting only the extant ^{14}C data, although Feniak and Unalakleet could be earlier and both may be evidence of a preceding Norton culture organizational imperative. Hypothetically, the regional spacing of Ipiutak *qargi* reflects several interacting politic-religious nodes for gift exchange, and for shamanic performances. In addition, commodities very likely traveled between *qargi*: iron and obsidian, ivory, caribou skins and seal oil. Symbolic exchange is indicated from the occurrence of pictograph-inscribed cobbles in *qargi* that are widely separated, geographically.

The *qargi* that are located on the coast are better documented archaeologically than those on the interior lakes, although neither of the Point Hope *qargi* are dated. A large oval *qargi* (a size over 120 square meters) was constructed between 1,509 years BP and 1,332 years BP along Golovnin Bay at Qitchauvik (Mason *et al.*, 2007: 62–3). The structure sheltered male activities such as stone-tool production (debitage and numerous end-blades) and wood-working (Alix, 2007). Its use as a theatrical space is in evidence from several wood carvings: human maskettes, a seated human in shamanic transformation, and of caribou, one with an x-ray skeletal motif (Mason *et al.*, 2007: 88). A considerable amount of wood-working and maintenance of arrow points was also undertaken by men within the Deering *qargi* (Larsen, 2001; Alix, 2009). The two undated Point Hope *qargi* differed in size but offered comparatively little in terms of artifacts; a few objects could be considered gender related, but the large space may have been of more use for performance or gatherings. Significantly, each Krusenstern community maintained a single large structure, with only sparse occupational debris. However, male-specific tool manufacture activities were rather more common at Deering (Larsen, 2001: 49–60) and possibly at Cape Krusenstern H-17 and H-30, while *qargi* were smaller, both only forty-two square meters, and were used between 1,450 years BP and 1,100 years BP (Giddings and Anderson, 1986: 30, 133, 136).

The sizable Deering and Feniak *qargi* also contained a sparse number of artifacts, although the inscribed stones in the latter are more indicative of shamanic practice. The Deering wood and birch-bark-roofed structure, over 108 square meters, is well preserved; the girth of the horizontal timbers in the Deering *qargi* was so substantial (approximately forty centimeters in diameter) that Larsen (2001: 18) inferred wood procurement from the forest by sled. Minimally, forty-five people, likely many more, could have occupied the Deering *qargi* (Mason, 1998) seated on the three benches placed around a central hearth (Larsen, 2001: 23ff). Men's labor is evident from the artifact inventory that had only a few decorated pieces (a harpoon socket and a flaker [Larsen, 2001: 32, 67]) and abundant stone, organic and wood objects, including bow parts, spoons, trays, and a very distinctive built-up "Athapaskan" sled (Larsen, 2001: 40ff). Five thick floors were deposited between 1,410 years BP and 970 years BP, based on ^{14}C ages ($n = 10$) on dog feces, birch bark or wood (Larsen, 2001: 30). The Deering *qargi* was one of the largest built prehistoric spaces known in northwest Alaska, and may betray the nature of social contacts into the boreal forest, and explain the peculiarities of Ipiutak.

Qargi occur in the Brooks Range, with two large structures constructed only twenty kilometers apart on the shores of two small lakes, Feniak Lake (Hall, 1974: 484) and Desperation Lake (Irving, 1962: 81), within the upper Noatak drainage (which itself lacked substantial Ipiutak populations [cf. Anderson, 1972]). Radiocarbon ages suggest that both of the *qargi* were used contemporaneously during two intervals: first possibly as early as between 1,850 and 1,750 years BP and subsequently between 1,350 years BP and 1,050 years BP (Lawn, 1975: 207–8; Gal, 1982: 170–1). The Feniak *qargi* was over 100 square meters in size, and, although incompletely reported (cf. Hall, 1973), contained nearly 15,000 waste flakes, as well as numerous pictographs and arrow points – presumably evidence of male, possibly, martial pursuits. The nearby twelve-meter-long oval structure at Desperation Lake, discovered by Irving (1962: 81), contained "Ipiutak-like" side blades and a discoid, as well as enigmatic pictographs and imposing, two-meter-wide stone walls; a single ^{14}C age of 1,830 ± 170 RCYBP (Gal, 1982: 172) offers only a broad age estimate between 203 BCE and 580 CE. The Desperation Lake structure may be among the earliest *qargi*, possibly from a Norton ("Near Ipiutak") occupation, and may precede the Feniak structure.

The origin of the *qargi* can be equated with the development of men's political and social power, manifested in cultic and shamanic activity. Along the eastern Norton Sound, Norton people constructed a ninety-six square meter *qargi* at Unalakleet using log-banked hearths ("in use for a rather long period" [Lutz, 1972: 67]) and similar to the Ipiutak *qargi* at Deering (Larsen, 2001; Lutz, 1972: 344). The Unalakleet *qargi* was possibly occupied by the mid first century CE, and possibly as late as the early seventh century, contemporaneous with early Ipiutak occupations within Golovin Bay and at Deering.

Warfare

The evidence for violent conflict ("warfare") during Ipiutak is unequivocal (see Mason, 2006, 2009a) and includes inferences from mortuary contexts, within households, and even battlefields. The quantity of bow and arrow technology is in itself an indication of the amount of warfare (see Mason, 2012), contrary to the caribou hunting attribution of Larsen (1952: 24) and Rainey (1942: 370). At Cape Krusentern, three partially burnt skeletons were uncovered, one wounded in the pelvis, recovered in a shallow pit with an adze that possibly was used in digging in an attempt to escape from the flames (Giddings and Anderson, 1986: 127). Several mixed burials occur at a presumptive battlefield locale, Battle Rock, and are notable for a lack of crania (Simmons, 1986: 356ff); however, a full taphonomic analysis is lacking. Deering Burial 1 offers far more definitive evidence of warfare: one vertebra bore a direct hit, and the individual was associated with fifteen arrow points and two fragmentary lances decorated with extra-regional designs, related to Cape Dezhneva, East Cape, who were possibly the enemy perpetrators (Mason, 2009b). The sternum of Point Hope Burial 89 was also impaled by an arrow point, and the chest was "literally shot full of arrows" (Larsen and Rainey, 1948: 61, 243); the cranium was apparently disarticulated prior to burial. Evidence of Ipiutak warfare also includes the artifactual: bone daggers and clubs, extremely thin barbed bone points, and the sheer number ($n > 500$) – especially within the houses – of "fishtail" points so useful in intrapersonal violence (Larsen and Rainey, 1948). Similar fishtail points are common in Cook Inlet and are considered evidence of warfare (Maschner, 2008). One piece of slat armor within House 7 was considered intrusive by Larsen and Rainey (1948: 192).

Specialized shamanistic and military expertise was also evident in the amount of grave goods and in the spatial arrangement of the cemetery (cf. Mason, 2006, re-interpreting Newton, 2002, and Larsen and Rainey, 1948). Shamanistic beliefs were revealed in a small fraction of the graves, with a preference for transformative imagery based on bears, the botfly, loons, seals and occasionally humans (Mason, 2009b). A pan-regional identity in Ipiutak intellectual culture is shown by the common motifs shared by East Cape, on Chukotka, as well as Deering and Point Spencer (Mason, 2009a).

Ideological origins of Ipiutak

Upon its discovery in 1939 at the end of the pre-radiocarbon era, the search for Ipiutak progenitors proceeded to employ the paradigm of anthropo-geography derived from nineteenth-century German and Scandinavian scholars (Larsen and Rainey, 1948; Larsen, 1952). A priori, it was assumed that its provocative animal art originated far beyond the Chukchi Sea or Bering Strait region based on broadly similar motifs and subjects of representation (Larsen, 1952). In this manner, Ipiutak was easily linked to the savage animal and open work carvings of shamanistic horsemen ("Scythians") across the entirety of north Asia, from as far west as the Yamal Peninsula near the Kara Sea, and as far as north China (Larsen, 1952: 28ff; Jettmar, 1964; Bunker *et al.*, 1970). Jenness (1952: 32ff) critiqued and summarized the argument, favoring the southern influences from the jade eye insets of the Han dynasty (200 BCE to 200 CE). A generation later, Collins (1971) offered a similar derivation for the composite maskoids; most Chinese forms are Shang dynasty (1100–700 BCE) and considerably predate the late first millennium CE. However, it is not the specific motif that is the question, it is rather the behavioral complex that spread the motif as part of its ritual labor.

The basis for a Eurasian origin for Ipiutak and/or OBS styles lies in deep time and through the psychic linkage with pan-Arctic shamanic beliefs and, possibly, psychotropic drug use (Mason, 1998; cf. Pearson, 2002). The converse is that shamanism, derived from Scythian roots, is a very late introduction into the Bering Sea world, a doubtful proposition (Sutherland, 2001a, b). If it were a recent entry or possibly a set of new or innovative beliefs, Ipiutak/OBS could represent the introduction of esoteric knowledge, through a cult of shamans, passed from shaman to mentor *across* ethnic and linguistic boundary lines. This passage of knowledge more likely resembles the spread of a new and powerful religion than of a *mere* esthetic style. This is why it confounds traditional taxonomic archaeology of the cultural historical mold that is so common in Alaska. However, does religion spread innocently and passively or does it inevitably follow at the point of a sword or arrow? Can we link the shaman's power with military power? In other words, this perspective would argue for an actual conquest of Point Hope by OBS peoples and the imposition of a new ideology on a preexisting Norton community. One might anticipate that the disorder of the Jabbertown burials might be explained as the chaotic impact of violent actions. Unfortunately, the data to resolve such a speculation are too slim at present.

Fate and significance of the Ipiutak culture

To conclude, the elaborate Ipiutak mortuary cult reflects an elaborate shamanic belief system, imbedded within an intercontinental sociopolitical network that obtained extra-regional commodities, principally iron and obsidian, from afar in Siberia. Neighborly interactions were not always peaceful and often devolved into violence, feuding, if not full-scale war. Ipiutak people aggregated seasonally at critical regional centers, gathering within community structures for cultic or theatric performances. Status was ascribed, in that young children were interred with sumptuous and spiritual objects. Unfortunately, most Ipiutak assemblages are comparatively late, between 1,400 years BP and 1,150 years BP, possibly at the termination of the culture/cult, a circumstance that does not inform us about its origin that appears, on present data, to precede 1,550 years BP, possibly by several centuries. On the other extreme, at its termination, virtually no Ipiutak components are younger than 1,050 years BP, with an abruptness that suggests a sudden collapse for the culture (Mason 1998, 2000, 2006, 2009c). Did a series of poor hunting years lead to famine, or did a single military defeat result in enslavement or population resettlement or relocation? In either case, these circumstances may reflect a catastrophic termination, with other possibilities including population replacement or migration. The development of the Ipiutak art style and mortuary pattern might be part of a cultural revitalization movement, possibly, resembling a "millennial" crisis cult that betokened the end of time, cosmic decline, an ideology developed in response to explain and to empower in adverse climates and political reversals. The *qargi* were virtually abandoned everywhere at the same time; across northwest Alaska, especially the Brooks Range, Ipiutak is succeeded by a hiatus of up to a century or more until the appearance of the Thule culture, associated with a northern group who were the ancestors of the modern Iñupiat (Mason and Bowers, 2009). Ipiutak collapsed so universally and completely that neither climate nor political change should be excluded from explanation – especially in view of the Medieval Climate Anomaly (MCA) that started around 1,050 years BP (Mason and Barber, 2003; Mason, 2009a).

The issue of a catastrophic, climatically driven termination to Ipiutak involves the detailed consideration of numerous proxy climate records that are regionally specific, including beach ridges and tree rings (Mason and Jordan, 1993; Mason and Gerlach, 1995; Mason and Barber, 2003), too diverse to be fully reviewed in this chapter (cf. Mason and Barber, 2003: 84ff; Mason, 2009a: 97). Further, new high-quality local datasets continue to revolutionize our understanding; for example, varved lake sediments from the central Brooks Range indicate that warmer conditions prevailed from 1,220 years BP to 1,100 years BP, with slightly higher temperatures in the following 150 years, until 980 CE, when temperatures cooled (Bird *et al.*, 2009). Extra-regional data show that warm weather prevailed in north China during the tenth century, albeit with considerable local variability (Ge and Wu, 2011: 25). The complex and varied responses involved in the MCA presented a series of possibly unprecedented challenges to Ipiutak peoples – involving rapid changes between warm and cold, often in the same decade (cf. various papers in Xoplaki *et al.*, 2011). Did sea mammal (walrus, seal) or salmonid populations decline? To resolve the issue of climatic forcing will require a considerable improvement in our state of knowledge about Ipiutak chronology, subsistence, and cosmology. Hopefully, this volume will serve as a motivation to supplement that database and answer the many outstanding questions raised herein.

Notes

1. The iconic image of the skull of Burial 51 with the ivory eyes, so crucial to the Ipiutak narrative, was presented at the end of the final report by Larsen and Rainey (1948), and is reproduced here (Figure 3.1). In addition, the grave goods from Burial 41, including those within the eye sockets, were briefly discussed and illustrated (*ibid.* p. 234–5, Plate 49).
2. While the estimate of 122 houses by explorer Knud Rasmussen (Ostermann, 1952: 47) is widely accepted (cf. Burch, 1981: 43), apparently no map was produced and recording methods, even definition standards, were likely imprecise. Writing of "Old Tigara," Larsen and Rainey (1948: 20) observe that the "... ruins now form four irregular mounds covered with tall grass ... for the most part it is impossible to recognize individual house mounds." Did Rasmussen count multi-room houses as one house or as a single house? As Burch (1981: 42) commented and illustrated, the surface configuration and dimensions of a Tikiġaġmiut house are difficult to discern even when in use. Lacking a precise map, the precision of any estimates should be questioned, especially those of Rasmussen and of Larsen and Rainey.
3. The first reliable ^{14}C dates from Ipiutak, acquired in the 1950s, supported the expectations of an early age for Ipiutak (Rainey and Ralph, 1959; Ralph and Ackerman, 1961).
4. The erosion rate along the north aspect of the Point Hope spit is estimated by two distance comparisons. First is that of the Reverend Driggs, who in 1909 reported 56.4 meters of loss in the prior eighteen years (1890–1908), 3.2 meters per year in relation to his church building at the Ipiutak site. This interval can be confused since Driggs' tenure is not mentioned by Larsen and Rainey (1948: 19). Several meters eroded in 1939–41 (Larsen and Rainey, 1948). Aerial photographic comparisons by Hosley (1972: 11) established that 73 meters had eroded from Tikiġaq between 1940 and 1967, at approximately 2.7 meters per year. The contribution to erosion by the extensive subsistence digging at Tikiġaq remains uncertain. In addition, the Tikiġaq mounds' high silt and ice content probably increases its rate of erosion as large blocks are undercut by waves, in contrast to Hosley's view. Erosion rates since 1967 are not documented, but it is unlikely that rates have continued at such high levels.
5. The amount of the "prize" money increases with the years, from $3.00 in 1942 (Rainey, 1942: 320), $5.00 in the 1970s (Rainey, 1971: 7), and in 1990 a reported $15.00 bonus for ivory eye burials (Kowanna in Hall, 1990: 38) – it is not known

if the post 1950 writers are correcting for inflation, although the Native informant does state that "$15 was a lot of money in those days."

6. The Point Hope mortuary data were assembled by Crass (1998), and also collated and further analyzed by Newton (2002: 15ff).
7. As enshrined on the Tikiġaq website: "Dating from approximately 600 BCE, Point Hope has one of the longest documented, continuous occupations of Iñupiaq marine mammal hunters in the Arctic. Layer upon layer of archeologic[al] remains have provided a window into the lifestyles and traditions of the region's people." (http://www.Tikigaq.com/inupiaq_people/our_roots.shtml)
8. The two samples from House 30 vary widely: The older age, *c.* 1,900 RCYBP, is less convincing than the 1,400 RCYBP age because unidentified charcoal inherently has a "whole tree" uncertainty. Hence the younger age seems preferable.
9. This assessment may require revision in light of the intensive ^{14}C dating by S. Anderson (2011: 162), which indicates that the Cape Krusenstern population was higher during the eleventh to fourteenth centuries CE.
10. The Kivalina site NOA-362 is a sizable assemblage, including four open work carvings, several fishtail points, an adze, and even lithic debitage, some associated with one of nine graves laid atop a wood frame, possibly the remains of a house depression. The grave is about forty centimeters below surface, typical for Ipiutak at Point Hope (Stern *et al.*, 2010).
11. The relationship of Ipiutak to the boreal forest, so evident to Larsen (2001), remains controversial; Mason (n.d. a) reviews this hoary and still controversial issue. However, only the very small Hahanudan Lake settlement of two structures is fully within the trees. Most interior Ipiutak sites lie either at the forest–tundra boundary or within the tundra. Fuel and wood supply issues would thus be critical for any winter occupation (Mason, 1998).
12. Newton's (2002) chi-square analayses were interpreted to exhibit no relationship statistically between grave goods and shamanism, or power. However, a closer inspection of her analyses reveals that the focus on group characteristics failed to analyze individual grave assemblages and minimized single grave assemblages. Further, Newton (2002: 24) reclassified wood burials without additional data, reducing sample size. Several grave and grave clusters do account for the majority of grave goods, and of grave goods of a particular type. For example, open work carvings occur in sizable numbers (>100) in only two clusters (Newton, 2002: 77), while arrowpoints (>100), indicative of war to Mason (1998), occur mostly in four clusters. While the relationship may not be statistically significant, sample size or other factors may be involved.
13. To quote Reuther and Bowers (2009: 216): "One obsidian artifact found in the upper portion of fill within . . . Ipiutak house area is geochemically similar to obsidian from Mt. Kankaren in Chukota, Siberia. This piece may relate to the Ipiutak occupation . . . the presence of this piece of Siberian obsidian and iron engraver tip . . . . within Deering . . . shows connections to East Asian trade networks."
14. However, it should be recalled that most, if not all OBS groups also had ready access to walrus, so that perhaps surplus walrus was accumulated by several OBS groups for trade with other non-OBS groups to the west and south of the Bering Strait, across Chukotka and Kamchatka.

References

Alix, C. (2007). Wood artifacts and technology at Qitchauvik. In *An Ipiutak Outlier: A Late 1st millennium AD Qarigi in Golovnin Bay*. Shared Beringian Heritage Program. Anchorage, AK: National Park Service, pp. 95–134.

Alix, C. (2009). Deering wood technology. In *The Archaeology of Deering, Alaska: Final Report on the Village Safe Water Program*, edited by P. M. Bowers. Fairbanks, AK: Northern Land Use Research, pp. 233–41.

Anderson, D. D. (1962). *Cape Krusenstern Ipiutak Economic and Settlement Patterns*. M.A. Brown University.

Anderson, D. D. (1972). An archaeological survey of the Noatak drainage, Alaska. *Arctic Anthropology*, 9 (1), 66–117.

Anderson, D. D. (1978). Tulaagiaq: A transitional Near Ipiutak-Ipiutak archaeological site from Kotzebue Sound, Alaska. *Anthropological Papers of the University of Alaska*, 19 (1), 45–57.

Anderson, D. D. (1983). Changing prehistoric Eskimo subsistence patterns: A working paper. In *Cultures of the Bering Sea Region*, papers from an International Symposium (Moscow), edited by H. N. Michael and J. W. VanStone. New York, NY: International Research and Exchanges Board, pp. 62–83.

Anderson, D. D. (1984). Prehistory of north Alaska. In *Handbook of North American Indians, Volume 5, Arctic*. Washington, D.C.: Smithsonian Institution Press, pp. 80–93.

Anderson, D. D. (1988). Onion Portage: The archaeology of a stratified site from the Kobuk River, northwest Alaska. *Anthropological Papers of the University of Alaska*, 22 (1–2), 1–163.

Anderson, S. (2011). *From Tundra to Forest: Ceramic Distribution and Social Interaction in Northwest Alaska*. Ph.D. University of Washington.

Arutiunov, S. A. and Sergeev, D. A. (2006a) [1969] *Ancient Cultures of the Asiatic Eskimos: The Uelen Cemetery*. Anchorage, AK: Shared Beringian Program, National Park Service. [Translation by R. L. Bland of *Drevnie kul'tury aziatskikh eskimosov (Uelenskii mogil'nik)*, Moscow: Akademiia Nauk SSSR.]

Arutiunov, S. A. and Sergeev, D. A. (2006b) [1975]. *Problems in the Ethnic History of the Bering Sea: Ekven Cemetery*. Anchorage, AK: Shared Beringian Program, National Park Service. [Translation by R. L. Bland of *Problemy etnishiskoi istorii Beringomoria: Ekven mogil'nik*, Akademiia Nauk SSSR, Institut Etografi Imeni, N.N. Moscow: Milkukho-Maklaia.]

Auger, E. E. (2005). *The Way of Inuit Art: Aesthetics and History in and beyond the Arctic*. Jefferson, NC, and London: McFarland and Co.

Bird, B. W., Abbott, M. B, Finney, B. P. and Kutchko, B. (2009). A 2000 year varve-based climate record from the central Brooks Range, Alaska. *Journal of Paleolimnology*, 41 (1), 25–41.

Bogorz, W. (1904–9). *The Chukchee, Part II. Religion*. Reprint from Vol. VII, Part II, Jesup North Pacific Expedition, edited by Franz Boas. Vol. XL, Memoirs of the American Museum of Natural History, New York.

Bowers, P. M. (Editor) (2009). *The Archaeology of Deering, Alaska: Final Report on the Village Safe Water Program*. Fairbanks, AK: Northern Land Use Research.

Monograph in preparation for the Alaska Department of Environmental Conservation Village Safe Water Office, the City of Deering, and the Deering IRA Council. Northern Land Use Research, Inc., Fairbanks.

Bunker, E. C., Chatwin, C. B. and Farkas, A. R. (1970). *"Animal Style" Art from East to West*. New York, NY: The Asia Society.

Burch, E. S. (1981). *Traditional Eskimo Hunters of Point Hope, Alaska: 1800–1875*. North Slope Borough.

Campbell, J. M. (1962). Cultural succession at Anaktuvuk Pass, arctic Alaska. In *Prehistoric Cultural Relations between the Arctic and Temperate Zones of North America*, edited by J. M. Campbell. Technical Paper No. 11, Arctic Institute of North America, Montreal, pp. 39–54,

Clark, D. W. (1977). *Hahanudan Lake: An Ipiutak Related Occupation of Western Interior Alaska*. Mercury Series Archaeological Survey Paper 71. Ottawa: National Museum of Man.

Collins, H. B. (1937). Archaeology of Saint Lawrence Island, Alaska. *Smithsonian Miscellaneous Collections*, 96 (1).

Collins, H. B. (1964). The arctic and subarctic. In *Prehistoric Man in the New World*. Chicago, IL: University of Chicago Press, pp. 85–114.

Collins, H. B. (1971). Composite masks: Chinese and Eskimo. *Anthropologica*, XIII, 271–8.

Collins, H. B. (1973). Eskimo art. In *The Far North: 2000 years of American Eskimo and Indian Art*. Washington, D.C.: National Gallery of Art, pp. 1–31.

Collins, H. B. (1976). The Okvik figurine: Madonna or Bear Mother. *Folk*, 17, 125–32.

Crass, B. (1998). *Pre-Christian Inuit Mortuary Practices: A Compendium of Archaeological and Ethnographical Sources*. Ph.D. University of Wisconsin.

de Laguna, F. (1934). *The Archaeology of Cook Inlet, Alaska*. Philadelphia, PA: University Museum, University of Pennsylvania.

Dumond, D. E. (1977). *Eskimos and Aleuts*. London: Thames and Hudson.

Dumond, D. E. (1978). Alaska and the Northwest coast. In *Ancient Native Americans*. San Francisco, CA: W. H. Freeman, pp. 45–93.

Dumond, D. E. (1982). Trends and traditions in Alaskan prehistory: The place of Norton culture. *Arctic Anthropology*, 19 (2), 39–51.

Dumond, D. E. (1987). *Eskimos and Aleuts*, Second Edition. London: Thames and Hudson.

Dumond, D. E. (2000). The Norton tradition. *Arctic Anthropology*, 37 (2), 1–22.

Dumond, D. E. (2009). A Note on labret use around the Bering and Chukchi Seas. *Alaska Journal of Anthropology*, 7 (2), 121–34.

Eliade, M. (1964). *Shamanism: Archaic Techniques of Ecstasy*. Princeton, NJ: Princeton University Press.

Gal, R. (1982). Appendix I: An annotated roster of archaeological radiocarbon dates from Alaska, north of 68° latitude. *Anthropological Papers of the University of Alaska*, 20 (1–2), 159–80.

Gannon, B. (1987). *1986 Archaeological survey along the proposed airport road relocation right of way, Kotzebue, Alaska*. Unpublished report, Alaska Department of Transportation, Fairbanks. Dated April 29, 1987.

Ge, Q., and Wu, W. (2011). Climate during the Medieval Climate Anomaly in China. In *Medieval Climate Anomaly. PAGES News*, 19 (1), 24–6.

Gerlach, S. C. (1989). *Models of Caribou Exploitation, Butchery, and Processing at the Croxton Site, Tukuto Lake, Alaska*. Ph.D., Department of Anthropology, Brown University.

Gerlach, S. C. and Mason, O. K. (1992). Calibrated radiocarbon dates and cultural interaction in the western Arctic. *Arctic Anthropology*, 29 (1), 54–81.

Giddings, J. L. (1964). *The Archaeology of Cape Denbigh*. Providence, RI: Brown University Press.

Giddings, J. L. (1967). *Ancient Men of the Arctic*. New York, NY: Alfred A. Knopf.

Giddings, J. L. and Anderson, D. D. (1986). *Beach Ridge Archaeology of Cape Krusenstern: Eskimo and Pre-Eskimo Settlements around Kotzebue Sound, Alaska*. Publications in Archeology 20. Anchorage, AK: National Park Service, U.S. Department of the Interior.

Hall, E. S. (1974). Archaeological investigations in the Noatak River valley, Summer 1973. *Contributions from the Center for Northern Studies*, No. 1, 460–523.

Hall, E. S. (1990). *A Cultural Resource Site Reconnaissance of Proposed Construction Areas in the Vicinity of Point Hope, Alaska*. Technical Memorandum 35, Brockport, NY: Edwin S. Hall and Associates.

Harritt, R. K. (1994). *Eskimo Prehistory on the Seward Peninsula*. Research Report AR 21. Anchorage, AK: National Park Service.

Harritt, R. K. (1995). The development and spread of the whale hunting complex in Bering Strait: Retrospective and prospects. In *Hunting the Largest Animals: Native Whaling in the Western Arctic and Subarctic*. Studies in Whaling No. 3, Occasional Paper No. 36, Circumpolar Institute. Edmonton: University of Alberta, pp. 33–51.

Hill, E. (2011). The historical ecology of walrus exploitation in the North Pacific. In *Human Impacts on Seals, Sea Lions and Sea Otters: Integrating Archaeology and Ecology in the Northeast Pacific*. Berkeley, CA: University of California Press, pp. 41–64.

Hoff, R., Thorsen, S. and Miraglia, R. (1991). Report of Section 106 Investigation. Historic Site Numbers: NOA-287, 288, Allotment FF 017606 Max Wilson, Sr. Archeology Section, Bureau of Indian Affairs, On file, Anchorage: State Historic Preservation Office, Office of History and Archaeology, State of Alaska.

Hoffecker, J. F. and Mason, O. K. (2010). Cape Espenberg Thule Origins Project. *Alaska Journal of Anthropology*, 8 (1), 143–4.

Hosley, E. (1967). *Archaeological Evaluation of Ancient Habitation Site and Surface and Submarine Geology, Point Hope, Alaska*. Report to the U.S. Corps of Engineers, Fairbanks, AK: University of Alaska.

Hosley, E. (1972). Archaeological evaluation of ancient habitation site, Point Hope, Alaska. In *Point Hope Beach Erosion, Point Hope, Alaska*, Appendix A, Survey Report, United States Army Corps of Engineers, Alaska District, Anchorage.

Irving, W. H. (1962). 1961 Field work in the western Brooks Range, Alaska. Preliminary Report. *Arctic Anthropology*, 1 (1), 76–83.

Jenness, D. (1952). Discussion of H. Larsen, "The Ipiutak Culture: Its Origin and Relationships." *Proceedings of the 29th International Congress of Americanists 3*. Chicago, IL: University of Chicago Press, pp. 30–4.
Jensen, A. M. (1997). *An Archaeological Field Survey in Connection with Proposed Construction at the Ipiutaq Site and Old Town Site, Point Hope, Alaska*. UIC Science Division Technical Report #1. Barrow, AK: UIC Cultural Resources, prepared for LCMF, Inc.
Jensen, A. M. (2009). *Nuvuk: Point Barrow, Alaska: The Thule Cemetery and Ipiutak Occupation*. Ph.D. Bryn Mawr College.
Jettmar, K. (1964). *Art of the Steppes*. New York, NY: Crown Publishers.
Larsen, H. (1952). The Ipiutak culture: Its origin and relationships. In *Proceedings of the 29th International Congress of Americanists*. Chicago, IL: University of Chicago Press, pp. 22–34.
Larsen, H. (1968a). Near Ipiutak and Uwelen-Okvik. *Folk*, 10, 81–90.
Larsen, H. (1968b). Trail Creek: Final report on the excavation of two caves on Seward Peninsula, Alaska. *Acta Arctica*, 15, 7–79.
Larsen, H. (1969/70). Some examples of bear cult among the Eskimos and other northern peoples. *Folk*, 11–12, 27–42.
Larsen, H. (1979/80). Examples of Ipiutak art from Point Spencer, Alaska. *Folk*, 21/22, 17–28.
Larsen, H. (1982). An artifactual comparison of finds of Norton and related cultures. *Arctic Anthropology*, 19 (2), 53–8.
Larsen, H. (2001). *Deering: A Men's House from Seward Peninsula, Alaska*. Publications of the [Danish] National Museum, Ethnographical Series, Volume 19. Department of Ethnography, SILA – Greenland Research Centre, Copenhagen: National Museum of Denmark.
Larsen, H. and Rainey, F. (1948). *Ipiutak and the Arctic Whale Hunting Culture*. Anthropological Papers of the American Museum of Natural History 42. New York, NY: American Museum of Natural History.
Lawn, B. (1975). University of Pennsylvania Radiocarbon Dates XVIII. Sect. F. Arctic. *Radiocarbon*, 7 (2), 207–15.
Lommel, A. (1967). *The World of the Early Hunters: Medicine Men, Shamans and Artists*. London: Evelyn, Adams & Mackay.
Lowenstein, T. (2008). *Ultimate Americans: Point Hope, Alaska: 1826–1909*. Fairbanks, AK: University of Alaska Press.
Lutz, B. J. (1972). *A Methodology for Determining Regional Intercultural Variation Within Norton, an Alaskan Archaeological Culture*. Unpublished Ph.D., University of Pennsylvania.
Lutz, B. (1973). An archaeological karigi at the site of UngLaqLiq, western Alaska. *Arctic Anthropology*, 10–1, 111–9.
Maschner, H. D. G. (2008). Fishtails, ancestors, and Old Islanders: Chirikof Island, the Alaska Peninsula and the dynamics of Western Alaska prehistory. *Alaska Journal of Anthropology*, 6 (1–2), 171–83.
Maschner, H. D. G. and Reedy-Maschner, K. (1998). Raid, retreat, defend (Repeat): The archaeology and ethnohistory of warfare on the North Pacific Rim. *Journal of Anthropological Archaeology*, 17, 19–51.

Mason, O. K. (1998). The contest between Ipiutak, Old Bering Sea and Birnirk polities and the origin of whaling during the first millennium A.D. along Bering Strait. *Journal of Anthropological Archaeology*, 17 (3), 240–325.

Mason, O. K. (2000). Archaeological Rorshach in delineating Ipiutak, Punuk and Birnirk in NW Alaska: Masters, slaves or partners in trade? In *Identities and Cultural Contacts in the Arctic*, Publication No. 8. Copenhagen: Danish Polar Center, pp. 229–51.

Mason, O. K. (2006). Ipiutak remains mysterious: A focal place still out of focus. In *Dynamics of Northern Societies*, Proceedings of a Symposium. Copenhagen: Danish National Museum and Danish Polar Center, pp. 106–20.

Mason, O. K. (2009a). The multiplication of forms: Bering Strait harpoon heads as a demic and macroevolutionary proxy. In *Macroevolution in Human Prehistory: Evolutionary Theory and Processual Archaeology*. New York, NY: Springer Verlag, pp. 73–107.

Mason, O. K. (2009b). Art, power, and cosmos in Bering Strait prehistory. In *Gifts from the Ancestors: Ancient Ivories from Bering Strait*. Exhibition catalog. Princeton, NJ: Princeton University Museum of Art, pp. 112–25.

Mason, O. K. (2009c). Mask affinities. In *The Archaeology of Deering, Alaska: Final Report on the Safe Water Archaeological Project*, edited by P. M. Bowers. Fairbanks, AK: Northern Land Use Research, pp. 249–66.

Mason, O. K. (2009d). The Ipiutak cult on the Seward Peninsula. In *The Archaeology of Deering, Alaska: Final Report on the Safe Water Archaeological Project*, edited by P. M. Bowers. Fairbanks, AK: Northern Land Use Research, pp. 267–72.

Mason, O. K. (2010). *An assessment of the Archives of Professor Helge Larsen held at the Danish National Museum, Copenhagen*. Report.

Mason, O. K. (2012). Memories of warfare: Archaeology and oral history in assessing the conflict and alliance model of Ernest S. Burch. *Arctic Anthropology* 49 (2), 72–91.

Mason, O. K. (n.d. a). From Norton to Ipiutak: The onset of sedentary salmonizers to its crisis. In *Handbook of Arctic Archaeology*. Oxford: Oxford University Press. In press.

Mason, O. K. (n.d. b). The Old Bering Sea florescence. In *Handbook of Arctic Archaeology*. Oxford: Oxford University Press. In press.

Mason, O. K. (n.d. c). Thule origins in the Old Bering Sea culture: The Inter-relationship of Punuk and Birnirk. In *Handbook of Arctic Archaeology*. Oxford: Oxford University Press. In press.

Mason, O. K. and Barber, V. (2003). A paleogeographic preface to the origins of whaling: Cold is better. In *Indigenous Ways to the Present: Native Whaling in the Western Arctic*, edited by A. P. McCartney. Circumpolar Institute, University of Alberta, Edmonton and Salt Lake City: University of Utah Press, pp. 69–108.

Mason, O. K. and Bowers, P. M. (2009). The origin of Thule is always elsewhere: Early Thule within Kotzebue Sound, cul de sac or nursery? In *The Thule Culture: New Perspectives in Inuit Prehistory: An International Symposium in Honor of Research Professor H.C. Gulløv*. Copenhagen: Danish National Museum, pp. 25–44.

Mason, O. K. and Gerlach, S. C. (1995). Chukchi sea hot spots, paleo-polynyas and caribou crashes: Climatic and ecological constraints on northern Alaska prehistory. *Arctic Anthropology*, 32 (1), 101–30.

Mason, O. K. and Jordan, J. W. (1993). Heightened North Pacific storminess and synchronous late Holocene erosion of northwest Alaska beach ridge complexes. *Quaternary Research*, 40 (1), 55–69.

Mason, O. K., Ganley, M. L., Sweeney, M., Alix, C. and Barber, V. (2007). *An Ipiutak Outlier: A Late 1st millennium AD Qarigi in Golovnin Bay*. Shared Beringian Heritage Program. Anchorage, AK: National Park Service.

McGhee, R. (1976). Differential artistic productivity in the Eskimo cultural tradition. *Current Anthropology*, 17 (2), 203–20.

Mills, R. O., Gerlach, S. C. and Bowers, P. M. (2005). Stability and change in the use of place at the Kame Terrace site, Anaktuvuk Pass, Alaska. *Anthropological Papers of the University of Alaska* (New Series), 4 (1), 27–58.

Morrow, P. and Volkman, T. A. (1975). The Loon with the ivory eyes: A study in symbolic archaeology. *Journal of American Folklore*, 88, 143–50.

Moss, M. L., and Bowers, P. M. (2007). Migratory bird harvest in Northwestern Alaska: A zooarchaeological analysis of Ipiutak and Thule occupations from the Deering archaeological district. *Arctic Anthropology*, 44 (1), 37–50.

Narroll, R. (1962). Floor area and settlement population. *American Antiquity*, 27 (4), 587–9.

Nelson, E. W. (1899). The Eskimo about Bering Strait. *18th Annual Report of the Bureau of American Ethnology for the Years 1896–1897*. Washington, D.C.: U.S. Government Printing Office.

Newton, J. I. M. (2002). *About Time: Chronological Variation as Seen in the Burial Features at Ipiutak, Point Hope*. M.A. University of Alaska.

Ostermann, H. (Editor) (1952 [1976]). The Alaskan Eskimos, as described in the posthumous notes of Dr. Knud Rasmussen. *Report of the 5th Thule Expedition 1921–1924, Vol X, No. 3*. Copenhagen: Gyldendalske Boghandel, Nordisk Forlag.

Pearson, J. L. (2002). *Shamanism and the Ancient Mind: A Cognitive Approach to Archaeology*. Walnut Creek, CA: Altamira Press.

Rainey, F. (1941). The Ipiutak culture at Point Hope, Alaska. *American Anthropologist*, 43 (3), 364–75.

Rainey, F. (1942). Discovering Alaska's oldest Arctic town. *The National Geographic Magazine*, 82 (3), 318–26.

Rainey, F. G. (1947). *The Whale Hunters of Tigara*. Anthropological Papers of the American Museum of Natural History 41(2). New York, NY: American Museum of Natural History.

Rainey, F. (1971). *The Ipiutak Culture: Excavations at Point Hope, Alaska*. Reading, MA: Addison-Wesley.

Rainey, F. and Ralph, E. (1959). Radiocarbon dating in the Arctic. *American Antiquity*, 24 (4), 365–74.

Ralph, E. and Ackerman, R. E. (1961). University of Pennsylvania radiocarbon dates IV. *Radiocarbon*, 3, 4–14.

Reanier, R. E. (1992). *Refinement of K Means Clustering: Spatial Analysis of the Bateman Site, Arctic Alaska*. Ph.D. University of Washington.

Reuther, J. D. (2009a). Lithic analysis. In *The Archaeology of Deering, Alaska: Final Report on the Village Safe Water Program*, edited by P. M. Bowers. Fairbank, AK: Northern Land Use Research, pp. 86–92.

Reuther, J. D. (2009b). Obsidian analysis data. In *The Archaeology of Deering, Alaska, Volume 3, Appendices. Final Report on the Village Safe Water Program*, edited by P. M. Bowers. Fairbanks, AK: Northern Land Use Research, pp. E-79–82.

Reuther, J. D. and Bowers, P. M. (2009). Geochronology. In *The Archaeology of Deering, Alaska: Final Report on the Village Safe Water Program*, edited by P. M. Bowers. Fairbanks, AK: Northern Land Use Research, pp. 201–17.

Rouget, G. (1985). *Music and Trance: A Theory of the Relations Between Music and Possession*. Chicago: University of Chicago Press.

Saleeby, B., Moss, M. O., Hays, J. M., Strathe, C. and Laybolt, D. L. (2009). Faunal analyses. In *The Archaeology of Deering, Alaska. Final Report on the Village Safe Water Program*, edited by P. M. Bowers. Fairbanks, AK: Northern Land Use Research, pp. 178–200.

Shepard, F. P. and Wanless, H. R. (1971). *Our Changing Coastlines*. New York, NY: McGraw-Hill.

Simmons, W. (1986). Human skeletal remains from Cape Krusenstern and Battle Rock. In *Beach Ridge Archaeology of Cape Krusenstern: Eskimo and Pre-Eskimo Settlements around Kotzebue Sound, Alaska*. Publications in Archeology 20. Anchorage, AK: National Park Service, pp. 356–61.

Smith, G. S. and Zimmerman, M. R. (1975). Tattooing found on a 1,600 year old frozen mummified body from St. Lawrence Island, Alaska. *American Antiquity*, 40, 434–7.

Stanford, D. J. (1976). *Walakpa: Its Place in the Birnirk and Thule Cultures*. Smithsonian Contributions to Anthropology 20. Washington, D.C.: Smithsonian Institution Press.

Steinacher, S. (1998). Mystery people of the arctic. *The Nome Nugget*, February 12, XCVIII (6), 1–6.

Speiss, A. (1979). *Reindeer and Caribou Hunters: An Archaeological Study*. New York, NY: Academic Press.

Stern, R., Reuther, J. D., Bowers, P. M., Gelvin-Reymiller, C. and Hays, J. M. (2010). *Cultural Resources Monitoring of Water Treatment Plant Construction (2009) in Kivalina, Alaska*. Report to Alaska Native Tribal Health Consortium, Fairbanks, AK: Northern Land Use Research.

Sutherland, P. (2001a). Shamanism and the iconography of Paleo-Eskimo art. In *The Archaeology of Shamanism*. London: Routledge, pp. 135–45.

Sutherland, P. (2001b). Bear imagery in Paleo-Eskimo art: Shamanism and traditional beliefs. *North Atlantic Studies*, 4 (1–2), 13–6.

Time magazine (1926). *Cephenemiya*. *Time*, XII [issue of April 5].

Time magazine (1941). Arctic metropolis. *Time*, XXXVII, 11, 58–9 [issue of March 17].

Time magazine (1948). Diggers. *Time*, LII, 16, 70–2 [issue of October 18].

Whitney, H. (2004). *Parasites of Caribou (2): Fly Larvae Infestations*. Publication APO10, dated July 27, Department of Natural Resources, Agriculture, Saint

Johns: Government of Newfoundland and Labrador [available at www.nr.gov.nl.ca/nr/agrifoods].

Willey, G. R. (1966). *An Introduction to American Archaeology, Volume 1: North and Middle America*. Englewood Cliffs, NJ: Prentice Hall.

Workman, W. B. (1982). Beyond the southern frontier: The Norton culture and the western Kenai Peninsula. *Arctic Anthropology*, 19 (2), 101–22.

Workman, W. B. (1992). Life and death in a first millennium AD Gulf of Alaska culture: The Kachemak tradition ceremonial complex. In *Ancient Images, Ancient Thought: The Archaeology of Ideology*, Proceedings of the 23rd Annual Chacmool Conference. Calgary: Archaeological Association of Calgary, pp. 19–25.

Xoplaki, E., Fleitmann, D., Diaz, H., von Gunten, L. and Kiefer, T. (Editors) (2011). *Medieval Climate Anomaly. PAGES News*, 19 (1), 1–32.

4 *Ancestor–descendant affinities between the Ipiutak and Tigara at Point Hope, Alaska, in the context of North American Arctic cranial variation*

BLAINE MALEY

Introduction

One of the long-standing questions in anthropology is the degree to which biological change and associated material culture are correlated over time. When presented with evidence for replacement or differentiation of associated technology and cultural attributes, to what extent are these cultural changes correlated with change in the human populations? Across the Arctic there is substantial archaeological evidence of a sweeping technological replacement event associated with advanced whale-hunting technology. This complex, broadly defined as the Thule, emerged out of the northern coast of Alaska around 1,100 years before present (BP), spreading rapidly across the Arctic (Dumond, 1987; McGhee, 2000).

The site of Point Hope, Alaska, is one of the prominent examples where archaeologists have documented this cultural transition. Sometime around 1,600 years BP, the first evidence of the Ipiutak culture at Point Hope emerges. Characterized by complex burial items, use of iron, and a distinctive ivory and wood carving tradition, the Ipiutak thrived at Point Hope for several centuries (Larsen and Rainey, 1948). While the Ipiutak culture seems to wane at Point Hope and other coastal sites around 1,300 years BP, there is continuity inland until roughly 700–800 years BP (Gerlach and Mason, 1992; cf. Mason, Chapter 3, this volume). When Point Hope is repopulated between 800–500 years BP by the Thule-affiliated Tigara population (Gerlach and

The Foragers of Point Hope: The Biology and Archaeology of Humans on the Edge of the Alaskan Arctic, eds. C. E. Hilton, B. M. Auerbach, L. W. Cowgill. Published by Cambridge University Press.

Mason, 1992; Mason, 2006), the question remains: to what extent are the Tigara descendant from the Ipiutak?

Cranial morphological variation from a geographical and temporal range of Alaskan Arctic samples has been used to examine population structure differentiation between pre-Thule and Thule-affiliated populations. To this end, biological variation patterns of these Alaskan Arctic populations have been examined to provide a comparative framework for testing the biological relationship between the Ipiutak and Tigara at Point Hope in the context of migration patterns and ancestor–descendant relationships.

The archaeology of Point Hope and the Thule transition

The site of Point Hope includes the Ipiutak and Tigara settlements, each associated with a distinct time period and material culture. A series of twenty-five ^{14}C dates from antler and wood has placed the Ipiutak culture at Point Hope securely between 1,600 and 1,300 years BP (Gerlach and Mason, 1992; Mason, 2006). After 1,300 years BP, occupation at Point Hope appears to diminish, with Ipiutak occupation ending by 1,050 years BP. There is some evidence of low-density Birnirk and early Thule occupations throughout this period until the subsequent Tigara occupation (Mason, 2009), beginning between 800 and 500 years BP (Gerlach and Mason, 1992; Mason, 2006; Jensen, Chapter 2, this volume). The material culture associated with the older Ipiutak settlement is characterized by distinct burial practices, worked ivory, and stone tool tradition. Although their subsistence strategy was aquatically based, there is also evidence of substantial reliance on caribou (Larsen and Rainey, 1948).

The Ipiutak material culture is regularly juxtaposed to that found at the later Tigara site, which fits into the more broadly defined Thule tradition. The Tigara consisted of a long continuous occupation representing the increasing complexity of the Arctic whale-hunting Thule culture. Associated with a more advanced blade technology, more modern artistic design, clay lamps, whalebone houses, and a whale- and ocean-based subsistence strategy, the material culture of the Tigara occupation clearly diverged from that of the Ipiutak (Larsen and Rainey, 1948; see Jensen, Chapter 2, this volume).

Biological anthropology research comparing the two populations is relatively scant. Dental dietary analysis comparing the Ipiutak and Tigara confirms they had different diets (Costa, 1980; El Zaatari, Chapter 6, this volume; Krueger, Chapter 5, this volume). Sexual differences in tooth wear between the two populations suggest a behavioral divergence in their division of labor (Madimenos, 2002). The only research directly comparing craniofacial typology between the Ipiutak and Tigara in an ancestor–descendant context is from 1960 (Debets, 1999; originally published in 1960 in Russian). Citing differences in cranial dimensions, body proportions, and long bone dimensions, this work claimed the Ipiutak were not direct ancestors of the Tigara, rather their similarity was a consequence of common ancestry for the two populations (Debets, 1999). Studies comparing the two populations using aDNA found similarity in haplogroups that are common among North American Arctic populations, although no exact overlap in haplotypes (Maley *et al.*, 2006; Maley, 2007).

Dating and material culture analysis of other regional sites place the Ipiutak material culture contemporaneously with the Birnirk of Northern Alaska, the Norton of the Bering Sea region south of Bering Strait, and the Old Bering Sea and Okvik cultures (Punuk) of Siberia and St. Lawrence Island (Giddings, 1960; Gerlach and Mason, 1992). These contemporaneous groups developed a set of advantageous technologies – seal-oil lamps, pottery, ground slate tools, sleds, and ice-hunting gear – not found among the Ipiutak. While there was some degree of overlap with neighboring cultures, the Ipiutak technological complex as a whole, as well as their artistic tradition, was maintained as distinct for the duration of their occupation at Point Hope.

Evidence of increased storm frequency and temperature fluctuation during the intervening period (between 1,200 and 600 years BP) has been used to suggest that the Ipiutak and other neighboring coastal cultures went inland during this period due to some combination of depleted aquatic resources (salmon and seal) and the increasingly devastating coastal weather (Hume, 1965; Mason and Gerlach, 1995; Mason, Chapter 3, this volume). After 1,200 years BP the Ipiutak material tradition was continued at the inland sites of Onion Portage, Itkillik Lake, South Meade, and Lake Tukuto, in some cases up until 500 years BP (Gerlach and Mason, 1992), although these attributions have more recently been called into question (Mason, Chapter 3, this volume). Sometime after 1,000 years BP (more commonly around 800–500 years BP), the coastal sites were repopulated, this time associated with Thule-like material culture containing technologies for more efficient exploitation of aquatic resources, including whale hunting (Mason and Gerlach, 1995). When the Tigara were established at Point Hope sometime after 600 years BP, their material culture was consistent with these repopulation events.

Evidence for Birnirk culture has been found at Point Barrow, dating between 1500 and 900 years BP. The Birnirk culture featured highly developed harpoon heads, decorative clay lamps and cooking pots, and distinct wood and ivory carving artistic traditions (Hollinger *et al.*, 2009; Mason, 2009). By 1,100 years BP, the Thule tradition appears to have emerged from the Birnirk, primarily defined by its association with advanced whale hunting and dog-pulled sleds. These allowed greater consistency in subsistence, greater mobility, and greater population size. The Thule material culture spread rapidly from the northern coast of Alaska, across northern Canada into Greenland, and eventually southwest along the Alaskan coast as far as the Aleutian Islands (Dumond, 1987;

Gerlach and Mason, 1992; Mason, 1998; Maley, 2011). Consistent with the spread of the Thule eastward, a previous study that examined cranial variation across the eastern Arctic found strong similarities between the Birnirk and the Thule peoples who had spread eastward, arguing the latter likely had little intermixing with the previous Dorset-affiliated occupants that they replaced (Maley, 2011). Because these central and eastern Arctic populations are likely derived from the Birnirk, having less similarity with the Ipiutak, they have been left out of this study.

Examining biological relationships in an ancestor–descendant context

Testing the hypothesis of ancestry and descent in modern human populations using morphological cranial variation is an exercise in degree. These relationships are universally complicated by a number of dynamic variables not limited to changes and differences in population size, local and distance admixture, and changes due to selective and environmental forces. The assumption that all Arctic people share common ancestry is based on genetic research that has shown that Arctic populations share a relatively unique common set of specific sub-haplogroups. Specific to Circum-Arctic populations, analysis of mtDNA diversity in Chukchi, northern Siberian, Native Alaskan, and Greenland populations contain only A and D haplogroups, with trace amounts of C (Shields *et al.*, 1993; Ward *et al.*, 1993; Merriwether *et al.*, 1995; Starikovskaya *et al.*, 1998; Saillard *et al.*, 2000; Derbeneva *et al.*, 2002; Rubicz *et al.*, 2003; Tamm *et al.*, 2007). This is in line with the archaeological record, which demonstrates evidence of historical migration-based population expansions with common material cultures and systems of niche construction. Despite human presence in Beringia during the early colonization of the Americas, archaeological evidence in the Arctic is sparse and equivocal until around 6,000–5,000 years BP (Dumond, 1987; see also Jensen, Chapter 2, this volume).

Based on the assumption of common ancestry, this research uses analytical methods that employ morphological variation to examine the biological relationships between the pre-Thule and Thule populations of western Alaska. This analysis is used to contextualize the relationships of the distinct cultural occupations at Point Hope. These relationships have been clarified using a comparative framework of Alaskan Arctic peoples employing a number of statistical methods that incorporate the variance/covariance of quantitative trait variation among the population samples.

To put this issue into a testable framework, the null hypothesis of similarity in variability patterns between the two pre-Thule population samples relative

Table 4.1. *Population list described by location (see Figure 4.1) according to museum acquisition logs with sample sizes used in this study*

ID	Sample	*n*	Affiliation	Museum	Region	Longitude	Latitude
1	Siberia	32	Thule	AMNH	Siberia	−172.25	64.407
2	Dillingham	40	Thule	SI	W Arctic	−155.262	59.877
3	Kuskoskwim	33	Thule	SI	W Arctic	−162.214	60.091
4	Bethel	37	Thule	SI	W Arctic	−161.834	60.77
5	Hooper Bay	25	Thule	SI	W Arctic	−166.097	61.531
6	Pastolik	53	Thule	SI	W Arctic	−163.304	62.998
7	Pilot Station	34	Thule	SI	W Arctic	−161.97	61.753
8	Holy Cross	42	Thule	SI	W Arctic	−159.562	62.682
9	Teller	35	Thule	SI	NW Arctic	−166.356	65.258
10	Wales	34	Thule	SI	NW Arctic	−168.101	65.656
11	Mitliktavik	33	Thule	SI	NW Arctic	−167.522	65.816
12	Sarichef	27	Thule	SI	NW Arctic	−166.123	66.234
13	Ipiutak, Point Hope	54	Ipiutak	AMNH	N Arctic	−166.72	68.351
14	Tigara, Point Hope	216	Thule	AMNH	N Arctic	−166.793	68.349
15	Point Barrow	38	Birnirk	SI	N Arctic	−156.715	71.29
16	Kittegazuit	42	Thule	CMC	N Arctic	−133.869	69.315

Cultural affiliations are presented to help define temporal affiliations. With the exception of the Tigara, Ipiutak, and Point Barrow samples, most of the other samples have little provenience and are commonly accepted as historic Thule.
Latitude and longitude are utilized for comparing the biological distances to geographical distances. Shading reflects different regional affiliations.
Museum key: American Museum of Natural History (AMNH), National Museum of Natural History – Smithsonian Institution (SI), Canadian Museum of Civilization (CMC).

to the Thule samples is tested. These variability patterns are used to quantify biological relationships by taking several statistical approaches, including principal coordinate analysis, biological distance analysis, Relethford and Blangero analysis, and Mantel testing of biological distance and geographic distance. The results of these analyses are then used to assess models for population movements among the different past cultures in western and northern Alaska.

Materials and measurements

Arctic population samples

To provide a comparative framework for testing hypotheses of ancestry and descent across the Alaskan Arctic, a number of discrete population samples were collected from an extensive geographic range, including an Arctic Siberian sample (Table 4.1 and Figure 4.1). See Maley (2011) for additional specific

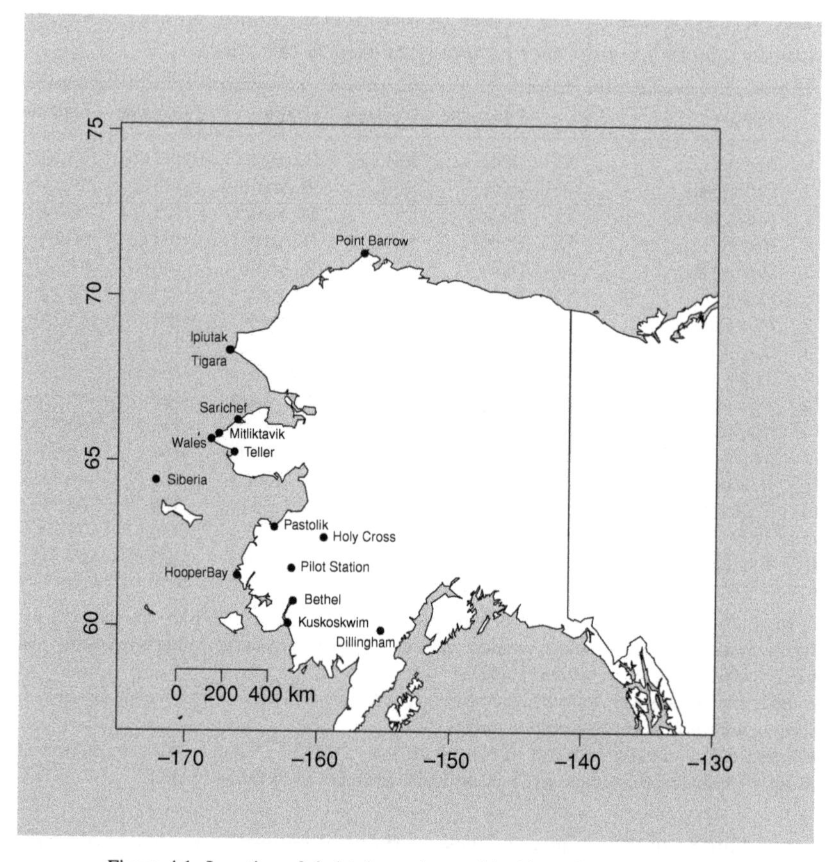

Figure 4.1. Location of skeletal samples used in this study (see Table 4.1 for additional information).

information about the sample composition. In addition to analysis of the samples separately, the population samples were affiliated with other geographically proximate populations to examine relevant patterns of regional variation. The samples were sorted into western Alaska, northwestern Alaska, and northern Alaska regions, based on their occupation of the three distinct Alaskan peninsulas, each separated by far-reaching inland water channels (Table 4.1). The Birnirk sample from Point Barrow is referred to as the Point Barrow sample. This needs to be clarified because an earlier study by Hollinger (2009) additionally utilized a later population sample from this area that has since been repatriated.

Locations of sites were defined by latitude and longitude in Google Earth. These were based on museum locality descriptions, historic maps, and published archaeological accounts. Pairwise geographic distances in kilometers were calculated from the latitude and longitude data in the R programming language (see "Data processing" below).

Cranial measurements

The cranium was divided into a series of traits (measurements listed in Table 4.2) that quantify global cranial shape according to Howells (1973), Moore-Jansen *et al.* (1994), and Maley (2011). The pattern of variance and covariance around the multivariate trait means was used to define relative differentiation among the samples being studied. This pattern provides the basis for calculating a number of statistics that define population structure and biological relationships.

Because of this relatively stable relationship between covariance structures, it is typical in studies that incorporate the patterns of morphological traits to utilize the phenotypic variance/covariance (V/Cov) matrix (**P**) as a proxy for the genetic V/Cov matrix (**G**) even though what is of interest is the genetic correlation among the cranial measurements. As long as the populations under examination have similar structural integration and trait heritabilities across populations, **P** will be proportional to **G**, a prediction that has been empirically supported among a wide range of studies between closely and distantly related taxa (Arnold, 1981; Atchley *et al.*, 1981; Cheverud, 1988, 1995, 1996; Kohn and Atchley, 1990; Venable and Burquez, 1990; Konigsberg and Ousley, 1995; Roff, 1995, 1996; Arnold and Phillips, 1999; Marroig and Cheverud, 2001; Marroig *et al.*, 2009; Porto *et al.*, 2009).

Data processing

All data handling and statistical analyses were performed in the R programming language (R Development Core Team, 2008) unless otherwise noted. Landmark data were collected using the Microscribe G2 digitizer and Microscribe Utility Software (Immersion, San Jose, CA). The three-dimensional landmark data were imported into R and converted into pairwise measurements for each sample. Additional information about measurement methods and errors are reported in Maley (2011). Missing data were filled using multiple imputation methods using the mice package in R (Horton and Kleinman, 2007).

Table 4.2. *Cranial measurements and their definitions according to Howells (1973), Moore-Jansen* et al. *(1994), and Maley (2011)*

ID	Measurements	Landmarks
1	Maxillo-alveolar length[a]	Orale to staphylion
2	Posterior segment of basion–nasion length[a]	Hormion to basion
3	Foramen magnum length	Basion to opisthion
4	Occipital cord length	Opisthion to lambda
5	Parietal cord length	Lambda to bregma
6	Frontal cord length	Nasion to bregma
7	Nasal height	Nasospinale to nasion
8	Basion–prosthion height – nasal height[a]	Orale to nasospinale
9	Nasio-occipital length[a]	Nasion to lambda
10	Basion–prosthion length[a]	Orale to basion
11	Basion–bregma height	Basion to bregma
12	Maxillo-alveolar breadth	Ectomolare (L) to ectomolare (R)
13	Upper facial breadth[a]	Frontomolare orb (L) to Frontomolare orb (R)
14	Bizygomatic diameter	Zygion (L) to zygion (R)
15	Biasterionic breadth	Asterion (L) to asterion (R)
16	Bimaxillary breadth	Zygomaxillare (L) to zygomaxillare (R)
17	Lateral cranial length[b]	Porion (R) to zygomaxillare (R)
18	Malar length, maximum	Zygomaxillare (R) to zygoorbitale (R)
19	Lateral facial length[b]	Frontomolare orb (R) to asterion (R)
20	Orbit height[a]	Frontomolare orb (R) to zygoorbitale (R)
21	Biauricular breadth (from sagittal plane)[a]	Bregma to porion (R)
22	Lamdoid suture cord length[b]	Lambda to asterion (R)
23	Coronal suture cord length[b]	Frontomolare orb (R) to bregma
24	Latero-posterior cranial length[b]	Porion (R) to asterion (R)
25	Posterior nasopharynx height[b]	Staphylion to hormion
26	Nasal breadth	Alare (L) to alare (R)
27	Mid-lateral nasopharynx length[b]	Sphenoid 8 to alare (R)
28	Interorbital breadth	Dacryon (L) to dacryon (R)
29	Anterior segment of basion–nasion length[a]	Hormion to nasion

[a] Used as a close proxy for Howells (1973) and Moore-Jansen (1994) measurements. These measurements use proxy landmarks of similar location, or substitute a subset of the measurements due to collection methods and consistency: orale for prosthion, staphylion for alveolon, frontomolare orbitale (Orb) for frontomolare temporale, lambda for opisthocranion, and porion for auriculare.
[b] From Maley (2011).

Sex and age

To control for the potentially confounding impact that sex differences have on measurements of population structure and differentiation, variation attributable to size-related sexual dimorphism has been removed using z-score standardization. The crania in each sample were placed into male (M) or female (F) groups based on a suite of dimorphic characteristics utilized for quantifying sexual differences in morphological studies (White and Folkens, 2000). These include general levels of robusticity specific to the nuchal crest, mastoid process, supra-orbital margin, supra-orbital ridge, and mental eminence of the mandible. After separating the population samples by sex (M and F), each sex category was pooled across all samples and standardized by dividing each individual's difference from the group mean by the group standard deviation for each trait. The M and F groups were then pooled and separated according to population sample.

Because age and developmental status can be confounding factors when comparing size and shape, only fully developed adult individuals were used for this research. Individuals were aged as adults based on dentition, requiring full eruption of the upper M3, or, in cases where it was clear no M3 was present or developing, of the M2.

Methods

All statistical analyses (unless otherwise noted) were calculated using the twenty-nine traits (Table 4.2) for all population samples (Table 4.1).

Principal coordinates analysis

Principal coordinates analysis (PCA) uses matrix decomposition of the V/Cov matrix to reduce datasets containing a large number of variables into several principal coordinate vectors that explain the greatest amount of variation in the dataset. For this research, the principal coordinates were plotted to explore the relationship among the different samples and Arctic regions in a graphical context. The eigenvalues and eigenvectors (scaled by dividing by the square root of the corresponding eigenvalues) were calculated from the **R**-matrix (see Table 4.3 for definition) output from RMET v5.0 (Relethford, 1997), then plotted for all samples between the first and second principal coordinates, as well as the first and third principal coordinates. Ninety-five percent confidence ellipses, calculated using the "car" package in R (Fox and Weisberg, 2011), were placed around the population samples included in each of the represented Arctic regions to demonstrate the differences in regional variation and overlaps in specific population affiliation.

Biological distance (D^2)

Biological distances and Relethford–Blangero statistics were calculated using RMET v5.0 (Relethford, 1997) for all pairwise samples based on their

Table 4.3. *Definitions of biological distance and Relethford–Blangero statistics used in this study*

Statistic/term	Definition
R-matrix	The standardized V/Cov matrix of sample relationships based on each sample's deviation from the total sample mean, the elements of the matrix represent the biological similarity between each pair of population samples.
r_{ii}	Phenotypic distance to the total population centroid for each sample and region (the diagonal elements of the R-matrix).
D^2	Biological pairwise distance between all samples and regions, calculated from the R-matrix elements: $d_{ii}^2 = r_{ii} + r_{jj} - 2r_{ij}$.
$\bar{v}_{Gi}$	Observed mean phenotypic variance for each sample and region over all traits.
$E(\bar{v}_{Gi})$	Expected mean phenotypic variance for each sample and region, which is the observed value weighted by its deviation from the average distance to the total sample centroid.
$\bar{v}_{Gi} - E(\bar{v}_{Gi})$	Residuals between observed and expected phenotypic variance for each sample and region.

Relethford (1994, 1997)

deviation from the total sample centroid. Mahalanobis D^2 is used to estimate the differences among the sample groups. The D^2, along with sample variance (presented in the form of a 95% confidence interval) for each population pairwise distance, provides the population biological relationship breakdown between the pre-Thule and Thule sample groups for examining ancestry and descent.

Relethford–Blangero analysis

Relethford–Blangero analysis provides a tool for examining population structure as a function of deviation from expected variation levels (Relethford and Blangero, 1990). This method compares observed within-sample phenotypic variance to total variance expected under panmixia of the entire population set. The observed phenotypic variance value was calculated using the diagonal of the phenotypic V/Cov matrix (**P**) for each sample, averaging over all twenty-nine traits. The expected variance for a sample is based on the average deviance among all samples, weighted by the sample's deviation from the total sample centroid (Relethford and Blangero, 1990). Although the analysis assumes the populations are contemporaneous, and panmixia is a theoretical possibility, the results still provide useful information for understanding the biological relationship in an ancestor–descendant context. Through an examination of deviations from expected variance relationships the impact of local versus distant gene flow can be assessed. Large positive residual values indicate greater than expected admixture from afar, while negative values suggest isolation or localized admixture from neighboring or closely related populations. In the context of the ancestor–descendant relationship, large positive deviations may indicate a possible population bottleneck event followed by population expansion.

Matrix comparison with other distance metrics

Biological distances were compared to geographic distances to examine whether specific pre-Thule affiliation improved the relationship between biological and geographical distance. To compare the structure of relationship matrices from different data sources, the Mantel test of matrix correspondence has been applied (Mantel, 1967; Sokal, 1979; Smouse *et al.*, 1986). The Mantel test statistics provide a correlation of the matrices based on random permutation of matrix elements and statistical support for matrix structure similarities between them. In this way, the geographical pairwise distance matrix of all populations was compared to the biological distance (D^2) matrix calculated from RMET. The expectation of this model is that populations that are close geographically will be biologically more similar, and those that are farther apart will be biologically less similar. In addition to an overall model, Mantel testing was done separately for each pre-Thule sample relative to the Thule "descendants." Mantel testing was done in R using the Mantel function in the "ecodist" package (Goslee and Urban, 2007).

Results and analysis

The results of the analyses performed are presented in the context of the comparative framework between the Ipiutak and Point Barrow population samples relative to the other Alaskan Arctic population samples and their affiliated regions.

Principal coordinate analysis

Along with each of the pairwise PCA plots, the regional group affiliations are shown in the context of the 95% confidence ellipse (CE) for each region (Figure 4.2). The 95% CE represents the area that 95% of the time would be expected to include the samples within the region. When specific samples lie in the 95% CE for another region, there are several ways this can be interpreted.

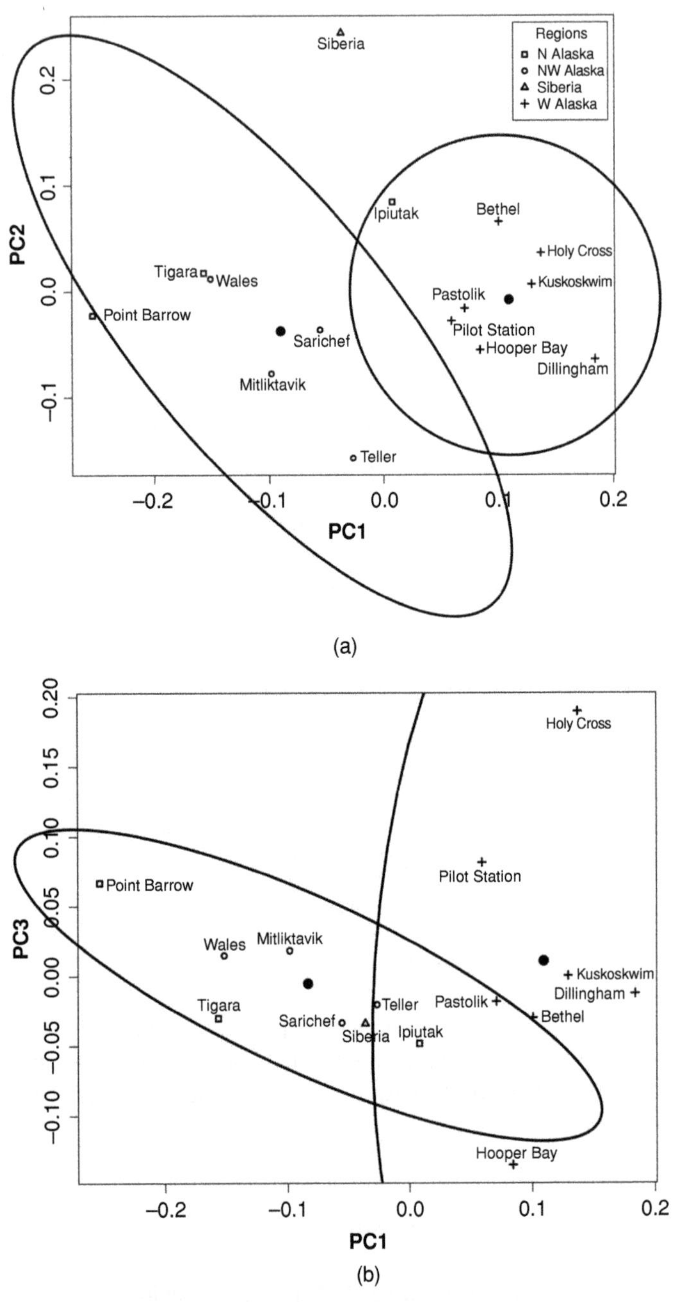

Figure 4.2. Scatter plots of the results from the PCA. (a) PC2 (20.6%) vs. PC1 (40.1%) for all Alaska samples. The first two principal components encompass 60.7% of the total variation. (b) PC3 (13.3%) vs. PC1 (40.1%) for all Alaska samples. The first and third principal components encompass 53.4% of the total variation. The 95% confidence ellipse for the N Alaska and NW Alaska regional groups is shown.

First, samples that fit within the 95% CE for another region fit within the expected pattern of variation of the other region with 95% confidence. This suggests a higher degree of shared ancestry or gene flow with the other samples in that region. Second, as in the case of northern Alaska, the within-region variance is so large as to suggest the samples are no more likely to belong in that region than they are to any other region. For the analysis and discussion of principal coordinates, northern Alaska was not analyzed as a region so that specific population affinities for the northern Alaskan population samples relative to other regions could be examined more closely. With a single sample, Siberia could not be assessed for regional variance.

PC2 vs. PC1 (Figure 4.2a) Encompassing 60.7% of the total variation in the relationship matrix, this plot shows the Ipiutak well within the 95% CE for the western Alaska region, but outside the CE for the northwestern Alaska region. The Tigara population sample resides only within the 95% CE for the northwestern Alaska region, while the Point Barrow population sample rests on the margin of this region. There is no overlap between the western Alaska and northwestern Alaska samples, suggesting a quantifiable differentiation between these geographic "isolates."

PC3 vs. PC1 (Figure 4.2b) Encompassing 53.4% of the total variation in the relationship matrix, this plot shows all three of the northern Alaska samples within the 95% CE of the northwestern Alaska region, while only the Ipiutak also aligns within the 95% CE of the western Alaska region.

This distribution provides support that the pre-Thule samples of the Ipiutak and Point Barrow may have had divergent ancestor–descendant pathways. Assuming greater similarity means a greater degree of shared ancestry, these plots suggest the Ipiutak may have played an ancestral role among the western Alaska populations, as well as to the northwestern Alaska and Tigara populations. In contrast, the Birnirk at Point Barrow appear to play a substantial ancestral role in northwestern Alaska, and little direct ancestral role in western Alaska. These plots also weakly suggest the Point Barrow Birnirk may possess a stronger ancestor–descendant relationship to the Tigara than the Ipiutak.

Biological distance (D^2)

Biological distances, including 95% confidence intervals (CI), between the Ipiutak, Tigara, and Point Barrow to all other samples are provided in Table 4.4 and Figure 4.3. By contrasting the two pre-Thule population samples in a regional context, a contrast in regional descent emerges between the Ipiutak and the Point Barrow population samples.

Table 4.4. *Biological distances (D^2) between the Ipiutak and Point Barrow samples and all other populations. Regional divisions are shown in alternating grey highlights, matching Table 4.1*

ID	Population	Ipiutak	95% CI	Tigara	95% CI	Barrow	95% CI
1	Siberia	0.069	(0.035–0.103)	0.091	(0.059–0.122)	0.153	(0.102–0.205)
2	Dillingham	0.070	(0.038–0.102)	0.132	(0.099–0.165)	0.212	(0.156–0.267)
3	Kuskoskwim	0.040	(0.012–0.068)	0.107	(0.074–0.140)	0.158	(0.107–0.210)
4	Bethel	0.024	(0.001–0.047)	0.089	(0.060–0.118)	0.150	(0.101–0.198)
5	Hooper Bay	0.058	(0.022–0.094)	0.094	(0.058–0.130)	0.159	(0.102–0.216)
6	Pastolik	0.049	(0.023–0.074)	0.079	(0.056–0.103)	0.134	(0.092–0.177)
7	Pilot Station	0.079	(0.044–0.114)	0.081	(0.052–0.110)	0.118	(0.073–0.163)
8	Holy Cross	0.085	(0.052–0.119)	0.152	(0.117–0.186)	0.186	(0.134–0.237)
9	Teller	0.081	(0.046–0.117)	0.069	(0.043–0.096)	0.105	(0.062–0.148)
10	Wales	0.056	(0.025–0.086)	0.036	(0.015–0.057)	0.041	(0.011–0.072)
11	Mitliktavik	0.090	(0.052–0.127)	0.059	(0.033–0.085)	0.065	(0.028–0.101)
12	Sarichef	0.049	(0.016–0.082)	0.057	(0.029–0.086)	0.084	(0.041–0.127)
13	Ipiutak	0.000	—	0.064	(0.043–0.085)	0.114	(0.075–0.153)
14	Tigara	0.064	(0.043–0.085)	0.000	—	0.049	(0.027–0.071)
15	Point Barrow	0.114	(0.075–0.153)	0.049	(0.027–0.071)	0.000	—

The Ipiutak, when compared directly to Point Barrow, have statistically significant lower D^2 values to all western Alaska samples, with the exception of Pilot Station, which although lower is not statistically significant when accounting for sampling variance. For the northwestern Alaska samples, including the Tigara, there are no statistically significant differences in the biological relationship between Ipiutak and Point Barrow populations.

Examining the pattern of relative D^2 values between the Ipiutak and Thule affiliated samples, the average D^2 relationship is lower to western Alaska (0.058) compared to the northwestern Alaska groups (0.069), although only several of the samples between the regions show statistically significant deviations. In contrast, Point Barrow shows a large deviation between average D^2 to western Alaska groups (0.160) relative to its average relationship to northwestern Alaska groups (0.073). In addition, several of the relationships between Point Barrow and northwestern regional groups, including the Tigara, are significantly lower than its D^2 to all but two of the western Alaska populations.

In terms of the D^2 relationships to the Tigara, values of 0.064 (0.043–0.085) and 0.049 (0.027–0.071) for the Ipiutak and Point Barrow population samples respectively, although lower for Point Barrow, are not statistically significantly different when accounting for sampling variance. When examining the relationship of the Tigara to the western Alaska and northwestern Alaska regions, although not similar in magnitude to the Point Barrow D^2 values, the

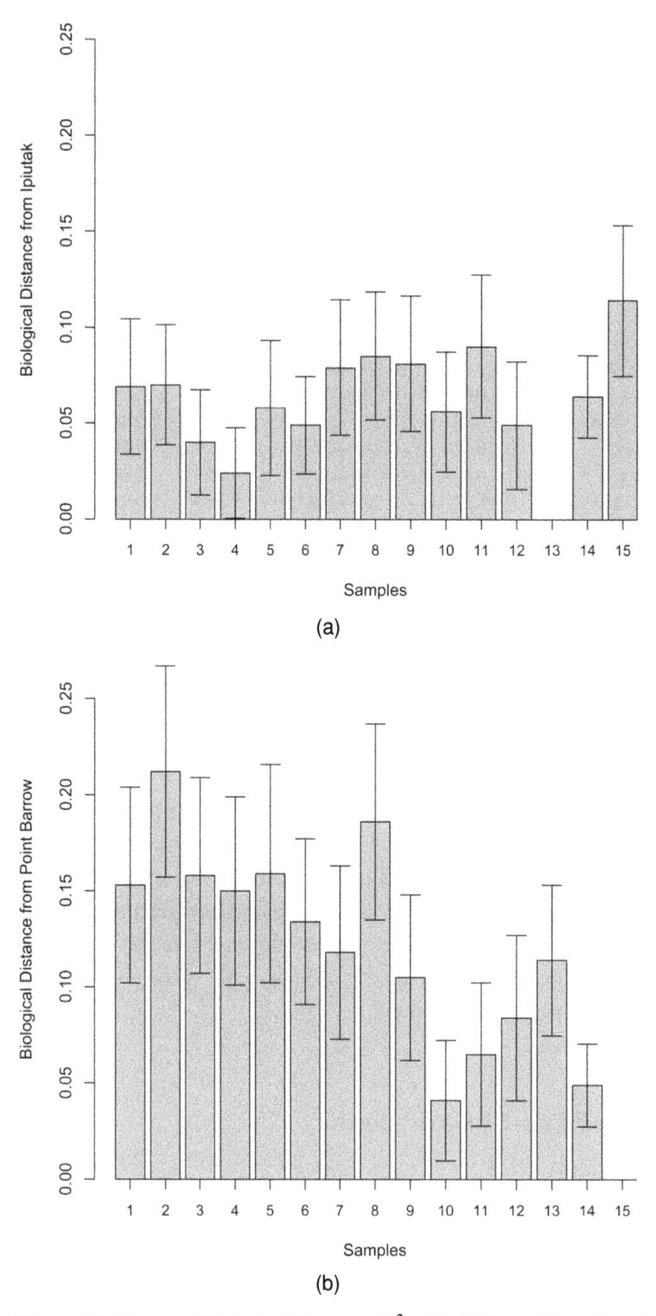

Figure 4.3. Barplot of biological distances (D^2) with 95% confidence bars. (a) The D^2 between the Ipiutak and all other Alaskan populations. (b) The D^2 between the Point Barrow sample and all other populations. The sample numbers refer to Table 4.4.

Table 4.5. *Relethford–Blangero analysis: Phenotypic distance to the centroid and 95% confidence interval, observed mean phenotypic variance, expected mean phenotypic variance, and residuals between observed and expected values*

ID	Sample	r_{ii}	95% CI	$\bar{v}_{Gi}$	$E(\bar{v}_{Gi})$	$\bar{v}_{Gi} - E(\bar{v}_{Gi})$
1	Siberia	0.067	(0.042–0.093)	0.846	0.872	−0.026
2	Dillingham	0.04	(0.022–0.058)	0.948	0.898	0.051
3	Kuskoskwim	0.018	(0.002–0.034)	0.928	0.918	0.009
4	Bethel	0.016	(0.002–0.031)	0.822	0.92	−0.098
5	Hooper Bay	0.028	(0.006–0.050)	0.803	0.909	−0.107
6	Pastolik	0.01	(0.000–0.019)	0.965	0.926	0.039
7	Pilot Station	0.018	(0.002–0.033)	0.96	0.919	0.041
8	Holy Cross	0.058	(0.038–0.079)	0.866	0.88	−0.015
9	Teller	0.038	(0.019–0.058)	0.828	0.899	−0.071
10	Wales	0.028	(0.010–0.045)	0.795	0.909	−0.115
11	Mitliktavik	0.027	(0.009–0.045)	0.874	0.91	−0.035
12	Sarichef	0.02	(0.001–0.040)	0.751	0.916	−0.165
13	Ipiutak	0.026	(0.013–0.039)	0.989	0.911	0.079
14	Tigara	0.042	(0.035–0.050)	0.967	0.895	0.072
15	Point Barrow	0.079	(0.054–0.104)	1.202	0.861	0.341

pattern is similar, showing the smallest average difference to the northwestern Alaska regional groups (0.055), and a larger average distance to the western Alaska regional groups (0.105).

This analysis suggests that the Ipiutak likely played a greater ancestral role among the western Alaska populations than the Birnirk at Point Barrow. The Point Barrow and Tigara show less similarity to the western Alaska populations, and appear to have more in common with the northwestern Alaska populations. Based on greater similarity and a similar regional relationship pattern across the Alaskan Arctic, the Tigara appear to be slightly more similar to the Birnirk at Point Barrow than to the Ipiutak.

Relethford–Blangero analysis

The Relethford–Blangero residuals are calculated as the difference between expected and observed biological variance (Table 4.5). The residuals are arranged according to region to reflect western to northern geographic position across the Alaskan Arctic (Figure 4.4).

A regional patterning among population samples becomes apparent, such that the western Alaska samples, while varying in sign, all have relatively small

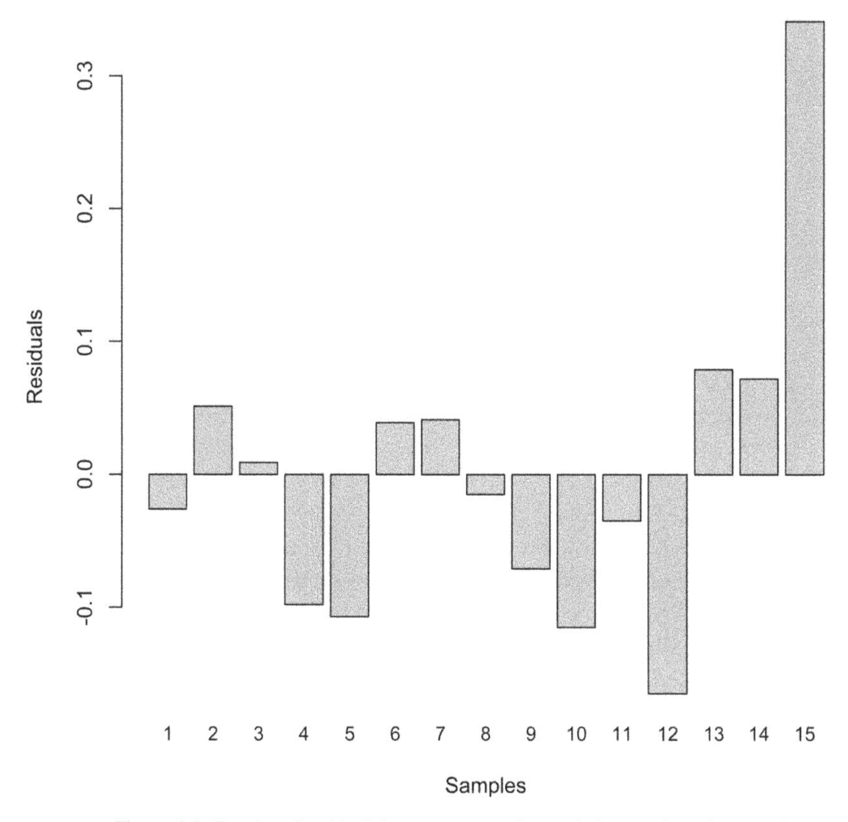

Figure 4.4. Barplot of residuals between mean observed phenotypic variance and expected phenotypic variance: $\bar{v}_{Gi} - E(\bar{v}_{Gi})$. The sample numbers refer to the sample list in Table 4.5.

deviations from the mean. The northwestern Alaska samples all have negative residual values, suggesting relative lack of substantial gene flow from afar, while maintaining regional intermixing and greater likelihood of common ancestry. The northern Alaska samples all show positive deviations, with magnitudes greater than all samples from the other Alaska regions. Positive deviations are generally suggestive of increased distant gene flow and greater isolation from local groups. However, in this case, the Point Barrow and Ipiutak are both pre-Thule affiliated populations, which also helps explain the increased observed variation among these samples. The high residual value of the Point Barrow suggests substantially more phenotypic variation in the population than would be expected, which is in line with previous research on this population (Ossenberg, 2005; Hollinger *et al.*, 2009).

Table 4.6. *Results of the Mantel test for biological distance (D^2) versus geographic distance for all samples, for all samples without Point Barrow, and for all samples without Ipiutak*

Test	Mantel r	95% CI	p-value
D^2–geodistance	0.670	(0.484–0.786)	<0.001
Without Point Barrow	0.542	(0.422–0.695)	<0.001
Without Ipiutak	0.769	(0.607–0.847)	<0.001

Note that the Siberian sample was removed for all tests.

Testing the relationship between biological and geographical distance

The Mantel statistics (Table 4.6) provide a test for correlation between biological distance and geographic distance. The significant Mantel r statistic for all samples (the Siberian sample was removed for all tests due to its deviant geographical relationship with the other samples) shows a statistically significant correlation of 0.670. To put this into a testable framework, the analysis was run separately for all samples without the Point Barrow sample, and for all samples without the Ipiutak sample to see if one of the pre-Thule samples provided a significantly better fit (higher correlation) to an isolation-by-distance model of differentiation. While the Mantel r was higher for the Point Barrow group 0.769 (0.607–0.847) than that containing the Ipiutak sample 0.542 (0.422–0.695), the slight overlap of the 95% confidence intervals failed to reject the hypothesis of stronger correlation for Point Barrow.

Discussion

As stated in the Introduction to this chapter, the question of continuity and biological relationship between Alaskan Arctic populations from different locations and different time periods is one of degree, because all of these Arctic populations share common ancestry in the not so distant past. The story of an ancestor–descendant relationship between the Ipiutak and Tigara must be evaluated in the context of their biological relationship to each other and to the pattern of their relationship to other Alaskan populations, including another potential source of Tigara ancestry, the Thule-deriving Birnirk at Point Barrow.

The Ipiutak occupied Point Hope between 1,600 and 1,300 years BP. As the population dwindled, there is evidence of increased coastal weather fluctuations and the possible movement inland into northwest Alaska (Mason, 2009; Mason,

Chapter 3, this volume). Sometime after 1,000 years BP (most dates are between 800 and 500 years BP), the coastal sites were repopulated with people associated with "Thule"-like material culture, which contained technologies for efficient exploitation of aquatic resources including whale hunting, generally considered to arise out from the Birnirk at Point Barrow (Mason and Gerlach, 1995; Hollinger *et al.*, 2009). The Tigara occupation at Point Hope began by 800 to 600 years BP. The question is: where did the Ipiutak go and to what extent do they play an ancestral role to the Tigara and other Thule-associated populations across the Alaskan Arctic?

The data presented in this chapter yield little substantive evidence tying the Ipiutak and the Tigara together in a direct ancestor–descendant relationship, due to the strong affinity between the Tigara and Birnirk at Point Hope and little comparative similarity between the Ipiutak and Point Barrow. According to results of the biological distance analysis and the PCA, the Ipiutak have the strongest association with the western Arctic populations, and a weaker association with the northwestern Alaska populations. In contrast, the Birnirk at Point Barrow show a significantly more distant relationship to the western Alaska populations, and a similar relationship to the northwestern Alaska and Tigara populations.

Figure 4.5 shows the scenario of population movement across the Alaskan Arctic that best summarizes the results of the quantitative analyses. These maps explain the ancestor–descendant relationship between the Ipiutak (Figure 4.5a) and Birnirk at Point Barrow (Figure 4.5b) to the Thule-affiliated populations strung along the rest of the Alaskan Arctic at the time of European contact. Figure 4.5a shows that sometime after 1,300 years BP, the Ipiutak left Point Hope and went inland during the period of intensified coastal weather. The Ipiutak archaeological complex has been excavated at inland sites of South Meade, Itkillik Lake, Lake Tukuto, and Onion Portage, dating up to 500 years BP (but see Mason, Chapter 3, this volume). As the weather improved along the coast, populations begin to re-occupy the coastal areas. The close biological relationship between the Ipiutak and many of these western and northwestern Alaska populations suggests the descendants of the Ipiutak at Point Hope migrated back to the coastal areas, this time moving south, deriving the western Alaska populations. There is also support from the analyses that the Ipiutak may have variably intermixed with Birnirk descendants to derive the northwestern Alaska populations and the Tigara.

Figure 4.5b shows a scenario of the pattern of descent from the Birnirk at Point Barrow. Previous work has suggested the Birnirk (who occupied Point Barrow shortly after the Ipiutak began to occupy Point Hope) were the originators of the Thule culture, which rapidly spread throughout the Arctic (Ossenberg, 2005; Hollinger *et al.*, 2009; Maley, 2011). The Birnirk have

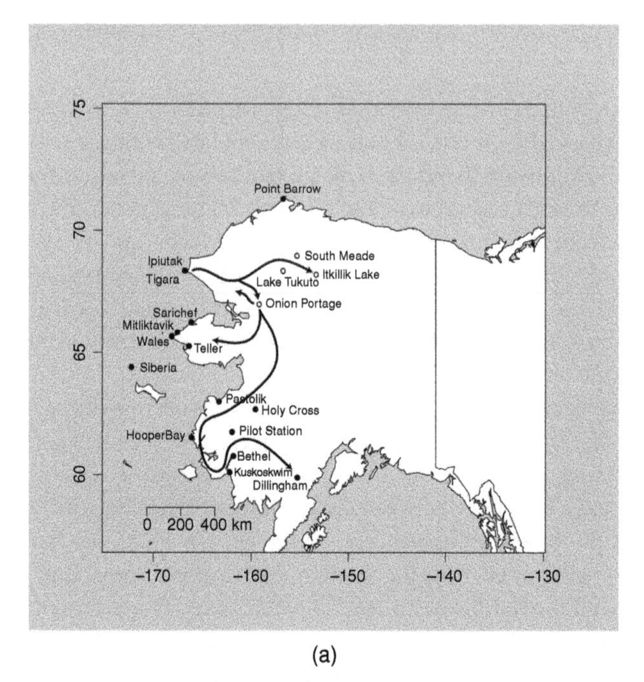

(a)

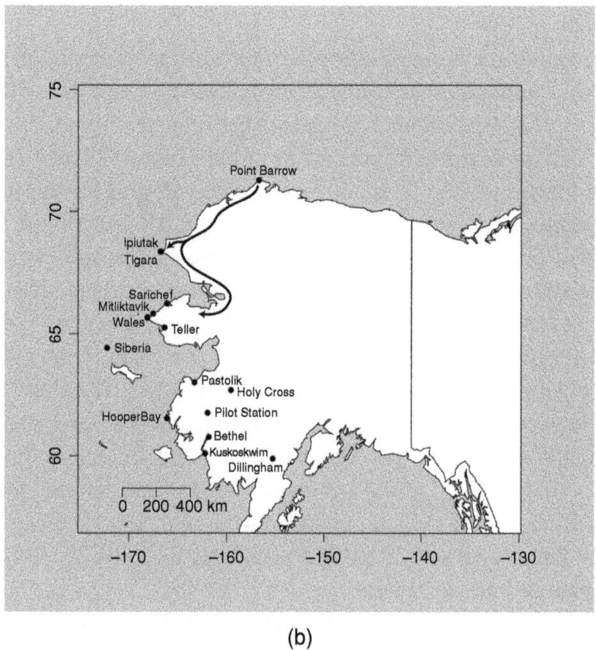

(b)

Figure 4.5. Maps showing the hypothetical movements of populations based on the results. (a) The proposed movement of the Ipiutak as a likely source of ancestry to the northwestern and western Arctic historic populations. (b) The proposed movement of the Birnirk at Point Barrow.

been described as culminating out of a number of previous traditions spanning the Arctic. Perhaps Point Barrow was a melting pot of other Arctic regions 1,100 years BP, with more distantly related populations coming together for cultural events or sharing technology. The high expected/observed variance residual of the Point Barrow sample in the Relethford–Blangero analysis helps explain Point Barrow as the origination of the Thule at this place and time, and the initiation of large-scale, technologically advanced whale hunting.

Conclusion

The pattern of population relationships between the pre-Thule and Thule populations spread over the large and challenging arctic environment is difficult to fully quantify due to the likelihood of a complicated mix of isolation, fragmentation, and gene flow after the initial founding event sometime prior to 5,000 years BP. The focus of this chapter has been to examine the relationship between the Ipiutak and the Tigara at Point Hope. However, to understand this relationship it is necessary to examine their relationship in the broader context of Alaskan Arctic variation patterns and ancestor–descendant relationships. Based on the analyses presented in this work, a pattern of ancestry and descent emerges from two primary sources, the Ipiutak at Point Hope and the Birnirk at Point Barrow. Overlapping in time, these two occupations along the northwest and north coast of Alaska appear to be fundamentally important in their contribution to the formation of human occupations across the Alaskan Arctic at the time of European contact.

Previous research suggests that as Point Hope became less hospitable the Ipiutak went inland and south. Perhaps, as the coastal climate improved, Ipiutak descendants returned to the coast, interacting to some extent with the Punuk and Birnirk derived populations. While the Ipiutak appear to have similarity to the Tigara, they also have a strong affinity to populations found throughout the western and northwestern Alaskan Arctic.

The Tigara have less in common with the western Alaska populations, likely intermixing more frequently with neighboring populations in northwestern Alaska. Although the Tigara likely have some ancestral relationship with the Ipiutak, they appear to have a slightly stronger affinity with the Birnirk population from Point Barrow, both in biological similarity and overall relationship patterns across the Alaska.

The overall scenario this research supports is that the Ipiutak from Point Hope and the Birnirk from Point Barrow both played a substantive ancestral role in the Tigara. In addition, the Ipiutak likely had a considerable ancestral role among the western and northwestern Alaska populations, while the Birnirk at Point Barrow appear to have had the greatest ancestral impact primarily in the Tigara and northwestern Alaskan populations. While these analyses do not point to a clean and direct lineage of descent from either pre-Thule population, the similarities and patterns of variation between them strongly suggest some degree of ancestry across the Alaskan Arctic from both groups.

Acknowledgements

I would like to thank the editors of this volume Charles Hilton, Libby Cowgill, and Benjamin Auerbach. For access to collections I would like to thank David Hunt at the Smithsonian Institution, and Ken Mowbray, Gisselle Garcia, and Ian Tattersall at the American Museum of Natural History. I would also like to extend a very special thanks to the Native American groups whose data I collected. Without their consent and support, this research would not have been possible. This work was funded by Washington University in St. Louis, Des Moines University, and the National Science Foundation (Doctoral Dissertation Improvement Grant # 0752134).

References

Arnold, S. J. (1981). Behavioral variation in natural populations. I. Phenotypic, genetic and environmental correlations between chemoreceptive responses to prey in the garter snake, *Thamnophis elegans. Evolution*, 35, 489–509.

Arnold, S. J. and Phillips, P. C. (1999). Hierarchical comparison of genetic variance-covariance matrices. II. Coastal-island divergence in the garter snake, *Thamnophis elegans. Evolution*, 53, 1516–27.

Atchley, W. R., Rutledge, J. J. and Cowley, D. E. (1981). Genetic components of size and shape. II. Multivariate covariance patterns in the rat and mouse skull. *Evolution*, 35, 1037–55.

Cheverud, J. M. (1988). A comparison of genetic and phenotypic correlations. *Evolution*, 42 (5), 958–68.

Cheverud, J. M. (1995). Morphological integration in the saddle-back tamarin (*Saguinus fuscicollis*) cranium. *The American Naturalist*, 145 (1), 63–89.

Cheverud, J. M. (1996). Quantitative genetic analysis of cranial morphology in the cotton-top (*Saguinus oedipus*) and saddle-back (*S. fuscicollis*) tamarins. *Journal of Evolutionary Biology*, 9 (1), 5–42.

Costa, R. L. (1980). Incidence of caries and abscesses in archeological Eskimo skeletal samples from Point Hope and Kodiak Island, Alaska. *American Journal of Physical Anthropology*, 52, 501–14.

Debets, G. (1999). *The Paleo-anthropology of the Early Eskimos (Ipiutak and Tigara).* Anchorage, AK: U.S. Department of Interior, National Park Service, Beringia Program, Anchorage, Alaska.

Derbeneva, O., Sukernik, R., Volodko, N. *et al.* (2002). Analysis of mitochondrial DNA diversity in the Aleuts of the Commander islands and its implications for the genetic history of Beringia. *The American Journal of Human Genetics*, 71 (2), 415–21.

Dumond, D. (1987). *The Eskimos and Aleuts*. London: Thames and Hudson.

Fox, J. and Weisberg, S. (2011). *An {R} Companion to Applied Regression*. Second Edition. Thousand Oaks, CA: Sage. URL: http://socserv.socsci.mcmaster.ca/jfox/Books/Companion.

Gerlach, C. and Mason, O. K. (1992). Calibrated radiocarbon dates and cultural interaction in the western Arctic. *Arctic Anthropology*, 29 (1), 54–81.

Giddings, J. (1960). The Archeology of Bering Strait. *Current Anthropology*, 1 (2), 121–38.

Goslee, S. C. and Urban, D. L. (2007). The ecodist package for dissimilarity-based analysis of ecological data. *Journal of Statistical Software*, 22 (7), 1–19.

Hollinger, R., Ousley, S. and Utermohle, C. (2009). The Thule migration: A new look at the archaeology and biology of the Point Barrow region populations. In *The Northern World, AD 900–1400*. Salt Lake City, UT: University of Utah Press, pp. 131–54.

Horton, N. and Kleinman, K. (2007). Much ado about nothing. *The American Statistician*, 61 (1), 79–90.

Howells, W. W. (1973). *Cranial Variation in Man: A Study by Multivariate Analysis of Patterns of Difference Among Recent Human Populations*. Cambridge, MA: Peabody Museum of Archaeology and Ethnology, Harvard University.

Hume, J. (1965). Sea-level changes during the last 2000 years at Point Barrow, Alaska. *Science*, 150 (3700), 1165–6.

Kohn, L. A. and Atchley, W. R. (1990). How similar are genetic correlation structures? Data from mice and rats. *Evolution*, 42, 467–81.

Konigsberg, L. W. and Ousley, S. D. (1995). Multivariate quantitative genetics of anthropometric traits from the Boas data. *Human Biology*, 67 (3), 481–98.

Larsen, H. and Rainey, F. (1948). *Ipiutak and the Arctic Whale Hunting Culture*. Anthropological Papers of the American Museum of Natural History 42. New York, NY: American Museum of Natural History.

Madimenos, F. (2002). *Dental Evidence for Division of Labor Among the Prehistoric Ipiutak and Tigara of Point Hope, Alaska*. M.A. Louisiana State University.

Maley, B. (2007). Using nested clade analysis to explore temporal change in ancient population structure with aDNA sequence data. *American Journal of Physical Anthropology*, S44, 163.

Maley, B. (2011). *Population Structure and Demographic History of Human Arctic Populations Using Quantitative Cranial Traits*. Ph.D. Washington University in St. Louis.

Maley, B., Doubleday, A., Kaestle, F. and Mowbray, K. (2006). Sorting out population structure and demographic history of the Tigara and Ipiutak cultures using ancient DNA analysis. *American Journal of Physical Anthropology*, S42, 124.

Mantel, N. (1967). The detection of disease clustering and a generalized regression approach. *Cancer Research*, 27 (2), 209–20.

Marroig, G. and Cheverud, J. M. (2001). A comparison of phenotypic variation and covariation patterns and the role of phylogeny, ecology, and ontogeny during cranial evolution of new world monkeys. *Evolution*, 55 (12), 2576–600.
Marroig, G., Shirai, L. T., Porto, A., de Oliveira, F. B. and De Conto, V. (2009). The evolution of modularity in the mammalian skull II: Evolutionary consequences. *Evolutionary Biology*, 36, 136–48.
Mason, O. K. (1998). The contest between the Ipiutak, Old Bering Sea, and Birnirk polities and the origin of whaling during the first millennium AD along Bering Strait. *Journal of Anthropological Archaeology*, 17 (3), 240–325.
Mason, O. K. (2006). Ipiutak remains mysterious: A focal place still out of focus. In *Dynamics of Northern Societies: Proceedings of the Sila/Nabo Conference on Arctic and North Atlantic Archaeology, Copenhagen, May 10th – 14th, 2004.* Copenhagen: Aarhus University Press, pp. 103–20.
Mason, O. K. (2009). Flight from the Bering Strait: Did Siberian Punuk/Thule military cadres conquer Northwest Alaska? In *The Northern World, AD 900–1400.* Salt Lake City, UT: University of Utah Press, pp. 76–130.
Mason, O. K. and Gerlach, S. (1995). Chukchi hot spots, paleo-polynyas, and caribou crashes: Climatic and ecological dimensions of north Alaska prehistory. *Arctic Anthropology*, 32 (1), 101–30.
McGhee, R. (2000). Radiocarbon dating and the timing of the Thule migration. In *Identities and Cultural Contacts in the Arctic: Proceedings from a Conference at the Danish National Museum, Copenhagen, November 30 to December 2, 1999.* Copenhagen: Danish Polar Center, pp. 181–91.
Merriwether, D., Rothhammer, F. and Ferrell, R. (1995). Distribution of the four founding lineage haplotypes in Native Americans suggests a single wave of migration for the New World. *American Journal of Physical Anthropology*, 98 (4), 411–30.
Moore-Jansen, P. M., Ousley, S. D. and Jantz, R. L. (1994). *Data Collection Procedures for Forensics Skeletal Material: Report of Investigations no. 48.* The University of Tennessee, Knoxville, TN: Department of Anthropology.
Ossenberg, N. (2005). Ethnogenesis in the central and eastern Arctic: A reconstruction based on cranial nonmetric traits. In *Contributions to the Study of the Dorset Palaeo-Eskimos*. Gatineau: Canadian Museum of Civilization, pp. 33–66.
Porto, A., de Oliveira, F., Shirai, L. and De Conto, V. (2009). The evolution of modularity in the mammalian skull I: Morphological integration patterns and magnitudes. *Evolutionary Biology*, 36, 118–35.
R Development Core Team (2008). *R: A Language and Environment for Statistical Computing*. R Foundation for Statistical Computing, Vienna, Austria.
Relethford, J. H. (1994). Craniometric variation among modern human populations. *American Journal of Physical Anthropology*, 95, 53–62.
Relethford, J. (1997). RMET v5.0: R matrix for METric data.
Relethford, J. and Blangero, J. (1990). Detection of differential gene flow from patterns of quantitative variation. *Human Biology*, 62 (1), 5–25.
Roff, D. A. (1995). The estimation of genetic correlations from phenotypic correlations: A test of Cheverud's conjecture. *Heredity*, 74, 481–90.

Roff, D. A. (1996). The evolution of genetic correlations: An analysis of patterns. *Evolution*, 50, 1392–403.
Rubicz, R., Schurr, T., Babb, P. and Crawford, M. (2003). Mitochondrial DNA variation and the origins of the Aleuts. *Human Biology*, 75 (6), 809–35.
Saillard, J., Forster, P., Lynnerup, N., Bandelt, H. and Nørby, S. (2000). mtDNA variation among Greenland Eskimos: The edge of the Beringian expansion. *The American Journal of Human Genetics*, 67 (3), 718–26.
Shields, G., Schmiechen, A., Frazier, B., *et al.* (1993). mtDNA sequences suggest a recent evolutionary divergence for Beringian and northern North American populations. *American Journal of Human Genetics*, 53 (3), 549–62.
Smouse, P., Long, J. and Sokal, R. (1986). Multiple regression and correlation extensions of the Mantel test of matrix correspondence. *Systematic Biology*, 35 (4), 627–32.
Sokal, R. R. (1979). Testing statistical significance of geographical variation patterns. *Systematic Zoology*, 28, 227–31.
Starikovskaya, Y., Sukernik, R., Schurr, T., Kogelnik, A. and Wallace, D. (1998). mtDNA diversity in Chukchi and Siberian Eskimos: implications for the genetic history of Ancient Beringia and the peopling of the New World. *The American Journal of Human Genetics*, 63 (5), 1473–91.
Tamm, E., Kivisild, T., Reidla, M., *et al.* (2007). Beringian standstill and spread of Native American founders. *PLoS One*, 2 (9), 1–6.
Venable, D. L. and Burquez, M. A. (1990). Quantitative genetics of size, shape, life-history, and fruit characteristics of the seed heteromorphic composite Heterosperma pinnatum. II. Correlation structure. *Evolution*, 44, 1748–63.
Ward, R., Redd, A., Valencia, D., Frazier, B., Paabo, S. (1993). Genetic and linguistic differentiation in the Americas. *Proceedings of the National Academy of Sciences USA*, 90 (22), 10663–7.
White, T. D. and Folkens, P. A. (2000). *Human Osteology*. Second Edition. San Diego, CA: Academic Press.

Part II

Biological variation among the foragers of Point Hope

5 *Contrasting the Ipiutak and Tigara: Evidence from incisor microwear texture analysis*

KRISTIN L. KRUEGER

Introduction

Point Hope provides an important and relevant model for the study of human adaptation in extreme marginal environments. Excavated in the first half of the twentieth century, the archaeological sites of Ipiutak and Tigara revealed themselves to be significant in understanding not only Arctic cultural affinities, but also migration patterns, subsistence strategies, and other aspects of Arctic forager adaptation (Larsen and Rainey, 1948). Considered one of the most extensive and influential archaeological sites in the Arctic, Point Hope proved to be a cache of settlement and burial data, with thousands of artifacts, 600 dwellings, and 500 burials associated with the two sites (Larsen and Rainey, 1948; Rainey, 1971).

Analyses of the vast array of artifacts and skeletal remains created more questions than answers. Over 10,000 cultural artifacts associated with the Ipiutak site were recovered; however, over 2,000 of them were related to land-mammal hunting, and very few were identified as whale-hunting implements (Larsen and Rainey, 1948; Rainey, 1971). This is in stark contrast to what would be expected at a coastal site, and led to the hypothesis that the Ipiutak followed the seasonal caribou migration to Point Hope, and relied predominantly on them for subsistence (Rainey, 1971). On the other hand, material artifacts found within the Tigara excavations included many harpoon blades, boats, and whale bone, which suggested subsistence strategies related to whaling (Larsen and Rainey, 1948).

These hypotheses have yet to be confirmed or disproved, as little bioarchaeological work has been completed on the Point Hope skeletal remains. Moreover, questions have also been raised surrounding the cultural adaptations of Arctic

The Foragers of Point Hope: The Biology and Archaeology of Humans on the Edge of the Alaskan Arctic, eds. C. E. Hilton, B. M. Auerbach, L. W. Cowgill. Published by Cambridge University Press.

peoples, including the use of teeth as tools, in relation to the Ipiutak and Tigara. Recent access to the collections has allowed for skeletal and dental analyses to develop and flourish for the first time since their excavation, and this study is one such example.

The analysis presented here provides supportive data not only to the hypotheses proposed from the archaeology, that the Ipiutak and Tigara relied on contrasting dietary strategies, but also sheds light on the different behavioral regimes of the Point Hope communities. Dental microwear texture analysis, a repeatable and objective technique for the collection of dental microwear, has been shown to be a valuable tool in distinguishing aspects of diet and behavior in recent and fossil humans, and is used here to provide a direct dietary and behavioral reconstruction of the Ipiutak and Tigara using the incisor teeth.

Background

Dental analyses of the Point Hope communities have been sporadic and somewhat conflicting in their results, with the majority of work addressing the overall health of the oral cavity, including carious lesions and abscesses (Costa, 1980a; Madimenos, 2005; Dabbs, 2011), antemortem tooth wear and loss (Costa, 1980b; Madimenos, 2005; Dabbs, 2011) and Arctic dentistry (Schwartz *et al.*, 1995). However, dietary reconstruction using molar microwear techniques has recently been completed, and provides a complementary analysis (El Zaatari, 2008; El Zaatari, Chapter 6, this volume).

The prevalence of carious lesions within the Ipiutak males and females has been examined, but results differed between studies (Costa, 1980a; Madimenos, 2005; Dabbs, 2011). While Dabbs (2011) found no carious lesions within the Ipiutak males or females, other researchers have reported them (Costa, 1980a; Madimenos, 2005). Costa (1980a) found that the Ipiutak males had a greater average number of lesions than the Ipiutak females (4.25 versus 2.20 lesions per individual, respectively). Costa (1980a) also reports that the Ipiutak males showed a general decrease in lesions over the course of the lifespan; however, these results have been questioned (Madimenos, 2005). However, Costa (1980a) and Madimenos (2005) agreed that the Ipiutak females had fewer carious lesions than the males, but the difference fluctuated between the two studies. Both studies also agreed that the average number of carious lesions remained the same over the course of the Ipiutak female lifespan (Costa, 1980a; Madimenos, 2005).

Studies also differed in the prevalence of carious lesions between the Tigara males and females. Dabbs (2011) and Costa (1980a) found that the Tigara males and females were more similar in carious lesion frequency than those of the Ipiutak, with an overall low occurrence and no differences over the course of the lifespan. For example, Costa (1980a) reported that the Tigara males had an average of 1.14 carious lesions per individual, while the Tigara females had a mean of 1.17 lesions. Contrastingly, Madimenos (2005) reported an overall decrease in carious lesions over the course of the lifespan for both the Tigara males and females. Finally, abscess frequency was found to be very low in both Point Hope communities (Costa, 1980a; Dabbs, 2011).

Antemortem loss of the anterior teeth differed between the sexes, but this difference remained steady between the communities. Both the Ipiutak and Tigara males demonstrated a low incidence of anterior tooth loss (Costa, 1980b; Madimenos, 2005). For example, Costa (1980b) reported an incisor loss rate of 5.3% and 8.8% for the Ipiutak and Tigara males, respectively. While the Ipiutak male loss rate was well correlated with chronological age, the rate of the Tigara males was not (Costa, 1980b). The Ipiutak and Tigara females, on the other hand, had a higher rate of antemortem anterior tooth loss (Costa, 1980b; Madimenos, 2005; Dabbs, 2011). Indeed, Costa (1980b) stated an average incisor loss percentage of 19.4% and 16.4% for the Ipiutak and Tigara females, respectively. In this case, the Ipiutak females did not have a progressive occurrence over the course of the lifespan, but the Tigara females did (Costa, 1980b; Dabbs, 2011). The female patterning of high antemortem anterior tooth loss was attributed to strenuous non-dietary anterior tooth use, or use of the teeth as tools, clamps, or third hands, in the Ipiutak, and progressively heavy occlusal wear in the Tigara (Costa, 1980b).

The Ipiutak and Tigara have been subjected to direct dietary reconstruction in the form of scanning electron microscopy-based dental microwear analyses. Significant differences were found between the two communities, with the Tigara sample having more microwear features, including pits and narrow scratches, than the Ipiutak (El Zaatari, 2008). These data support the idea that the Ipiutak and Tigara maintained different dietary strategies, and differences in microwear features may indicate a more abrasive diet for the Tigara (El Zaatari, 2008).

Although measurements and counts of microwear features are valuable in determining associations between feature patterns and diet/tooth use, three-dimensional analyses of surface textures are becoming a progressively favorable alternative to these feature-based microwear studies (Ungar *et al.*, 2003, 2007, 2008a, b; R. Scott *et al.*, 2006; Krueger *et al.*, 2008; Merceron *et al.*, 2009; J. Scott *et al.*, 2009; Krueger and Ungar, 2010, 2012). Dental microwear texture analysis employs a combination of white-light confocal profilometry and scale-sensitive fractal analysis, an engineering protocol rooted in fractal geometry (Scott *et al.*, 2006). This technique generates several distinct texture attributes that have proven useful in distinguishing groups based on different microwear-producing factors.

Incisor microwear texture analyses have been extremely useful for characterizing bioarchaeological/ethnographic groups based on the three factors recognized to produce incisor microwear: diet, non-dietary anterior tooth use behaviors, and abrasive loads (Krueger and Ungar, 2010; Krueger, 2011). For example, anisotropy, a measure of feature texture orientation, is an accurate indicator of dietary or non-dietary use of the anterior teeth. Textural fill volume reveals the magnitude of the loading regime among those groups that utilized non-dietary practices. Heterogeneity, a measure of texture variation, denotes abrasive loads, but is intensified by non-dietary behaviors (Krueger and Ungar, 2010; Krueger, 2011).

To date, incisor microwear texture data have been collected representing twelve recent human populations with known or inferred diets, types of non-dietary anterior tooth use behaviors, and abrasive loads. Results are clear that this approach is valuable for testing hypotheses concerning the etiology of anterior tooth wear in both modern and fossil humans (Krueger and Ungar, 2010, 2012; Krueger, 2011). Five of these twelve samples range from the high to low Arctic, including the Ipiutak and Tigara from Point Hope, the Aleut from various Aleutian Islands, the Sadlermiut from the Nunavut Territory, and the Coast Tsimshian from the Prince Rupert Harbour area of British Columbia. These five samples collectively provide data that suggest fundamental differences in Arctic behavioral strategies and Arctic adaptation as a whole. The three latter samples are used to provide the best comparative analogue to those of the two Point Hope samples. Lastly, the data and analyses presented from these samples are consistent with previous results, and offer a unique perspective on the contrasting dietary and behavioral strategies of the Ipiutak and Tigara of Point Hope.

Materials and methods

Sampled groups

Ipiutak. The coastal Ipiutak site is located 125 miles north of the Arctic Circle on the gravel and sandy Tigara (Iñupiat: Tikiġaq) peninsula on the northwest coast of Alaska (Rainey, 1941). The site dates from before 1,600 to approximately 1,100 years BP, and is both arctic and arid (Larsen and Rainey, 1948). Helge Larsen and Froelich Rainey excavated the site in the late 1930s and early 1940s (see Jensen, Chapter 2, this volume). The Ipiutak culture was identified as distinctive due to specialized ivory carvings unknown in previous Arctic excavations and research (Larsen and Rainey, 1948).

Archaeological evidence indirectly suggested the Ipiutak were heavily reliant on caribou resources, with thousands of land-mammal hunting implements, including antler arrowheads, recovered in excavations (Rainey, 1941; Larsen and Rainey, 1948). It was hypothesized that the Ipiutak came to Point Hope to follow the caribou migration, and, while there, also subsisted on sea mammals such as seals and walrus (but see Mason, Chapter 3, this volume); however, whales were not a significant resource, as no boats and few harpoons and associated implements were found (Rainey, 1941; Larsen and Rainey, 1948; Lester and Shapiro, 1968). Direct dietary reconstructions of the Ipiutak buttress the archaeological evidence (El Zaatari, 2008).

High rates of antemortem anterior tooth loss and wear have led to hypotheses concerning non-dietary anterior tooth use behaviors in the Ipiutak (Costa, 1980b, 1982; Madimenos, 2005). Indeed, this sample reached their greatest tooth wear early in life, with the majority of individuals showing their maximum during the 26–30 years age range (Costa, 1982). No differences in wear rates between sexes were found in the Ipiutak sample, and striae, associated with hide or sinew pulling across the anterior teeth, were found on both male and female dentitions (Madimenos, 2005). Tooth wear may also have been exacerbated by heavy environmental and dietary abrasive loads associated with the landscape of Point Hope itself, as well as underground food preparation techniques.

Tigara. The later Tigara site was excavated at the same time as that of the Ipiutak, and dates from as early as 800 years BP, to *c*. 300 years BP. Neoglaciation-induced climate change at 750 years BP limited whale-hunting areas, but Tigara remained intact (Dabbs, 2009). This led to the Tigara Peninsula being inundated with settlers and becoming permanently settled at this time (Dabbs, 2009). Indeed, Point Hope is still regarded as one of the best whale-hunting areas in the world.

As with the Ipiutak, analyses of the Tigara archaeological excavations indirectly indicated dietary strategies. The assemblage was considerably different from that of the Ipiutak, with the majority of artifacts related to sea-mammal hunting, particularly whales (Larsen and Rainey, 1948). It was suggested that the Tigara were similar in their diet to the post-contact Point Hope occupants, who relied on whale, fish, seal, and walrus either eaten fresh or dried on open racks (Foote, 1992). Differences in dietary strategies between the Ipiutak and Tigara were supported with direct dietary reconstructions, including dental microwear analyses (El Zaatari, 2008).

Non-dietary anterior tooth use behaviors have also been suggested for the Tigara; however, patterns of tooth wear and antemortem tooth loss were different than those of the Ipiutak (Costa, 1980b, 1982; Madimenos, 2005). The Tigara demonstrated a more gradual tooth wear and antemortem tooth loss pattern, with maximums not reached until the oldest age group examined (40 years and older; Costa, 1980b, 1982). Moreover, while Tigara males had a higher frequency of tooth chipping and notching, females displayed a greater number of striae (Madimenos, 2005). While this suggests non-dietary anterior tooth use behaviors in both sexes, it may indicate different tasks associated with a gendered division of labor. Tooth wear patterns also indicate that while environmental and dietary abrasives also likely played a role, loads may not have been as heavy as those of the Ipiutak (Krueger, 2011).

Aleut. Ales Hrdlička collected the Aleut sample in the 1930s from Agattu, Amaknak, Kagamil, Unmak, and Unalaska Islands, all of which form part of the oceanic and rainy eastern archipelago of the Aleutian Islands (Hrdlička, 1945). The skeletal sample is dated from 3,400 to 400 years BP, suggesting Paleo- and Neo-Aleut affiliation (Coltrain, 2010). Stable isotope analyses indicate a marine diet likely composed of raw and dried fish, sea mammals, and shellfish, supplemented by foxes, rodents, birds, and tubers, as the average Paleo-Aleut and Neo-Aleut $\delta^{14}C$ and $\delta^{15}N$ values are −12.3 and 19.5, and −12.7 and 20.3, respectively (Hrdlička, 1945; Moorrees, 1957; Hoffman, 1993; Coltrain, 2010; Krueger and Ungar, 2010).

Ethnographic and historic reports document non-dietary anterior tooth use behaviors in the Aleut (Moorrees, 1957; Merbs, 1968), including softening wood for boat frames and grasping hides in clothing preparation (Moorrees, 1957; Campbell, 1967; Oliver, 1988). Landscapes across the Aleutian Islands are predominantly covered with mosses and shrubs, and sand beaches are limited in number; moreover, archaeological evidence suggests habitations were well protected from high winds (Hrdlička, 1945; Hoffman, 1993). Therefore, environmental and/or dietary abrasives are not predicted to have been a source of dental wear.

Nunavut Territory. The Nunavut Territory skeletal remains are from the polar Arctic sites of Native Point, Kamarvik, and Silumiut, located in the northwest Hudson Bay area. The Native Point site, excavated by Henry Collins in the 1950s, is attributed to the Sadlermiut, an Inuit population of both Thule and Dorset genetic ancestry (Coltrain, 2009). The skeletal remains have been dated to 650 to 100 years BP (Coltrain *et al.*, 2004). The average $\delta^{14}C$ and $\delta^{15}N$ values for this sample were −13.2 and 20.3, respectively, consistent with a diet dominated by caribou, walrus, ringed seal, and seabirds (Coltrain *et al.*, 2004; Coltrain, 2009). Faunal analyses support an emphasis on caribou, with nearly 25% of the assemblage identified as such (Collins, 1962; Ryan, 2011). The Thule burials from Kamarvik and Silumiut were excavated by Charles Merbs in the late 1960s, and have been radiocarbon dated to 950 to 350 years BP (Coltrain, 2009). Their average $\delta^{14}C$ and $\delta^{15}N$ values were −14.3 and 17.5, indicating a heavy reliance on terrestrial mammals, such as caribou, as well as lower-level marine taxa (Coltrain, 2009).

The Nunavut Territory dental remains show extensive evidence of trauma (Turner and Cadien, 1969; Merbs, 1983; Wood, 1992). Chipping, flaking, fracturing of the tooth crowns, labial wear, and antemortem tooth loss have been attributed to using the anterior dentition as tools for tasks such as preparing caribou skins for clothing, tents, and boats (Wood, 1992). Frequency of antemortem tooth loss and wear was also high; this has been recognized as additional evidence for use of the anterior dentition in non-dietary behaviors (Wood, 1992).

Postcanine tooth gross wear is also extensive, and most likely indicates heavy abrasive loads (Wood, 1992). Ethnographic accounts of Inuit in the area detail the practice of drying meat on the ground, as well as storing it in houses covered by sand, moss, and peat (Marsh, 1976). Moreover, sand was often used to clean houses (Mathiassen, 1927). These practices would no doubt have contributed considerable dietary and environmental abrasives to the Nunavut Territory Inuit lifestyle.

Coast Tsimshian. The Boardwalk and Reservoir sites are two of eleven sites found among the Digby and Kaien Islands in the oceanic, temperate rainforest areas of Prince Rupert Harbour. George MacDonald excavated the area during the late 1960s and early 1970s as part of the North Coast Prehistory Project (Stewart *et al.*, 2009). The 200 individuals excavated were determined to be of Coast Tsimshian ancestry, and have been dated to between 4,000 and 700 years BP (Cybulski, 1974, 1978; Stewart *et al.*, 2009).

Faunal analyses of the sites indicated that the Coast Tsimshian diet was dominated by fish, especially salmon, though Pacific herring, Pacific tomcod, sculpin, sole, flounder, and other fish were all present at the site in fewer frequencies (Stewart *et al.*, 2009). Land and sea mammals, such as mule deer, blacktail deer, sea otters, beaver, and seals, as well as birds, were also consumed on rare occasion (Stewart *et al.*, 2009).

Non-dietary anterior tooth use behaviors have also been suggested for the Coast Tsimshian (Cybulski, 1974). Gross wear patterns found on the anterior teeth have been attributed to weaving practices; that is, root fibers for blanket and basketry production were moistened and softened using the anterior dentition, which, over time, produced linear grooves on the incisal edges (Cybulski, 1974). This connection was proposed from ethnographic records of the Tsimshian and nearby Tlingit, a group that shared elements of material culture practices with the Tsimshian. While blanket and basket weaving practices are found in the Tsimshian records, the importance of the mouth and teeth in these activities is found in those from the Tlingit (Cybulski, 1974).

Data collection

This study used high-resolution replicas of the maxillary central incisors of the anterior teeth of individuals from each of the groups. Each tooth was gently cleaned using acetone and cotton swabs. President's Jet (Coltène-Whaledent), a high-resolution, regular body impression material, was used to create a negative of each tooth surface. High-resolution epoxy casts were produced with Epotek 301 base and hardener (Epoxy Technologies). This standard procedure has been shown to reproduce microwear features to a fraction of a micrometer (Beynon, 1987).

The labial surface, closest to the incisal edge of each tooth, was inspected for dental microwear using a light microscope at low magnification. Those observed with antemortem wear were scanned using a Sensofar Plμ white-light confocal profiler (Solarius Development Inc., Sunnyvale, California), with a lateral sampling of 0.18 μm and a vertical resolution of 0.005 μm (Scott *et al.*, 2006). Four adjacent scans of each incisor labial surface were taken with a 100× objective lens, for a total work area of 276 × 204 μm (Scott *et al.*, 2006). The point clouds for each individual were leveled, and any defects, such as dust or preservative, were deleted using Solarmap Universal software (Solarius Development Inc., Sunnyvale, California). Resultant data were then imported into Toothfrax and SFrax scale-sensitive fractal analysis (SSFA) software packages (Surfract; http://www.surfract.com) for surface texture characterization.

Four texture variables, complexity (*Asfc*), anisotropy (*epLsar*), textural fill volume (*Tfv*), and heterogeneity (*HAsfc*), are used here to characterize the microwear surfaces. While these are detailed in Scott *et al.* (2006), a brief description is provided here. Complexity is a measure of change in surface roughness with the scale of observation. That is, overlapping features of varying sizes typically have high complexity values. Anisotropy is a measure of surface texture directionality; a surface dominated by fine parallel striations is highly anisotropic. Textural fill volume is a measure of the average diameter and depth of features removed from a microwear surface. Since this measures the difference between volumes of large (10 μm) and small (2 μm) square cuboids that "fill" a surface, features of intermediate size show high values. Lastly, heterogeneity measures texture variation across a scanned surface. In this case, sampled areas were divided into 3 × 3 grids ($HAsfc_9$) and 9 × 9 grids ($HAsfc_{81}$).

Statistical analyses

Median values for every texture attribute were calculated for each of the four scans that represented an individual (Scott *et al.*, 2006). A multivariate analysis of variance (MANOVA) was calculated on rank-transformed data (Conover and Iman, 1981) with the sampled groups as independent variables and the texture attributes as dependent variables. Analyses of variance (ANOVAs) for individual texture attributes and pairwise comparisons for the samples were completed as needed to determine the sources of significant variation. Both Tukey's HSD and Fisher's LSD tests were used to balance risks of Type I and Type II errors (Cook and Farewell, 1996). It is important to note that Tukey's HSD test results were used as the benchmark for significance, while Fisher's LSD test results were considered suggestive or of marginal significance.

Results

The Ipiutak and Tigara samples demonstrated remarkable incisor microwear texture differences between each other, while also showing affinities with the other Arctic groups. The MANOVA test results indicate significant variation in incisor microwear textures among all the samples, and individual ANOVAs revealed variation in all texture attributes. In particular, the Ipiutak and Tigara differed significantly from each other in all four texture characteristics. Results are illustrated in Figures 5.1 and 5.2, and descriptive and analytical statistics are presented in Tables 5.1, 5.2, and 5.3.

Complexity

The Ipiutak had the highest complexity average of all the samples, while the Tigara were moderately low in their mean value. Tukey's HSD pairwise comparisons indicate the Ipiutak, Nunavut Territory, and Coast Tsimshian had significantly higher average *Asfc* values than did the Tigara and Aleut. Fisher's LSD tests suggest that the Nunavut Territory sample had higher average complexity values than those of the Coast Tsimshian. The Ipiutak and Nunavut Territory were similar in average complexity values, while the Tigara and Aleut were comparable.

Anisotropy

In addition to the highest average complexity, the Ipiutak demonstrated the lowest mean anisotropy value. The Tigara anisotropy mean, on the other hand, was situated on the opposite end of the spectrum, and had the highest value of the Arctic samples. Tukey's HSD results indicate the Ipiutak and Nunavut

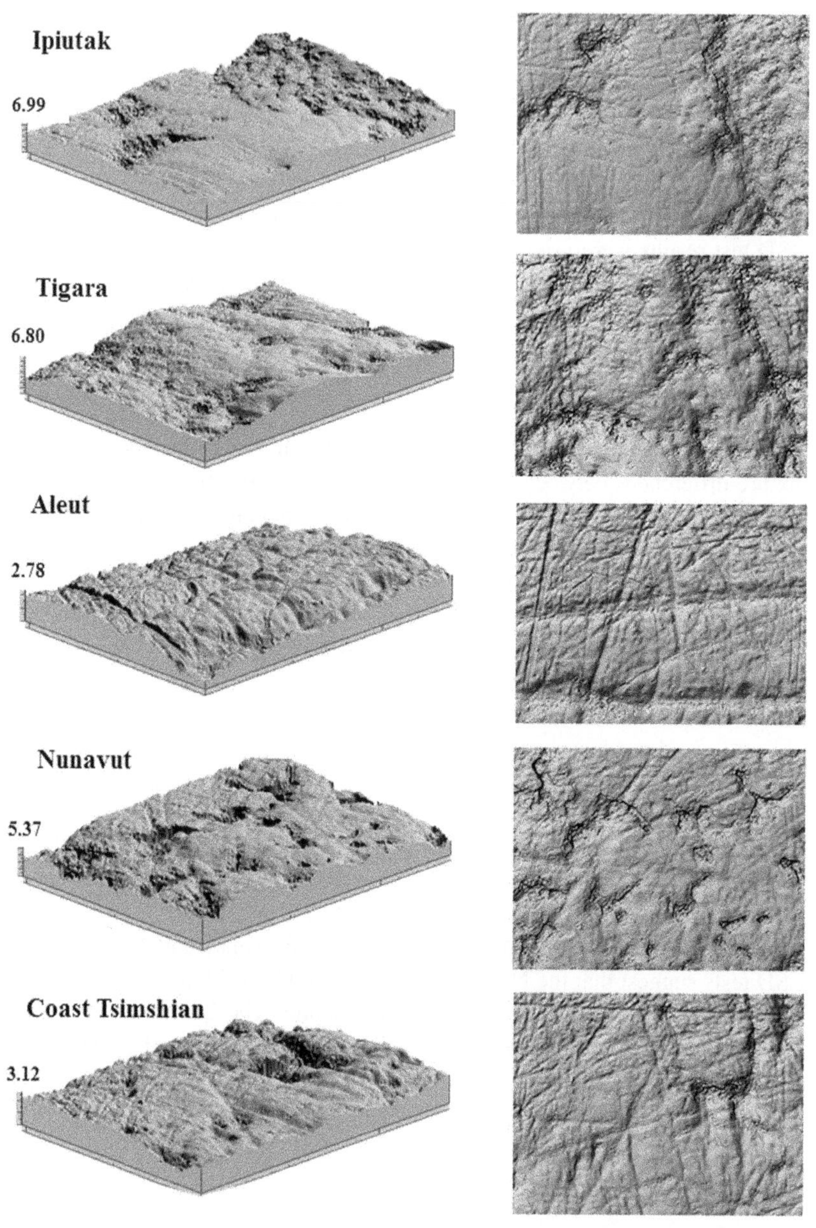

Figure 5.1. Two- and three-dimensional photo simulations of the samples. These images represent one of four adjacent scans for each individual, and measure an area of 102 × 138 μm.

Table 5.1. *Descriptive statistics of dental texture analyses performed in this study*

Group	n		*Asfc*	*epLsar*	*Tfv*	$HAsfc_9$	$HAsfc_{81}$
Ipiutak	22	Mean	3.4291	0.0020	12143.0159	0.6636	1.3582
		Median	2.2650	0.0018	12842.5900	0.5950	1.2550
		SD	3.0264	0.0008	4253.4876	0.3110	0.8006
		SEM	0.6452	0.0002	906.8466	0.0663	0.1707
Tigara	34	Mean	1.2041	0.0032	7296.0188	0.5309	0.8918
		Median	0.8150	0.0029	6269.7050	0.4750	0.7200
		SD	0.9862	0.0015	5391.2044	0.2610	0.4584
		SEM	0.1691	0.0003	924.5839	0.0448	0.0786
Aleut	24	Mean	0.9346	0.0030	7434.4971	0.3838	0.6058
		Median	0.9000	0.0030	7334.6650	0.3800	0.5500
		SD	0.4501	0.0011	5272.1881	0.0723	0.1186
		SEM	0.0919	0.0002	1076.1809	0.0148	0.0242
Nunavut	27	Mean	3.2767	0.0020	12449.2719	0.5974	1.0919
		Median	2.6000	0.0018	12905.6500	0.4700	1.0500
		SD	2.1260	0.0010	3464.0371	0.3154	0.3762
		SEM	0.4092	0.0002	666.6543	0.0607	0.0724
Tsimshian	25	Mean	1.8536	0.0024	5766.6396	0.4568	0.7004
		Median	1.8100	0.0019	3079.7100	0.4600	0.6800
		SD	0.7771	0.0013	5196.3965	0.1184	0.2016
		SEM	0.1554	0.0003	1039.2793	0.0237	0.0403

Territory were significantly lower in feature texture orientation than the Tigara and Aleut. Fisher's pairwise comparisons also suggested the Tigara and Aleut were marginally higher in anisotropy than the Coast Tsimshian. As with the complexity texture attribute, the Ipiutak and Nunavut Territory samples were congruent in anisotropy values, and those of the Tigara and Aleut were also consistent.

Textural fill volume

The Ipiutak sample demonstrated the second highest textural fill volume average of the Arctic samples, with the Nunavut Territory marginally exceeding it. The Tigara, alternatively, was moderate in its *Tfv* mean. Tukey's HSD test results once again revealed the shared values of the Ipiutak and Nunavut, with both these samples showing significantly higher values that those of the Tigara, Aleut, and Coast Tsimshian. The Tigara and Aleut were both moderate in their values.

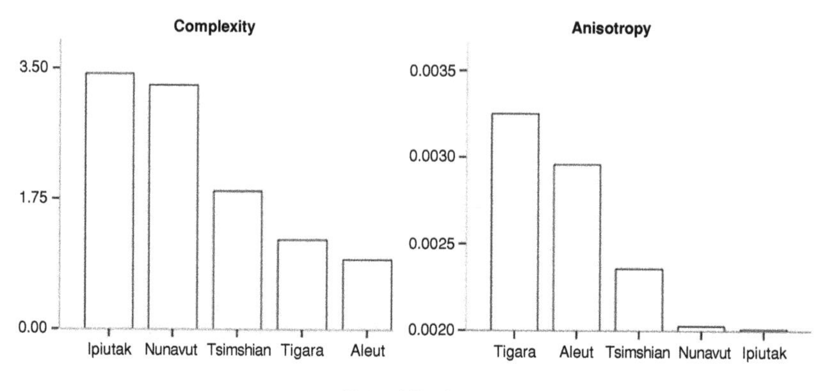

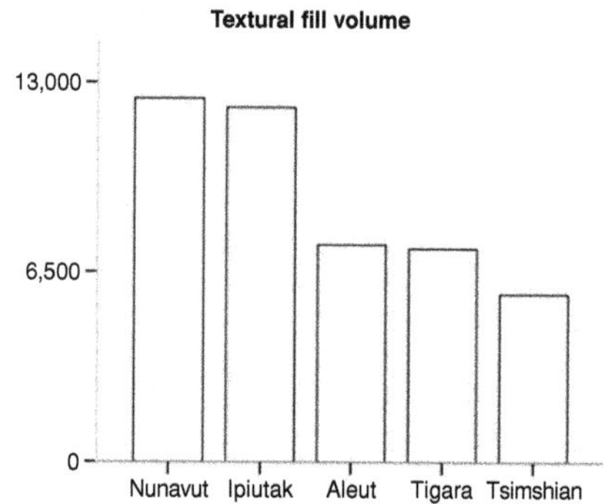

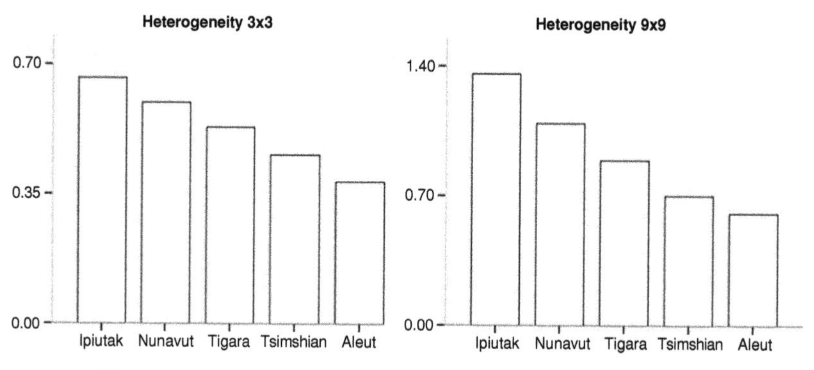

Figure 5.2. Bar charts showing the mean texture values of the samples.

Heterogeneity

Lastly, the Ipiutak had the highest heterogeneity 3 × 3 and 9 × 9 average values of the Arctic samples examined here, while the Tigara was moderate in its mean. The Ipiutak, Nunavut Territory, and Tigara samples were significantly

Table 5.2. *MANOVA and ANOVA results of ranked microwear texture data*

	Value	F	df	p
MANOVA				
Test statistic				
Wilk's lambda	0.056	413.818	5123	0.00
Pallai trace	0.944	413.818	5123	0.00
Hotelling's trace	16.822	413.818	5123	0.00
ANOVAs				
Measurement				
Asfc		20.501	4	0.00
epLsar		5.972	4	0.00
Tfv		9.873	4	0.00
HAsfc9		5.766	4	0.00
HAsfc81		12.869	4	0.00

Table 5.3. *Pairwise comparisons with significant Tukey's and Fisher's results*

	Ipiutak	Tigara	Aleut	Nunavut
Asfc				
Tigara	−44.812**			
Aleut	−51.153**	−6.342		
Nunavut	7.965	52.776**	59.119**	
Tsimshian	−12.551	32.261**	38.603**	−20.516*
epLsar				
Tigara	34.440**			
Aleut	31.288**	−3.152		
Nunavut	0.547	−33.892**	−30.741**	
Tsimshian	9.415	−25.025*	−21.873*	8.867
Tfv				
Tigara	−34.988**			
Aleut	−34.945**	0.043		
Nunavut	0.354	35.342**	35.299**	
Tsimshian	−43.811**	−8.823	−8.866	−44.164**
HAsfc9				
Tigara	−19.730*			
Aleut	−46.777**	−27.047**		
Nunavut	−10.614	9.115	36.162**	
Tsimshian	−26.718*	−6.988	20.058	−16.104
HAsfc81				
Tigara	−27.352**			
Aleut	−55.413**	−28.061**		
Nunavut	−3.066	24.286**	52.347**	
Tsimshian	−40.695**	−13.343	14.718	−37.629**

* Fisher's LSD $p < 0.05$

** Tukey's HSD $p < 0.05$

higher in $HAsfc_9$ than the Aleut according to Tukey's HSD tests. However, Fisher's tests suggest the Ipiutak was marginally higher in $HAsfc_9$ than the Tigara and Coast Tsimshian. The fine-scale variant of heterogeneity showed variation among the groups as well, with Tukey's HSD tests indicating the Ipiutak and Nunavut Territory were significantly higher than the Tigara, Aleut, and Coast Tsimshian. Additionally, the Tigara was higher in $HAsfc_{81}$ than the Aleut.

Discussion

These data show that the incisor microwear texture signatures of the Ipiutak and Tigara lie on opposite ends of the texture spectrum, but each one demonstrated affinities to one or more of the comparative groups. Although the Ipiutak and Tigara lived at the same location, data presented here suggest that their dietary and behavioral strategies differed radically. We can better understand these differences in anterior tooth use behaviors by comparing their microwear texture values with a variety of other Arctic groups that vary in their observed or inferred diets and other behaviors, as well as in their abrasive environments. Each texture attribute will be addressed separately.

Complexity (Asfc)

Complexity values in previous incisor microwear texture analyses indicate that *Asfc* represents a balance between level of non-dietary anterior tooth use behaviors and level of abrasive exposure (Krueger, 2011). If both non-dietary tasks (i.e., use of teeth as tools or as a third hand) and abrasive exposure are high, a high complexity value will result; the effects are cumulative. If one effect is high, or if both are moderate, this will result in moderate complexity values.

The Ipiutak sample has a mean *Asfc* value of 3.43. This value represents not only the highest complexity value of all the groups, but it is also nearly identical to that of the Nunavut Territory sample. The Nunavut groups were described as participating in non-dietary anterior tooth use behaviors in the form of clamping and processing caribou hide, and also as being exposed to heavy environmental and dietary abrasive loads; the high complexity value reflects this. These results suggest that the Ipiutak extensively used their incisors in non-dietary tasks, perhaps related to the clamping and processing of caribou hides, and were also exposed to high levels of environmental and/or dietary abrasives.

Both the Ipiutak and Nunavut samples, along with the Coast Tsimshian, have significantly higher *Asfc* values than those of the Tigara and Aleut. The Tigara average complexity value was 1.20, and is most similar to that of the Aleut, a group that employed some non-dietary behaviors, and was exposed to low abrasive loads. The moderate complexity values of the Tigara may indicate use of the anterior dentition less frequently, or in non-dietary tasks with a lower magnitude of loading, along with moderate abrasive loads.

Anisotropy (epLsar)

Anisotropy has been extremely reliable in distinguishing dietary from non-dietary anterior tooth use in previous analyses of incisor microwear textures (Krueger and Ungar, 2010, 2012; Krueger, 2011). High anisotropy values on the labial surfaces of incisor teeth seem to reflect abrasives being dragged apico-cervically, as would be expected during the incising of food items. However, low anisotropy values, or a lack of feature directionality across the labial surface, would suggest the incisor teeth being used in a variety of tasks, including non-dietary anterior tooth use (e.g., clamping, grasping, tool production, etc.).

All the samples examined here have been reported or inferred to have employed non-dietary uses of the anterior teeth (Krueger and Ungar, 2010, 2012; Krueger, 2011). Their low anisotropy values support this claim, and the range of anisotropy values can help to distinguish different uses of the incisor teeth in non-dietary behaviors. For example, the Ipiutak and Nunavut Territory samples have identical, extremely low *epLsar* values, which were significantly lower than those of the Tigara and Aleut. This suggests that the Ipiutak and Nunavut groups engaged in similar non-dietary activities, perhaps related to intense and frequent processing of caribou hide.

The similar values of the Aleut and Tigara, found on the higher end of the non-dietary spectrum, indicate these two groups participated in comparable behaviors. The Aleut are known to have used their anterior dentition for hide preparation, softening wood fibers, and other tasks, at least on occasion. This suggests the Tigara engaged in similar behaviors, or at least those that were less frequent or involved more moderate levels of non-dietary anterior tooth use than those of the Ipiutak and Nunavut Territory. Lastly, the moderately low *epLsar* value of the Coast Tsimshian is congruent with reported use of the anterior dentition in softening fibers for basketry and blanket weaving (Cybulski, 1974).

Textural fill volume (Tfv)

Textural fill volume is associated with loading regime for those groups that participated in non-dietary anterior tooth use behaviors (Krueger and Ungar, 2010, 2012; Krueger, 2011). High *Tfv* values are typically found in those groups that are reported or inferred to engage in high stress or heavy loading during non-dietary anterior tooth use behaviors.

The Ipiutak mean *Tfv* value is 12,143, an extremely high value that is bested only slightly by that of the Nunavut Territory (12,449). Predictably, these samples are significantly higher in this attribute than the Tigara, Aleut, and Coast Tsimshian samples. These extremely high values indicate that both the Ipiutak and Nunavut Territory individuals sampled here engaged in a very high loading regime. These values, coupled with the extremely low anisotropy of these two samples, strongly supports behavioral strategies involving a continuous, intense, non-dietary use of the anterior dentition, possibly in association with heavy use of the incisors in caribou hide preparation.

The Tigara and Aleut samples are, once again, extremely congruent in their textural fill volumes, with mean values of 7,296 and 7,434, respectively. These similar values, along with the anisotropy measures, suggest a moderate anterior loading regime, perhaps related to moderate hide preparation and wood softening. Lastly, the Coast Tsimshian has the lowest *Tfv* mean value, indicating lower magnitudes of anterior tooth loading most likely related to the processing of soft plant material for weaving practices.

Heterogeneity (HAsfc)

Heterogeneity of incisor microwear surfaces has been associated in the past with exposure to abrasives. Groups living in areas with heavy abrasive loads tend to have more heterogeneous microwear textures, resulting in high heterogeneity values (Krueger and Ungar, 2010, 2012; Krueger, 2011). However, non-dietary anterior tooth use behaviors may intensify this signal (Krueger, 2011). The Ipiutak have the highest $HAsfc_9$ and $HAsfc_{81}$ values of the samples, with the Nunavut Territory close behind. However, the Tigara and Aleut are not as similar as in the other texture attributes. In fact, the Ipiutak, Nunavut, and Tigara samples were all significantly higher in 3×3 values than the Aleut, suggesting minor coarse abrasive exposure for the latter. However, since all these groups participated in non-dietary anterior tooth use, these behaviors may be intensifying the heterogeneity signal. Consequently, the proposed intense non-dietary behaviors of the Ipiutak could account for the marginally higher 3×3 value of this sample over that of the Tigara.

Analyses of the fine-scale variant of this attribute also demonstrated differences within the samples. The Ipiutak and Nunavut Territory individuals were significantly greater in 9 × 9 values than the Tigara, Aleut, and Coast Tsimshian, and the Tigara was significantly greater than the Aleut. These differences may indicate differences in fine environmental and/or dietary abrasive loads among the groups, but the extremely high values of the Ipiutak and Nunavut Territory suggest the intense non-dietary anterior tooth use behaviors may be escalating the signal.

Conclusion

The Ipiutak are equivalent to the Nunavut Territory sample in all texture attributes. While the extremely low anisotropy of the two groups indicates non-dietary anterior tooth use behaviors, the high textural fill volume suggests a heavy anterior loading regime. When taken together, these texture variables indicate the Nunavut and Ipiutak were engaging in intense clamping and grasping activities with their anterior teeth. This signal is consistent with an intense use of the incisors in non-dietary behaviors, most likely in relation to a continuous and heavy regimen of caribou hide preparation. The similarly high complexity and heterogeneity signals of the Ipiutak and Nunavut indicate that both of these samples were exposed to high levels of environmental and/or dietary abrasives. In summary, the incisor microwear textures of the Point Hope Ipiutak indicate a dietary reliance on caribou, intense non-dietary anterior tooth use behaviors, and high abrasive loads.

On the other hand, the Tigara individuals are comparable to the Aleut in all texture attributes with the exception of heterogeneity. The marginally low anisotropy indicates non-dietary anterior tooth use behaviors in these two samples, and the similar *Tfv* suggests a moderate anterior loading regime. Taken together, this indicates a more moderate non-dietary anterior tooth use regimen, possibly consisting of hide preparation and wood softening. The complexity values buttress the moderate anterior tooth loads, but may also suggest moderate abrasive loads. Finally, the heterogeneity values of the Tigara signal that this sample was exposed to higher abrasive loads than that of the Aleut. Thus, the incisor microwear textures of the Point Hope Tigara suggest a dietary reliance on sea mammals, moderate non-dietary anterior tooth use behaviors, and exposure to moderate levels of environmental and/or dietary abrasive loads.

In conclusion, this study suggests that the Ipiutak and Tigara of Point Hope, Alaska, used their incisor teeth in dietary and behavioral strategies, but in contrasting ways. Indeed, the microwear texture signals of the two Point Hope communities were utterly distinctive, providing evidence to support the archaeological claim that these two groups relied on different dietary and behavioral regimes.

Acknowledgements

I would like to express my gratitude to Charles Hilton for introducing me to Point Hope many years ago, and to thank him, Libby Cowgill, and Benjamin Auerbach for inviting me to be a part of this book. I also thank curators at the American Museum of Natural History, U.S. National Museum of Natural History, and Canadian Museum of Civilization, as well as the Inuit Heritage Trust for their permission to study specimens in their care. I acknowledge Peter Ungar for permission to use the Aleut casts in his care, and for his microwear texture input. I also thank three reviewers for their thoughtful feedback and comments. This study was funded by the U.S. National Science Foundation DDIG program (BCS-0925818).

References

Beynon, A. D. (1987). Replication technique for studying microstructure in fossil enamel. *Scanning Microscopy*, 1, 663–9.

Campbell, A. (1967). *A Voyage Round the World from 1806 to 1812*. Honolulu, HI: University of Hawaii Press.

Collins, H. B. (1962). Bering Strait to Greenland. In *Prehistoric Cultural Relations Between the Arctic and Temperate Zones of North America*. Montreal: Arctic Institute of North America, pp. 126–39.

Coltrain, J. B. (2009). Sealing whaling and caribou revisited: Additional insights from the skeletal isotope chemistry of eastern Arctic foragers. *Journal of Archaeological Science*, 36, 764–75.

Coltrain, J. B. (2010). Temporal and dietary reconstruction of past Aleut populations: Stable- and radio-isotope evidence revisited. *Arctic*, 63, 391–8.

Coltrain, J. B., Hayes, M. G. and O'Rourke, D. H. (2004). Sealing, whaling and caribou: The skeletal isotope chemistry of Eastern Arctic foragers. *Journal of Archaeological Science*, 31, 39–57.

Conover, W. J. and Iman, R. L. (1981). Rank transformations as a bridge between parametric and nonparametric statistics. *American Statistician*, 35, 124–9.

Cook, R. J. and Farewell, V. T. (1996). Multiplicity considerations in the design and analysis of clinical trials. *Journal of the Royal Statistical Society: Series A*, 159, 93–110.

Costa, R. L. (1980a). Incidence of caries and abscesses in archeological Eskimo skeletal samples from Point Hope and Kodiak Island, Alaska. *American Journal of Physical Anthropology*, 52, 501–14.

Costa, R. L. (1980b). Age, sex, and antemortem loss of teeth in prehistoric Eskimo samples from Point Hope and Kodiak Island, Alaska. *American Journal of Physical Anthropology*, 53, 579–87.

Costa, R. L. (1982). Periodontal disease in the prehistoric Ipiutak and Tigara skeletal remains from Point Hope, Alaska. *American Journal of Physical Anthropology*, 59, 97–110.

Cybulski, J. S. (1974). Tooth wear and material culture: Precontact patterns in the Tsimshian area British Columbia. *Syesis*, 7, 31–5.

Cybulski, J. S. (1978). Modified human bones and skulls from Prince Rupert Harbour, British Columbia. *Canadian Journal of Archaeology*, 2, 15–32.

Dabbs, G. (2009). *Health and Nutrition at Prehistoric Point Hope, Alaska: Application and Critique of the Western Hemisphere Health Index*. Ph.D. University of Arkansas.

Dabbs, G. (2011). Health status among prehistoric Eskimos from Point Hope, Alaska. *American Journal of Physical Anthropology*, 146, 94–103.

El Zaatari, S. (2008). Occlusal molar microwear and the diets of the Ipiutak and Tigara populations (Point Hope) with comparisons to the Aleut and Arikara. *Journal of Archaeological Sciences*, 35, 2517–22.

Foote, B. A. (1992). *The Tigara Eskimos and Their Environment*. Point Hope: North Slope Borough Commission on Iñupiat History Language and Culture.

Hoffman, K. L. (1993). *Unalaska Aleut Subsistence Adaptations at the time of Early Russian Contact as Represented in the Reese Bay Artifact Assemblage*. M.A. University of Arkansas.

Hrdlička, A. (1945). *The Aleutian and Commander Islands and their Inhabitants*. Philadelphia, PA: Wistar Institute of Anatomy and Biology.

Krueger, K. L. (2011). *Dietary and behavioral Strategies of Neandertals and Anatomically Modern Humans: Evidence from Anterior Dental Microwear Texture Analysis*. Ph.D. University of Arkansas.

Krueger, K. L. and Ungar, P. S. (2010). Incisor microwear textures of five bioarchaeological groups. *International Journal of Osteoarchaeology*, 20, 549–60.

Krueger, K. L. and Ungar, P. S. (2012). Anterior dental microwear texture analysis of the Krapina Neandertals. *Central European Journal of Geosciences*, 4, 651–62.

Krueger, K. L., Scott, J. R., Kay, R. F. and Ungar, P. S. (2008). Technical note: Dental microwear textures of "Phase I" and "Phase II" facets. *American Journal of Physical Anthropology*, 137, 485–90.

Larsen, H. and Rainey, F. (1948). *Ipiutak and the Arctic Whale Hunting Culture*. Anthropological Papers of the American Museum of Natural History 42. New York, NY: American Museum of Natural History.

Lester, C. W. and Shapiro, H. L. (1968). Vertebral arch defects in the lumbar vertebrae of pre-historic American Eskimos. *American Journal of Physical Anthropology*, 28, 43–8.

Madimenos, F. (2005). *Dental Evidence for Division of Labor Among the Prehistoric Ipiutak and Tigara of Point Hope, Alaska*. M.A. Louisiana State University.

Marsh, D. B. (1976). The stone winter houses of the Sadlermiut. *The Beaver*, 307, 36–9.

Mathiassen, T. (1927). *Archaeology of the Central Eskimos part I: Descriptive part.* Report of the Fifth Thule Expedition 1921–1924, Volume 4. Copenhagen: Reitzels.

Merbs, C. (1968). Anterior tooth loss in arctic populations. *Southwest Journal of Anthropology*, 24, 20–32.

Merbs, C. (1983). *Patterns of Activity-induced Pathology in a Canadian Inuit Population.* Ottawa: National Museum of Man Mercury Series Archaeological Survey of Canada Paper No. 19.

Merceron, G. M., Scott, J. R., Scott, R. S. *et al.* (2009). Seed predation for an early Colobine as a link between frugivory and folivory? Evidence from dental microwear texture analysis of *Mesopithecus* (Late Miocene of Eurasia). *Journal of Human Evolution*, 57, 732–8.

Moorrees, C. F. A. (1957). *The Aleut Dentition: A Correlative Study of Dental Characteristics in an Eskimoid People.* Cambridge, MA: Harvard University Press.

Oliver, E. R. (1988). *Journal of an Aleutian Year.* Seattle, WA: University of Washington Press.

Rainey, F. (1941). The Ipiutak culture at Point Hope, Alaska. *American Anthropologist*, 43, 364–75.

Rainey, F. (1971). *The Ipiutak Culture: Excavations at Point Hope, Alaska.* Reading, MA: Addison-Wesley.

Ryan, K. (2011). Comments on Coltrain et al., *Journal of Archaeological Science* 31, 2004 "Sealing, whaling and caribou: the skeletal isotope chemistry of eastern Arctic foragers", and Coltrain, *Journal of Archaeological Science* 36, 2009 "Sealing, whaling and caribou revisited: additional insights from the skeletal isotope chemistry of eastern Arctic foragers". *Journal of Archaeological Science*, 38, 2858–65.

Schwartz, J. H., Brauer, J. and Gordon-Larsen, P. (1995). Brief communication: Tigaran (Point Hope, Alaska) tooth drilling. *American Journal of Physical Anthropology*, 97, 77–82.

Scott, J. R., Ungar, P. S., Jungers, W. L. *et al.* (2009). Dental microwear texture analysis of the archaeolemurids and megaladapids, two families of subfossil lemurs from Madagascar. *Journal of Human Evolution*, 56, 405–16.

Scott, R. S., Ungar, P. S., Bergstrom, T. S. *et al.* (2006). Dental microwear texture analysis: Technical considerations. *Journal of Human Evolution*, 51, 339–49.

Stewart, K. M., Stewart, F. L. and Coupland, G. (2009). Boardwalk northern northwest coast Canada: a new face to an old site. *Canadian Journal of Archaeology*, 33, 205–33.

Turner, C. and Cadien, J. (1969). Dental chipping in Aleuts, Eskimo, and Indians. *American Journal of Physical Anthropology*, 31, 303–10.

Ungar, P. S., Brown, C. A., Bergstrom, T. S. and Walker, A. (2003). Quantification of dental microwear by tandem scanning confocal microscopy and scale-sensitive fractal analyses. *Scanning*, 25, 185–93.

Ungar, P. S., Merceron, G. and Scott, R. S. (2007). Dental microwear texture analysis of varswater bovids and early Pliocene paleoenvironments of Langebaanweg

Western Cape Province South Africa. *Journal of Mammalian Evolution*, 14, 163–81.
Ungar, P. S., Scott, R. S., Scott, J. R. and Teaford, M. (2008a). Dental microwear analysis: Historical perspectives and new approaches. In *Technique and Application in Dental Anthropology*. Cambridge: Cambridge University Press, pp. 389–425.
Ungar, P. S., Grine, F. E. and Teaford, M. F. (2008b). Dental microwear and diet of the Plio-Pleistocene hominin *Paranthropus boisei*. *PLoS One*, 3, 1–6.
Wood, S. R. (1992). *Tooth Wear and the Sexual Division of Labour in an Inuit Population*. M.A. Simon Fraser University.

6 *The diets of the Ipiutak and Tigara (Point Hope, Alaska): Evidence from occlusal molar microwear texture analysis*

SIREEN EL ZAATARI

Introduction

Point Hope, Alaska, lies in an arctic tundra landscape that lacks trees, but supports other kinds of vegetation such as mosses, lichens, grasses, and small flowering plants. During the winter season, hard-packed snow covers the landscape (Larsen and Rainey, 1948). Yet the ocean currents always leave a strip of water close to the shore relatively free from pack ice. This strip supports the wide variety of sea mammals for which Point Hope is famous. Some of these sea mammals pass by the area seasonally, while others are abundant almost year-round (Larsen and Rainey, 1948). Particularly, the migration of bowhead whales close to Point Hope makes the area a prime location for whaling. Not far inland, terrestrial animals such as polar bear and caribou are present. The caribou herds also come to the shore in summer to graze (Lester and Shapiro, 1968).

The archaeological record shows that distinct cultural groups have occupied Point Hope on an almost continuous basis since at least 2,400 years BP (Rainey, 1971; Dumond, 1987). For their subsistence, all the ancient inhabitants of Point Hope would have had to rely on the same resources available in the area. However, several lines of evidence suggest that the distinct populations differed in their choice of their main dietary resources (e.g., Rainey, 1947; Larsen and Rainey, 1948; Lester and Shapiro, 1968; Costa, 1980; El Zaatari, 2008). Occlusal molar microwear textures of dental samples of the two archaeologically defined cultural groups of Point Hope, the Ipiutak (*c.* 1,600 years BP to 1,100 years BP) and Tigara (*c.* 800 years BP to 300 years BP), are analyzed in this study to identify any differences in their diets.

The Foragers of Point Hope: The Biology and Archaeology of Humans on the Edge of the Alaskan Arctic, eds. C. E. Hilton, B. M. Auerbach, L. W. Cowgill. Published by Cambridge University Press.

Previous reconstructions of the Ipiutak and Tigara diets

Archaeological and faunal records

Reconstructions of diets and subsistence patterns of both the Ipiutak and Tigara are based mainly on archaeological finds. Excavations at Point Hope showed clear technological distinctions between the old Tigara and the Ipiutak people, indicating that the subsistence patterns of these two groups differed significantly (see contributions by Jensen, Chapter 2, this volume, and by Mason, Chapter 3, this volume). The archaeological and faunal remains show that the diets of the Ipiutak and Tigara were substantially different such that, although both groups exploited fish and seal, the Ipiutak people utilized caribou, whereas the Tigara were whalers who utilized whales rather than caribou for meat.

The archaeological data suggest that the Tigara was a coastal population whose diet was similar to that of the early contact period residents of Point Hope, i.e., a marine-based diet composed of 35–60% fats, 35–65% proteins, and very little carbohydrates (Waugh, 1930; Costa, 1980). The presence of arrowheads in the material culture assemblage in relatively low quantities in relation to the artifacts used for sea hunting, including whaling, implies that, although terrestrial animals, mainly caribou, formed a part of the meat component of the Tigara diet, the major part of this component came mostly from fish (Lester and Shapiro, 1968). Therefore, it can be concluded that the Tigara subsistence was almost exclusively based on the exploitation of sea mammals, mostly whales, walrus, and seal (Rainey, 1947). The limited carbohydrates would have come mostly from raw roots (Waugh, 1930; Costa, 1980). Just like the early contact inhabitants of Point Hope, the Tigara most likely ate their meat fresh, dried it on open racks, or kept it frozen underground as a year-round staple (de Poncins, 1941; Giddings, 1967; Costa, 1980). Uncooked seal skin together with its attached subcutaneous fat would have often been chewed for prolonged periods of time (Balikci, 1970).

On the other hand, the Ipiutak material culture assemblage is dominated by arrowheads relative to artifacts used for coastal hunting (Rainey, 1941). In fact, archery-related artifacts comprise 22% of the entire assemblage while marine hunting artifacts (harpoons and harpoon parts) constitute around 3% only (Larsen and Rainey, 1948). In addition, whaling artifacts and boat remains are lacking from the Ipiutak assemblages, indicating that the Ipiutak did not practice any kind of whaling (Larsen and Rainey, 1948). It has been argued that the Ipiutak population was a migratory hunting culture that occupied the coast only on a seasonal basis (Rainey, 1941; Larsen and Rainey, 1948; Lester and Shapiro, 1968; Rainey, 1971). When they inhabited the coast, the Ipiutak lived mainly on sea mammals, including seals and walruses, but not whales, and terrestrial animals, mainly caribou (Lester and Shapiro, 1968). Remains of these animals constitute around 98% of the faunal elements excavated at the Ipiutak settlement. The remaining 2% are those of small animals, such as birds and fresh-water fish (Lester and Shapiro, 1968). As expected based on the Ipiutak material culture assemblage, whale bones are basically nonexistent in the faunal record left by the Ipiutak (Larsen and Rainey, 1948).

Dental diseases

Studies of periodontal diseases show that both the Ipiutak and Tigara suffered from such diseases, but at distinct rates probably reflecting dietary differences between the two populations (Costa, 1982). The percentage of carious teeth and the rate of decayed, missing, or filled (DMF) teeth are higher among the Ipiutak compared to the Tigara (Costa, 1980). However, severe periodontal disease was greater among the Tigara (Costa, 1982).

Macroscopic dental wear

An examination of the macroscopic dental wear shows that the majority of the individuals of both the Ipiutak and Tigara populations exhibit an extreme level of occlusal wear (Costa, 1982). The variation between the two Point Hope populations appears in the average age at which the individuals achieved maximum tooth wear. For the Ipiutak, tooth wear reached its maximum in individuals between 26 and 30 years of age, whereas for the Tigara maximum tooth wear was not achieved until an age of 40 years or more (Costa, 1982). The level of macroscopic dental wear for the Point Hope individuals was found to be much higher than that observed for the dentitions of individuals of the Kodiak Islands (Costa, 1980). This is most likely related to the large amount of grit included in the diet of Point Hope inhabitants (Costa, 1980). Point Hope inhabitants relied on dried meat (fish, whale, seal, and caribou), which was most likely dried on open racks and was therefore subject to prevailing winds. Since Point Hope consists of mostly sandy beaches, it would be easy for wind-blown sand to become attached to the drying meat and ingested with it (Costa, 1977). Another indication that grit was ingested with the food is the fact that the worn molar teeth from Point Hope were often chipped, cracked, or broken antemortem in a manner indicative of the mastication of highly abrasive particles (Costa, 1977).

Dental microwear

A more direct documentation of dietary differences between the Ipiutak and Tigara is available through the study of occlusal molar microwear using feature-based scanning electron microscopy (SEM) analyses (El Zaatari, 2008). The results of this analysis reveal significant differences in the microwear signatures, and therefore diets, of the two Point Hope populations, with the Tigara showing a more abrasive diet shortly before death compared to the Ipiutak (El Zaatari, 2008). In comparison to the microwear fabrics of two other modern human groups, the Aleut and Arikara, the results of this feature-based microwear study show that the two Point Hope populations had significantly more abrasive diets (El Zaatari, 2008).

New developments in the field of dental microwear analysis have led to the establishment of microwear texture analysis, a fully automated approach to the study of dental microwear where the scanning confocal profilometry replaces the scanning electron microscopy and scale-sensitive fractal analysis is introduced as a tool for three-dimensional analysis of microwear features (Ungar *et al.*, 2003; Scott *et al.*, 2006). In comparison with the more traditional techniques of microwear analysis, microwear texture analysis increases consistency and repeatability of surface measurements through the elimination of the observer subjectivity involved in the process of data collection (Scott *et al.*, 2006). This subjectivity has been shown to result in high levels of inter-observer and intra-observer error rates, making a direct comparison and combination of dental microwear data collected by different observers impossible (Grine *et al.*, 2002). Microwear texture analysis further enhances repeatability of measurements through offering more realistic and reliable three-dimensional representations of tooth surfaces (Scott *et al.*, 2006). These representations, unlike the ones collected with scanning electron microscopy, are not affected by instrumental settings that can lead to different representations of the same surface (Scott *et al.*, 2006). In this current study, the method of microwear texture analysis is used for the examination of the occlusal molar microwear signatures of individuals belonging to the Ipiutak, Tigara, Aleut, and Arikara populations.

Material

Point Hope samples

Occlusal molar microwear textures of a total of forty-four adult individuals from Point Hope were analyzed (see Table 6.1 for sample details). Of these,

Table 6.1. *Microwear data for the Ipiutak and Tigara specimens*

Specimen	Tooth	Sex	Age group	*Asfc*	*epLsar*	*Smc*	*Tfv*	*HAsfc*
Ipiutak								
PH_99_1-99	RM^1	Female	16–25	2.786	0.0010	0.153	15173.4	0.055
PH_99_1-199	LM_2	Female	26–35	2.146	0.0024	0.208	10322.6	0.370
PH_99_1-169	RM^1	Female	26–35	1.470	0.0014	0.267	5957.0	0.148
PH_99_1-111	RM^1	Female	26–35	14.256	0.0028	0.267	13194.0	0.095
PH_99_1-182	LM^2	Female	Unavailable	8.478	0.0022	0.150	17739.3	0.229
PH_99_1-166	LM^1	Male	16–25	19.641	0.0020	0.150	15671.7	0.243
PH_99_1-89B	LM^1	Male	16–25	9.577	0.0027	0.156	17411.2	0.411
PH_99_1-84A	LM_1	Male	16–25	3.729	0.0029	0.150	16569.8	0.126
PH_99_1-105	RM^1	Male	16–25	26.759	0.0012	0.150	14982.5	0.150
PH_99_1-98	RM^1	Male	26–35	11.326	0.0014	0.267	15135.6	0.191
PH_99_1-80	LM_2	Male	26–35	8.400	0.0026	0.151	14541.0	0.457
PH_99_1-94	LM^1	Male	26–35	23.113	0.0013	0.208	15960.2	0.255
PH_99_1-96A	LM^2	Male	36–40	1.666	0.0023	0.151	8476.8	0.112
PH_99_1-181	RM_2	Male	46+	10.596	0.0017	0.150	15054.2	0.328
PH_99_1-88	RM^1	Unavailable	16–25	13.797	0.0022	0.150	12269.7	0.081
PH_99_1-195	RM_1	Unavailable	Unavailable	9.485	0.0009	0.159	14783.1	0.067
PH_99_1-95	RM^1	Unavailable	Unavailable	3.330	0.0020	0.213	14489.3	0.053
Tigara								
PH_99_1-337	LM_1	Female	16–25	1.738	0.0061	0.150	1271.3	0.112
PH_99_1-450	LM_1	Female	16–25	2.375	0.0037	0.267	17559.1	0.152
PH_99_1-482	LM_1	Female	16–25	3.121	0.0021	0.267	13957.5	0.442
PH_99_1-510	RM^1	Female	16–25	7.605	0.0038	0.209	14116.9	0.664
PH_99_1-489	LM_1	Female	16–25	5.006	0.0016	0.150	6546.3	0.461
PH_99_1-507	RM^2	Female	16–25	4.221	0.0029	0.267	18675.8	0.632
PH_99_1-359	LM^1	Female	16–25	7.686	0.0009	0.150	8646.6	0.315
PH_99_1-240	LM_2	Female	26–35	1.320	0.0027	0.267	5411.4	0.184
PH_99_1-316	RM_1	Female	26–35	4.263	0.0026	0.150	14370.1	0.154
PH_99_1-302	LM^2	Female	26–35	3.333	0.0009	0.150	14198.5	0.171
PH_99_1-514	LM^2	Female	26–35	2.753	0.0021	0.266	3525.6	0.050
PH_326	LM^2	Female	26–35	11.330	0.0030	0.152	14932.7	0.305
PH_99_1-429	RM^1	Female	26–35	5.812	0.0021	0.151	15087.5	0.295
PH_99_1-294	RM^2	Female	26–35	14.470	0.0026	0.266	14366.3	0.496
PH_99_1-377	RM^2	Female	26–35	6.735	0.0011	0.267	15466.8	0.263
PH_99_1-364	LM_2	Female	36–40	5.646	0.0048	0.433	14512.4	0.745
PH_99_1-546	RM^2	Female	36–40	3.488	0.0023	0.150	11755.1	0.062
PH_99_1-256	LM^2	Male	16–25	1.241	0.0034	0.266	4959.7	0.070
PH_99_1-228	LM_1	Male	16–25	5.218	0.0032	0.267	22411.8	0.508
PH_99_1-392	LM^2	Male	16–25	3.514	0.0062	0.210	15502.9	0.521
PH_99_1-303	LM_1	Male	16–25	3.659	0.0035	0.151	13863.8	0.189
PH_99_1-481	RM^2	Male	16–25	16.259	0.0029	0.267	16078.4	0.093
PH_99_1-386	RM^2	Male	26–35	21.881	0.0045	0.152	15202.2	0.334
PH_99_1-511	RM^2	Male	26–35	5.089	0.0017	0.150	8717.2	0.153
PH_99_1-259	LM_1	Unavailable	Unavailable	0.754	0.0072	0.267	7753.2	0.135
PH_99_1-325	LM_2	Unavailable	Unavailable	4.239	0.0028	0.150	13733.5	0.073
PH_99_1-440	LM_1	Unavailable	Unavailable	20.193	0.0021	0.209	13097.7	0.130

seventeen are associated with the Ipiutak culture and twenty-seven with the Tigara culture. Although the complete Point Hope skeletal collection consists of a large number of individuals, the excessive amount of occlusal wear on the molars of these individuals, to the point of almost complete loss of enamel on that surface, greatly reduced the potential sample.

Comparative samples

For comparative purposes, occlusal molar microwear data were collected from twenty-one individuals from the Aleutian Islands and sixteen individuals from the Mobridge Arikara Site in South Dakota. Each individual was represented by one molar (either upper or lower, M1 or M2). Unfortunately no demographic information was available for these samples. It should be noted that the majority of the specimens belonging to the Point Hope and comparative samples are the same as those used in the feature-based SEM microwear analyses reported by El Zaatari (2008).

The Aleut skeletal collection consists of protohistoric specimens dating to as recently as 250 years BP (Ungar and Spencer, 1999; but see Coltrain, 2010). Hrdlička collected these skeletons between 1936 and 1938 from the Aleutian Islands of Agattu, Amaknak, Kagamil, Umnak, and Unalaska (Hrdlička, 1945). Ethnohistoric accounts indicate that the Aleutian Islanders relied almost exclusively on marine animals for their subsistence, including fresh and dried fish, mollusks, and sea mammals (Hrdlička, 1945; Laughlin, 1963). They also supplemented their diet with land resources such as edible tubers. Rodents and foxes were occasionally eaten as well (Hrdlička, 1945; Laughlin, 1963). The Aleut were also known for their extensive chewing of frozen and dried animal meat and skin.

Mathew Stirling excavated the Arikara skeletal remains from the Mobridge Site (39WWI), South Dakota, which dates to between 1600 and 1700 CE (Wedel, 1955; Jantz, 1973). The diet of this group consisted of a mix of wild and cultivated plant and animal foods. They hunted bison as well as deer, antelope, and jackrabbit (Meyer, 1977). Meat was cut into strips and dried (Meyer, 1977). Concerning plant foods, they collected black cherries, peppers, grapes, pumpkin, and goosefoot, and cultivated maize, beans, squash, and sunflowers (Hurt, 1969; Meyer, 1977; Blakeslee, 1994; Tuross and Fogel, 1994).

Methods

This study was based on the analysis of dental casts. These casts were prepared following established procedures (Teaford and Oyen, 1989). The occlusal surfaces of the molars were cleaned with cotton swabs soaked in water. Alcohol and/or acetone were used in the cases when the dental remains required more thorough cleaning to remove adherent dirt or preservatives. Molds of these dentitions were then made using President Microsystem (polysiloxane vinyl) impression material (Colténe-Whaledent Corp.). Casts were produced from these molds using Epotek 301 epoxy and hardener (Epoxy Technologies Inc.). Dental casts for the Point Hope material were prepared for this study, whereas Professor Peter Ungar provided those for the comparative groups, the Arikara and Aleut. It should be noted that these latter casts were prepared following the same procedure described above.

The casts were examined with light microscopy to determine their suitability for microwear analyses. Those that were free of postmortem damage and were thus suitable for further analysis (i.e., a total of forty-four individuals from Point Hope and thirty-seven from the comparative samples) were examined using a Sensofar Plμ white-light confocal profiler (Solarius, Inc.) following Scott *et al.* (2005, 2006). One molar (either upper or lower, M1 or M2) was used to represent each individual. Including data from both molars is possible since it has been demonstrated that there are no significant differences in microwear patterns between these molars in recent hunter-gatherer groups (El Zaatari, 2010). Using a magnification of 100×, lateral sampling intervals of 0.18 μm, and vertical resolution of 0.005 μm, four adjacent scans were taken of a "Phase II" facet for a total sampling area of $276 \times 204\ \mu m^2$. Data were analyzed using Toothfrax and Sfrax scale-sensitive fractal analysis (SSFA) software (Surfract Corp.) (Scott *et al.*, 2005, 2006). Five variables were measured: area-scale fractal complexity (*Asfc*), anisotropy (*epLsar*), scale of maximum complexity (*Smc*), texture fill volume (*Tfv*), and heterogeneity of complexity (*HAsfc*). Values for individual specimens are reported as medians of the four fields sampled for each specimen.

Detailed descriptions of these variables and their computations can be found elsewhere (e.g., Scott *et al.*, 2006; El Zaatari, 2010; see also discussion in Krueger, Chapter 5, this volume). Briefly, area-scale fractal complexity (*Asfc*) is a measure of change in roughness across the different scales of observation. The faster a measured surface area increases with resolution, the more complex the surface. A surface dominated by many overlying features of varying sizes would have a high *Asfc* value. Anisotropy (*epLsar*) is measured as the Mean vector of relative lengths (i.e., the sums of line segments divided by straight-line distances between the endpoints) of the sampled profiles across a surface at specific scale (1.8 μm) and orientation intervals (5°). A more anisotropic surface would be a surface dominated by parallel striations. The scale of maximum complexity (*Smc*) measures the fine-scale limit of the steepest part of the curve over which Asfc is calculated. High Smc values should correspond to more complex coarse features. Texture fill volume (*Tfv*) is calculated as the difference in summed volume of fine and large cuboids (with 2 μm and 10 μm diameters, respectively) that would fill a surface. A surface with a high *Tfv* value would have many large and/or deep features. Heterogeneity (*HAsfc*) is measured as the median absolute deviation of *Asfc* divided by the median of *Asfc* for the four scans representing each individual specimen. The individual *HAsfc* values used in this study were calculated using the four scans for each specimen without splitting single scans into smaller sub-regions. A more heterogeneous surface would have high degree of variability in complexity across a facet.

Statistical analyses focused first on assessing sex- and age-related differences in the five variables for the Ipiutak and Tigara samples. Sex designations for the Point Hope specimens were determined by Professor C. Hilton following standard osteological analyses of the ossa coxarum outlined by Buikstra and Ublaker (1994) (C. Hilton, personal communication). However, some specimens lacked pelves and their sex designations remain indeterminate. Age data are from Costa (1977). The second set of statistical analyses focused on comparisons of microwear signatures between the two Point Hope populations and among these populations and the Aleut and Arikara. All statistical analyses were performed on rank-transformed data in order to reduce the possible effects of violating assumptions associated with parametric statistical tests (Conover and Iman, 1981). Data for the five variables were compared among the different groups using a multivariate analysis of variance model (MANOVA) (Neff and Marcus, 1980). Single classification ANOVAs on each variable, along with multiple comparisons tests, were used to determine sources of significant differences when present (Sokal and Rohlf, 1995). Both Fisher's least significant difference (LSD) and Tukey's honestly significant difference (HSD) post-hoc tests were used to balance Type I and II errors (Cook and Farewell, 1996).

Results

Sex- and age-related differences for the Ipiutak and Tigara groups

Results reveal significant differences in the microwear patterns between two age groups of the Ipiutak sample. Ipiutak individuals belonging to the 16–25 years age group have, on average, significantly lower mean *Smc* values compared to those belonging to the 26–35 years age group (Tables 6.2 and 6.3). The MANOVA results indicate no significant age-related differences in microwear patterns for the Tigara sample and no sex-related differences in microwear signatures for either the Ipiutak or the Tigara populations (Table 6.2).

Table 6.2. *Summary statistics for the Ipiutak and Tigara samples divided by sex and age*

		Asfc			*epLsar*			*Smc*			*Tfv*			*HAsfc*		
	n	Mean	SD	SE	Mean	SD	SE	Mean	SD	SE	Mean	SD	SE	Mean	SD	SE
Ipiutak[a]																
Division by sex																
Male	9	12.756	8.595	2.865	0.0020	0.0006	0.0002	0.170	0.041	0.014	14867.0	2556.3	852.1	0.253	0.124	0.041
Female	5	5.827	5.474	2.448	0.0020	0.0007	0.0003	0.209	0.058	0.026	12477.2	4545.0	2032.6	0.179	0.125	0.056
Division by age																
16–25	6	12.715	9.333	3.810	0.0020	0.0008	0.0003	0.152	0.003	0.001	15346.4	1760.4	718.7	0.178	0.132	0.054
26–35	6	10.119	8.109	3.310	0.0020	0.0007	0.0003	0.228	0.048	0.019	12518.4	3771.4	1539.7	0.253	0.138	0.056
36–40	1	1.666	—	—	0.0023	—	—	0.151	—	—	8476.8	—	—	0.112	—	—
40+	1	10.596	—	—	0.0017	—	—	0.150	—	—	15054.2	—	—	0.328	—	—
Tigara[b]																
Division by sex																
Male	7	8.123	7.764	2.934	0.0036	0.0014	0.0005	0.209	0.058	0.021	13819.4	5600.7	2116.9	0.267	0.189	0.072
Female	17	5.347	3.461	0.839	0.0027	0.0014	0.0003	0.218	0.079	0.019	12023.5	5047.7	1224.2	0.324	0.216	0.052
Division by age																
16–25	12	5.137	4.050	1.169	0.0034	0.0016	0.0005	0.218	0.055	0.016	12799.2	6197.9	1789.2	0.347	0.218	0.063
26–35	10	7.699	6.390	2.020	0.0023	0.0010	0.0003	0.197	0.060	0.019	12127.8	4500.6	1423.2	0.241	0.125	0.040
36–40	2	4.567	1.526	1.079	0.0036	0.0018	0.0013	0.292	0.200	0.142	13133.8	1949.7	1378.7	0.404	0.483	0.342

[a] Three Ipiutak specimens are excluded due to the lack of sex and age data.
[b] Three Tigara specimens are excluded due to the lack of sex and age data.

Table 6.3. *Results of statistical analyses for the Ipiutak and Tigara age and sex divisions. Significant differences are in bold.*

	Value	*F*	Hypothesis *df*	Error *df*	*p*-value
(A) MANOVA results					
Ipiutak					
Sex groups comparisons					
Wilks' lambda	0.647	0.874	5	8	0.538
Pillai's trace	0.353	0.874	5	8	0.538
Hotelling's trace	0.546	0.874	5	8	0.538
Age groups comparisons[a]					
Wilks' lambda	0.187	5.221	5	6	**0.034**
Pillai's trace	0.813	5.221	5	6	**0.034**
Hotelling's trace	4.351	5.221	5	6	**0.034**
Tigara					
Sex groups comparisons					
Wilks' lambda	0.777	1.034	5	18	0.428
Pillai's trace	0.223	1.034	5	18	0.428
Hotelling's trace	0.287	1.034	5	18	0.428
Age groups comparisons[b]					
Wilks' lambda	0.735	1.155	5	16	0.373
Pillai's trace	0.265	1.155	5	16	0.373
Hotelling's trace	0.361	1.155	5	16	0.373
(B) Individual ANOVAs for the Ipiutak age groups comparisons					
	Asfc	*epLsar*	*Smc*	*Tfv*	*HAsfc*
F	0.387	0.023	17.010	3.040	1.289
df	1	1	1	1	1
p-value	0.548	0.881	**0.002**	0.112	0.283

[a] The Ipiutak 36–40 and 40+ age groups are excluded from this analysis due to their small sample sizes ($n = 1$ each).
[b] The Tigara 36–40 age group is excluded from this analysis due to its small sample size ($n = 2$).

Comparisons between Ipiutak and Tigara

Results show that the occlusal molar microwear textures differed significantly between the two Point Hope populations (Tables 6.4 and 6.5). Specifically, the Ipiutak have a significantly lower mean *epLsar* value compared to the Tigara (Tables 6.4, and 6.5C). No significant differences were detected in the remaining four variables (Table 6.5).

Table 6.4. *Summary statistics for the Ipiutak, Tigara, Aleut, and Arikara samples*

	n:	Aleut 21	Arikara 16	Tigara 27	Ipiutak 17
Measurement					
Asfc					
	Mean	2.590	1.152	6.406	10.033
	SD	1.600	1.087	5.615	7.605
	SE	0.349	0.272	1.081	1.845
epLsar					
	Mean	0.0028	0.0027	0.0031	0.0019
	SD	0.0011	0.0015	0.0016	0.0006
	SE	0.0002	0.0003	0.0003	0.0002
Smc					
	Mean	0.454	0.820	0.215	0.182
	SD	0.734	0.970	0.070	0.046
	SE	0.160	0.242	0.135	0.011
Tfv					
	Mean	10383.2	5672.8	12434.1	13984.2
	SD	5419.6	5997.8	4946.7	3133.6
	SE	1182.7	1499.5	952.0	760.0
HAsfc					
	Mean	0.221	0.195	0.286	0.198
	SD	0.136	0.146	0.204	0.129
	SE	0.030	0.037	0.039	0.031

Comparison among the two Point Hope groups, Aleut, and Arikara

In comparison to the Aleut and Arikara, the two Point Hope populations have, on average, significantly higher *Asfc* than the Aleut and Arikara (Tables 6.4 and 6.5). In addition, both the Ipiutak and Tigara were found to have significantly lower mean *Smc* and higher mean *Tfv* compared to the Arikara (Tables 6.4 and 6.5C). Finally, the Ipiutak's average *epLsar* value is only marginally significantly lower than that of the Aleut, whereas the Ipiutak's average *Tfv* is marginally significantly higher than that of the Aleut (Table 6.5).

Discussion and conclusions

Occlusal molar microwear analysis has the potential to detect slight differences in the diets of closely related species (e.g., Teaford, 1986, 1993). The occlusal molar microwear signature captures primarily the effects of the mechanical properties of the food itself. However, this signature is greatly influenced by

Table 6.5. *Results of statistical analysis for the Ipiutak, Tigara, Aleut, and Arikara. Significant differences are in bold.*

	Value	*F*	Hypothesis *df*	Error *df*	*p*-value
(A) MANOVA results					
Wilks' lambda	0.426	4.870	15	201.922	**<0.001**
Pillai's trace	0.675	4.357	15	225.000	**<0.001**
Hotelling's trace	1.118	5.341	15	215.000	**<0.001**

(B) Individual ANOVAs

ANOVA	*Asfc*	*epLsar*	*Smc*	*Tfv*	*HAsfc*
F	20.248	2.856	5.129	8.658	1.065
df	3	3	3	3	3
p-value	**<0.001**	**0.043**	**0.003**	**<0.001**	0.369

(C) Multiple comparisons tests (matrices of pairwise mean differences)

	Aleut			Arikara			Tigara		
		p			*p*			*p*	
	Value	Tukey's	Fisher's	Value	Tukey's	Fisher's	Value	Tukey's	Fisher's
Arikara									
Asfc	17.5	**0.022**	**0.004**						
epLsar	6.2	0.846	0.416						
Smc	19.8	**0.035**	**0.007**						
Tfv	20.2	**0.022**	**0.004**						
Tigara									
Asfc	17.9	**0.005**	**0.001**	35.3	**<0.001**	**<0.001**			
epLsar	1.4	0.996	0.829	7.6	0.715	0.293			
Smc	0.7	0.999	0.909	20.5	**0.018**	**0.003**			
Tfv	6.1	0.743	0.315	26.3	**0.001**	**<0.001**			
Ipiutak									
Asfc	25.4	**<0.001**	**<0.001**	42.9	**<0.001**	**<0.001**	7.6	0.528	0.178
epLsar	17.8	0.086	**0.019**	11.6	0.461	0.146	19.2	**0.038**	**0.008**
Smc	8.3	0.642	0.242	28.0	**0.002**	**<0.001**	7.6	0.670	0.260
Tfv	14.8	0.136	**0.032**	35.0	**<0.001**	**<0.001**	8.7	0.533	0.180

extraneous abrasives ingested with the food items, since these abrasives usually leave marks on the enamel surface as well (e.g., Teaford and Glander, 1991; Daegling and Grine, 1999). Concerning the two Point Hope populations, it is believed that their diets consisted mostly of meat. Meat by itself is not hard enough to scratch the enamel surface. The microwear signature caused by eating meat free of extraneous abrasive particles is generally an indirect one, resulting from the effects of tooth-on-tooth abrasion during mastication (e.g., Puech *et al.*, 1981; Teaford and Runestad, 1992). Such abrasion creates large numbers of small pits on the occlusal surface (e.g., Puech *et al.*, 1981; Teaford and Runestad, 1992). Also, studies of microwear signatures among human groups whose diets consist mostly of meat show that differences in the signatures observed were mostly the result of variations in the amount and physical properties of the extraneous abrasive particles ingested with the meat (e.g., Teaford *et al.*, 2001; Organ *et al.*, 2005; El Zaatari, 2010). These particles are usually present in the environment and are attached to the meat as part of the food preparation techniques and then ingested with it. The consumption of relatively large amounts of these abrasive particles results in the formation of numerous microscopic pits on the occlusal molar surface and the formation of surfaces with highly complex textures (e.g., Teaford *et al.*, 2001; Organ *et al.*, 2005; El Zaatari, 2010).

Results show that both the Ipiutak and Tigara have significantly higher levels of occlusal molar surface complexity compared to the Aleut and Arikara. Even though the ethnohistoric reports indicate that the Aleut also relied almost exclusively on meat (consumed raw, dried, or frozen) for their subsistence, the higher levels of surface complexity observed for the two Point Hope populations is not surprising considering the greater potential of sand and grit particles being attached to the meat at Point Hope compared to the Aleutian Islands (Waugh, 1930; de Poncins, 1941; Giddings, 1967; Balikci, 1970; Draper, 1977). Point Hope consists mainly of sandy beaches. In addition, the archaeological evidence suggests that the Tigara, like the prehistoric coastal Inuit, consumed high amounts of frozen meat that was either dried on open racks or stored underground. At Point Hope, this process of freezing meat would have facilitated the ingestion of sand particles, thus subjecting the Tigara to the consumption of high amounts of grit attached to their food. Not much is known about the Ipiutak food preparation techniques. Yet they too, like the Tigara, would have had to rely on substantial amounts of frozen meat, considering the extreme climatic conditions at Point Hope. Therefore, the Ipiutak probably ingested high amounts of extraneous sand particles present in their environment, thus leading to the observed high levels of surface complexity.

In contrast to the area of Point Hope, Alaska, the Aleutian Islands consist of volcanic landscapes that lack sandy beaches. This reduced the possibility of the attachment of small grit or sand particles to the Aleut's meat during its preparation (i.e., its drying or freezing). In terms of microwear signatures, this is reflected in the significantly lower levels of surface complexity observed for the Aleut individuals compared to those belonging to the two Point Hope populations. Therefore, the differences in microwear signatures between the Aleuts and the two Point Hope populations are most likely a manifestation of differences in the amount of extraneous abrasive particles being ingested by these populations.

Archaeological data indicate the Arikara had a mixed diet including both animal and plant foods (Hurt, 1969; Meyer, 1977; Blakeslee, 1994; Tuross and

Fogel, 1994). As expected, the Arikara show a microwear pattern that sets it apart from the two Point Hope and the Aleut groups. Microwear data indicate that the Arikara had a diet that included significantly fewer abrasive particles compared to the diets of the three Alaskan groups. This is reflected in the significantly lower complexity values for the individuals belonging to the Arikara sample relative to the Alaskan groups. The Arikara sample is also differentiated in having ingested, on average, larger abrasive particles, as is evident from its significantly high level of scale of maximum complexity, and few large and/or deep features, as is evident from its significantly lower texture fill volume value compared to the Ipiutak, Tigara, and Aleut populations. It should be noted that the findings of the microwear texture analyses reported here in relation to the amount of abrasives included in the diets of the populations studied, i.e., the two Point Hope populations having the highest and the Arikara having the lowest, are consistent with those of the feature-based microwear analysis for the same populations reported by El Zaatari (2008). Additional direct comparisons of the results obtained by the two microwear analysis methods are not possible since none of the other feature attributes reported by El Zaatari (2008) are analogous to the surface texture attributes.

All lines of dietary evidence show that even though the Ipiutak and Tigara were highly reliant on meat for their subsistence, their diets differed. The current analysis of the Ipiutak and Tigara occlusal molar microwear textures detects significant differences in the signatures of these two groups, specifically in their level of anisotropy. These differences are most likely linked to dietary differences. Ipiutak have significantly less anisotropic surfaces compared to the Tigara. Other researchers have found that the level of anisotropy is related to the directionality of jaw movements; the higher the value the more the surface is dominated by parallel striations (e.g., Scott *et al.*, 2006; Ungar *et al.*, 2007). Higher anisotropy values most likely result from the ingestion of more tough rather than hard items (e.g., Scott *et al.*, 2006; Ungar *et al.*, 2007). The higher value observed for the Tigara probably suggests that they were eating larger amounts of tough foods compared to the Ipiutak. This might be linked to higher levels of chewing of uncooked tough seal skin or eating higher levels of dried meat, both of which have been argued to be possible Tigara behaviors. The Ipiutak might have relied less on frozen meat, since whale meat was not included in their diet, but, rather, they supplemented their diet with smaller sea mammals and caribou. No other significant differences were detected in the microwear textures between the Ipiutak and Tigara.

Finally, sex-related and age-related variations in the microwear signatures of the Ipiutak and Tigara have been assessed. No sex-related differences have been detected in either of the two Point Hope groups. Age-related differences were observed for the Ipiutak sample. Specifically, individuals in the 16–25 years age group have on average a significantly lower scale of maximum complexity value compared to those in the 26–35 years age group, indicating that the latter group was ingesting larger abrasive particles. This might be related to the age at which Ipiutak individuals attained maximum tooth wear recorded by Costa (1982). For the Ipiutak, tooth wear reached its maximum in individuals between 26 and 30 years of age, whereas for the Tigara maximum tooth wear was not achieved until an age of 40 years or more (Costa, 1982). Unfortunately, none of Tigara specimens sampled for this study were over 40 years of age. However, due to the small sample sizes, the results of the sex- and age-related microwear differences, or their lack of, should be considered with caution.

In sum, material culture, faunal remains, and dental macrowear all suggest that the two Point Hope populations had significantly different diets. These indications were confirmed by the results of feature-based microwear analyses (El Zaatari, 2008) and are further confirmed with the results of microwear texture analyses.

Acknowledgements

I would like to thank the editors, Charles Hilton, Libby Cowgill, and Benjamin Auerbach for inviting me to contribute to this volume. I would also like to thank Ken Mowbray and Ian Tattersall for allowing access to the Point Hope specimens under their care. I am grateful to Peter Ungar for providing casts of the Aleut and Arikara specimens, as well as providing access to the microscope facilities at the Department of Anthropology at the University of Arkansas at Fayetteville. I thank Dr. G. Richard Scott and two anonymous reviewers for their comments on an earlier version of this chapter. The microwear data collection was supported by the National Science Foundation (0452155 to E. Grine and S. El Zaatari and 0315157 to P.S. Ungar) and the LSB Leakey Foundation.

References

Balikci, A. (1970). *The Netsilik Eskimo*. Prospect Heights, NY: Natural History Press.

Blakeslee, D. J. (1994). The archaeological context of human skeletons in the northern and central Plains. In *Skeletal Biology in the Great Plain: Migration, Warfare, Health and Subsistence*. Washington, D.C.: Smithsonian Institution Press, pp. 9–32.

Buikstra, J. E. and Ubelaker, D. H. (1994). *Standards for Data Collection from Human Skeletal Remains*. Fayetteville, AK: Arkansas Archaeological Survey Report Number 44.

Coltrain, J. B. (2010). Temporal and dietary reconstruction of past Aleut populations: stable- and radio-isotope evidence revisited. *Arctic*, 63, 391–8.

Conover, W. J. and Iman, R. L. (1981). Rank transformations as a bridge between parametric and nonparametric statistics. *The American Statistician*, 35, 124–9.
Cook, R. J. and Farewell, V. T. (1996). Multiplicity considerations in the design and analysis of clinical trials. *Journal of the Royal Statistical Society Series A*, 159, 93–110.
Costa, R. L. (1977). *Dental Pathology and Related Factors in Archaeological Eskimo Skeletal Samples from Point Hope and Kodiak Island, Alaska*. Ph.D. University of Pennsylvania.
Costa, R. L. (1980). Incidence of caries and abscesses in Archaeological Eskimo skeletal samples from Point Hope and Kodiak Island, Alaska. *American Journal of Physical Anthropology*, 52, 501–14.
Costa, R. L. (1982). Periodontal disease in the prehistoric Ipiutak and Tigara skeletal remains from Point Hope, Alaska. *American Journal of Physical Anthropology*, 59, 97–110.
Daegling, D. V. and Grine, F. E. (1999). Terrestrial foraging and dental microwear in Papio ursinus. *Primates*, 40, 559–72.
de Poncins, G. (1941). *Kabloona*. New York, NY: Reynal & Hichcock Inc.
Draper, H. H. (1977). The aboriginal Eskimo diet. *American Anthropologist*, 79, 309–16.
Dumond, D. E. (1987). *The Eskimos and Aleuts*. London: Thames and Hudson.
El Zaatari, S. (2008). Occlusal molar microwear and the diets of the Ipiutak and Tigara populations (Point Hope) with comparisons to the Aleut and Arikara. *Journal of Archaeological Science*, 35, 2517–22.
El Zaatari, S. (2010). Occlusal microwear texture analysis and the diets of historical/prehistoric hunter-gatherers. *International Journal of Osteoarchaeology*, 20, 67–87.
Giddings, J. L. (1967). *Ancient Men of the Arctic*. London: Secker and Warburg.
Grine, F. E., Ungar, P. S. and Teaford, M. F. (2002). Error rates in dental microwear quantification using scanning electron microscopy. *Scanning*, 24, 144–53.
Hrdlička, A. (1945). *The Aleutian and Commander Islands and their Inhabitants*. Philadelphia, PA: Wistar Institute of Anatomy and Biology.
Hurt, W. R. (1969). Seasonal economic and settlement patterns of the Arikara. *Plains Anthropologist*, 14, 32–7.
Jantz, R. L. (1973). Microevolutionary change in Arikara crania: Multivariate analysis. *American Journal of Physical Anthropology*, 38, 15–26.
Larsen, H. and Rainey, F. (1948). *Ipiutak and the Arctic Whale Hunting Culture*. Anthropological Papers of the American Museum of Natural History 42. New York, NY: American Museum of Natural History.
Laughlin, W. S. (1963). Eskimos and Aleuts: Their origins and evolution. *Science*, 142, 633–45.
Lester, C. W. and Shapiro, H. L. (1968). Vertebral arch defects in the lumbar vertebrae of pre-historic American Eskimos. *American Journal of Physical Anthropology*, 28, 43–8.
Meyer, R. W. (1977). *The Village Indians of the Upper Missouri*. Lincoln, NE: University of Nebraska Press.

Neff, N. A. and Marcus, L. F. (1980). *A Survey of Multivariate Methods for Systematics*. New York, NY, American Museum of Natural History (manuscript).

Organ, J. M., Teaford, M. F. and Larsen, C. S. (2005). Dietary inferences from dental occlusal microwear at Mission San Luis de Apalachee. *American Journal of Physical Anthropology*, 128, 801–11.

Puech, P. F., Prone, A. and Albertini, H. (1981). Reproduction expérimental des processus d'altération de la surface dentaire par friction non abrasive et non adhésive: Application à l'étude de alimentation de l'homme fossile. *Comptes Rendus de l'Académie des Sciences, Série II*, 293, 729–34.

Rainey, F. G. (1941). The Ipiutak culture at Point Hope, Alaska. *American Anthropologist*, 43, 364–75.

Rainey, F. G. (1947). *The Whale Hunters of Tigara*. Anthropological Papers of the American Museum of Natural History 41(2). New York, NY: American Museum of Natural History.

Rainey, F. (1971). *The Ipiutak Culture: Excavations at Point Hope, Alaska*. Reading, MA: Addison-Wesley.

Scott, R. S., Ungar, P. S., Bergstrom, T. S., *et al.* (2005). Dental microwear texture analysis shows within-species diet variability in fossil hominins. *Nature*, 436, 693–5.

Scott, R. S., Ungar, P. S., Bergstrom, T. S., *et al.* (2006). Dental microwear texture analysis: Technical considerations. *Journal of Human Evolution*, 51, 339–49.

Sokal, R. R. and Rohlf, F. J. (1995). *Biometry*. New York, NY: W.H. Freeman and Company.

Teaford, M. F. (1986). Dental microwear and diet in two species of Colobus. In *Proceedings of the Tenth Annual International Primatological Conference, Volume 2: Primate Ecology and Conservation*. Cambridge: Cambridge University Press, pp. 63–6.

Teaford, M. F. (1993). Dental microwear and diet in extant and extinct Theropithecus: Preliminary analyses. In *Theropithecus: The Rise and Fall of a Primate Genus*. Cambridge: Cambridge University Press, pp. 331–49.

Teaford, M. F. and Glander, K. E. (1991). Dental microwear in live, wild-trapped *Alouatta palliata* from Costa Rica. *American Journal of Physical Anthropology*, 85, 313–19.

Teaford, M. F. and Oyen, O. J. (1989). Live primates and dental replication: New problems and new techniques. *American Journal of Physical Anthropology*, 80, 73–81.

Teaford, M. F. and Runestad, J. A. (1992). Dental microwear and diet in Venezuelan primates. *American Journal of Physical Anthropology*, 88, 347–64.

Teaford, M. F., Larsen, C. S., Pastor, R. and Noble, V. E. (2001). Pits and scratches: Microscopic evidence of tooth use and masticatory behavior in La Florida. In *Bioarchaeology of Spanish Florida: The Impact of Colonialism*. Gainesville, FL: University Press of Florida, pp. 82–112.

Tuross, N. and Fogel, M. L. (1994). Stable isotope analysis and subsistence patterns at the Sully Site. In *Skeletal Biology of the Great Plain: Migration, Warfare, Health and Subsistence*. Washington, D.C.: Smithsonian Insitution Press, pp. 283–9.

Ungar, P. S. and Spencer, M. A. (1999). Incisor microwear, diet, and tooth use in three Amerindian populations. *American Journal of Physical Anthropology*, 109, 387–96.

Ungar, P. S., Brown, C. A., Bergstrom, T. S. and Walker, A. (2003). Quantification of dental microwear by tandem scanning confocal microscopy and scale-sensitive fractal analyses. *Scanning*, 25, 185–93.

Ungar, P. S., Merceron, G. and Scott, R. S. (2007). Dental microwear texture analysis of Varswater bovids and early Pliocene paleoenvironments of Langebaanweg, Western Cape Province, South Africa. *Journal of Mammalian Evolution*, 14, 163–81.

Waugh, L. M. (1930). A study of the nutrition and teeth of the Eskimo of North Bering Sea and Arctic Alaska. *Journal of Dental Research*, 10, 387–93.

Wedel, W. (1955). Archaeological materials from the vicinity of Mobridge, South Dakota. *Smithsonian Institution, Bureau of American Ethnology*, 154, 85–167.

7 *Postcranial pathological lesions in precontact Ipiutak and Tigara skeletal remains of Point Hope, Alaska*

CHARLES E. HILTON, MARSHA D. OGILVIE, MEGAN LATCHAW CZARNIECKI, AND SARAH GOSSETT

Introduction

Disease burdens are common conditions seen throughout all human societies. Documenting their manifestation and prevalence patterns cross-culturally can provide greater anthropological understandings of their impact within different human cultural systems, especially for those cultural communities living within harsh environments such as the Arctic. In recent traditional communities of the North American Arctic, ethnohistoric records and ethnographic studies provide important, though limited information on the type of morbidities seen at the time of contact. Additionally, a lively discussion exists regarding the prevalence of endemic diseases and their impact on cultural behaviors as well as the levels of comorbidities seen within the precontact northwest coastal Alaska human foraging communities (Costa, 1980a, b, 1982; Fortuine, 1986, 1989, 2005; Keenleyside, 1998; Gossett and Hilton, 2004; Latchaw and Hilton, 2004, 2006; Dabbs, 2009b, 2011). In contrast, although researchers (Merbs, 1983; Hart Hansen *et al.*, 1991; Arriaza, 1995; Larsen and Kelly, 1995; Webb, 1995; Larsen, 1997) have investigated patterns of morbidity associated with forager skeletal remains worldwide, few studies have examined the skeletal evidence for disease burdens within the prehistoric adult postcranial remains of the Arctic foragers from Point Hope, Alaska (Gossett and Hilton, 2004; Latchaw and Hilton, 2004, 2006; Ogilvie, 2006; Dabbs, 2009b, 2011). Previous paleopathological studies examining human remains from the Point Hope archaeological sites have focused primarily on cranial and dental lesions

The Foragers of Point Hope: The Biology and Archaeology of Humans on the Edge of the Alaskan Arctic, eds. C. E. Hilton, B. M. Auerbach, L. W. Cowgill. Published by Cambridge University Press.

in order to evaluate developmental or periodontal diseases (Costa, 1980a, b, 1982; Keenleyside, 1998; Madimenos, 2005). Given that different foraging strategies may be associated with unique sex- and age-specific morbidity risks and exposures for individuals (Larsen, 1997), in this chapter we ask whether the contrasts in resource-targeting foci and the requisite infrastructure for such subistence strategies produce significantly more pervasive adult postcranial skeletal lesions (PCSLs) in the whale-centric Tigara (800–300 years BP) or the earlier non-whaling Ipiutak (*c.* 1,600–1,100 years BP) Arctic foragers of Point Hope.

Our chapter investigates specifically the prevalence patterns of select adult PCSLs as indirect indicators of morbidities and physical workloads among the women and men of the precontact Ipiutak and Tigara cultural periods of Point Hope, Alaska. We apply epidemiological principles (Rothman, 2002; Gordis, 2009) to paleopathological examinations in conjunction with available archaeological and ethnographic evidence related to foragers. This combination may provide a valuable approximation for reconstructing past lifeways of prehistoric human groups and assessing endemic and chronic diseases within groups that utilized different subsistence and cultural behavior regimes. Based on these methods, the findings presented herein point to important differences in disease burdens with a shift to a more whale-centric subsistence livelihood in our later, Tigara sample, and we suggest that this shift affected Tigara women more substantially than men compared to their Ipiutak counterparts.

Materials and methods

Sample background

Our research was designed to investigate patterns of adult disease burdens seen within the Ipiutak and Tigara samples from Point Hope, Alaska. Human burials encompassing the Point Hope skeletal collection were originally excavated between 1939 and 1941 by Froehlich Rainey and Helge Larsen as part of the archaeological excavations sponsored by the University of Alaska, the Danish Government, and the American Museum of Natural History (Larsen and Rainey, 1948; Giddings, 1968; Rainey, 1971; Anderson, 1984; Dumond, 1984; see Mason, Chapter 3, this volume). The majority of archaeological artifacts as well as the human remains are currently curated at the American Museum of Natural History (Larsen and Rainey, 1948; Giddings, 1968; Rainey, 1971). The prevalence of adult PCSLs, indicative of fractures, spondylolysis, developmental defects, and infectious diseases, seen within the samples are compared and used here as a framework for addressing the overall levels of disease burden and to assess the effects of subsistence-related physical activities on males and females.

Ipiutak foragers

As described by Mason (Chapter 3, this volume), Ipiutak foragers lived on the Point Hope spit for roughly 500 years (*c.* 1,600–1,100 years BP). According to Mason, Ipiutak house construction and placement suggests that only a small percentage of houses were occupied concurrently at any given time and that the Ipiutak community size was small with only 125–200 people. In contrast to previous researchers suggesting a large population size at Point Hope (Rainey, 1941, 1971; Larsen and Rainey, 1948), Mason's smaller estimated population sizes are more in agreement with ethnographic and ethnohistoric descriptions of small settlement sizes for historic northwest coastal Alaska Iñupiaq communities (Murdoch, 1892; Nelson, 1899; Stefansson, 1914; Spencer, 1959; Nelson, 1969; Burch, 1981, 1988; Worl, 1999). The high frequency of caribou (*Rangifer tarandus grantz*) derived artifacts led Larsen and Rainey (1948) to conclude that caribou was a primary terrestrial resource target, used both as a source of food and as raw material for manufacturing tools. However, the wide range of faunal remains seen in the Ipiutak cultural horizons indicate a subsistence base that utilized sea mammals, particularly seals and walruses (Mason, Chapter 3, this volume). Giddings (1968) noted that the Ipiutak did not appear to have watercraft of any type. Although Larsen and Rainey (1948) concluded that the Ipiutak were only seasonal residents of Point Hope, who in the winter months hunted terrestrial game within the interior, Mason proposes that Ipiutak residency at Point Hope was more permanent.

Tigara foragers

The Tigara lived on the Point Hope spit between 800 and 300 years BP (Larsen and Rainey, 1948; Rainey, 1947, 1971; Anderson, 1984; Schwartz *et al.*, 1995; Jensen, Chapter 2, this volume). Archaeologists consider the Tigara to have engaged in less seasonally based residential mobility in comparison to their Ipiutak counterparts (Larsen and Rainey, 1948; Rainey, 1971; Anderson, 1984; Dumond, 1984). Instead, the Tigara are thought to have maintained permanent year-round residency on the Point Hope spit and to have incorporated the hunting of bowhead whales (*Balaena mysticetus*) into their subsistence economy. However, faunal remains indicate a reliance on a wide variety of other maritime animal resources, such as walruses and various species of seals, as well as a variety of fish and bird species. In contemporary Iñupiaq communities, whale hunting is seasonal and depends upon the creation of an offshore ice-free corridor or lead, like that seen at Point Hope; whales pass through the lead during their annual migration (Burch, 1981, 1988; Sakakibara, 2007). Ethnographic observations (Murdoch, 1892; Nelson, 1899; Spencer, 1959; Nelson, 1969; Spencer, 1977; Burch, 1981, 1988; Sheehan, 1985; Chance, 1990; Worl, 1999; Sakakibara, 2007) indicate that, in targeting large-bodied migrating marine mammals such as bowhead whales and walruses, traditional Iñupiaq hunting involves high levels of planning depth in preparing and maintaining specialized hunting technology. Whale hunting generally begins in the spring (April–May) and continues through the summer months (August–September) (Rainey, 1947; Spencer, 1977; Worl, 1999; Sakakibara, 2007). Iñupiaq whale hunters use an *umiak*, a paddle-propelled large-sized open-air boat generally made with walrus skin stretched over a driftwood frame, which could carry a crew of approximately eight individuals, including the captain and the harpooner (Spencer, 1977; Sheehan, 1985; Sakakibara, 2007). Contemporary indigenous whale hunting is a highly conspicuous large-scale coordinated subsistence activity with a high risk for bodily harm (Spencer, 1977; Sheehan, 1985; Alvard and Nolin, 2002; Sakakibara, 2007).

When successful, whale hunting can produce a high payoff (Alvard and Nolin, 2002). However, hunting success rates in contemporary traditional whale hunts (Spencer, 1977; Alvard and Nolin, 2002) suggest that prehistoric Tigara households most likely relied on other animal food resources, such as walrus, that could be procured by a single individual or several individuals hunting cooperatively (Murdoch, 1892; Nelson, 1899; Spencer, 1959; Nelson, 1969; Burch, 1981, 1988; Sheehan, 1985; Chance, 1990; Mason, Chapter 3, this volume).

Study specimens

The Ipiutak and Tigara human skeletal samples in this study were selected from the more than 500 archaeologically derived human burials from Point Hope dating from *c.* 1,600 years BP to the historic period (Larsen and Rainey, 1948; Schwartz *et al.*, 1995; Holliday and Hilton, 2010). The prehistoric cultural designation (i.e., Near Ipiutak, Norton, Ipiutak, Birnirk, and/or Tigara) for each burial is based on the archaeological context and the associated artifacts (Larsen and Rainey, 1948). However, the exact number of excavated burials for each cultural period remains unclear due to a lack of burial integrity, preservation bias, some misattribution to specific cultural periods, and commingling (Hilton, personal observation). Most excavated Point Hope burials belong to either the

Ipiutak and Tigara cultural periods (Larsen and Rainey, 1948), with the Tigara burials comprising the largest component.

The adult Point Hope skeletal samples in this study include forty-three Ipiutak skeletons (twenty-five males, eighteen females) and one hundred and fifty-five Tigara skeletons (eighty-two males, seventy-three females). Not all skeletons preserve all postcranial regions of interest, and, thus, sample size varies for several analyses of pathological lesions.

Several criteria were used to select skeletons for inclusion in the current study. Like many archaeological skeletal series, the state of preservation for the Point Hope burials varies from excellent to extremely poor. All Point Hope skeletons available for analysis were examined for their overall level of postcranial completeness and developmental maturity. Skeletons included in the study exhibited complete fusion of the epiphyses of major long bones. Sex and age designations were determined using standard osteological procedures as outlined by Buikstra and Ubelaker (1994) and these designations were verified by at least two observers. Our adult age estimations are based on evaluations of auricular surface and pubic symphysis morphology of the *os coxae* (Buisktra and Ubelaker, 1994). The age estimations for each adult skeleton were collapsed into two primary age categories: "Young" adults (18–39 years of age) and "Old" adults (≥40 years of age). The "cut-off" age between "Young" and "Old" adults for both males and females is based on the documented age ranges described for "reproductive" and "post-reproductive" women in traditional foraging societies (Howell, 1979; Hill and Hurtado, 1996). A third age category, "Adult," was created to include those skeletally mature individuals for which an age estimation was undetermined due to the lack of key diagnostic regions in the *os coxae*. All Point Hope skeletons in the current study were examined by Charles Hilton.

Postcranial skeletal lesions

Each postcranial skeleton included in the current analysis was examined macroscopically for evidence of pathological lesions indicative of trauma (e.g., healed fractures, pseudoarthroses, osteomyelitis, dislocations, Schmorl's node, vertebral collapse, etc.), non-specific infections (e.g., periostitis), infectious disease, developmental anomalies/defects (i.e., vertebral border shifting, spina bifida), and extreme degenerative joint disease (i.e., eburnation). The presence of such lesions was recorded for each individual. While access to radiographic equipment was limited, noninvasive radiographic examination aided our analysis of one individual. As hunter-gatherers habitually utilize strenuous subsistence-related physical activities in their daily lives, we expect adult skeletal remains of Arctic foragers to exhibit minor to moderate degenerative joint disease (DJD) regardless of adult age. Therefore, the less severe but more pervasive forms of DJD were excluded in our lesion frequency counts. However, we tabulate the presence of eburnation, an extreme form of DJD. Additional descriptions are provided below for several types of skeletal lesions and defects, including parry fractures, spondylolysis, and vertebral border shifting.

Bones of the forearm were closely examined for evidence of parry fractures, i.e., mid-shaft fracturing of the ulna and sometimes the radius (Ferguson, 1997). Parry fractures are considered a "classic" fracture that occurs as part of defensive posturing when an individual uses his or her forearm to ward off a blow to the face (Ferguson, 1997). Although other types of trauma may produce similar lesions, parry fractures are considered indicative of interpersonal physical violence and may appear at higher frequency rates in human societies with high levels of interpersonal physical violence (Ferguson, 1997), such as the warfare and blood feuds associated with recent Eskimo societies (Murdoch, 1892; Nelson, 1899; Spencer, 1959; Nelson, 1969; Burch, 1981, 1988). All other major long bones of the upper and lower limbs as well as the vertebral column were examined in this study for evidence of healed fractures. Ribs were not examined given their frequently poor state of preservation in a prehistoric context.

Spondylolysis affects the neural arch of a vertebra and manifests as bilateral spatial separations at both the left and right *pars interarticularis* (Aufderheide and Rodriguez-Martin, 1998). Complete bilateral spondylolysis produces a vertebra composed of two separate pieces, a ventral piece consisting of the body, the pedicles, the superior articular processes, and the transverse processes as well as a dorsal piece consisting of the laminae, inferior articular processes, and the spinous process (Aufderheide and Rodriguez-Martin, 1998). Unilateral spondylolysis is a single spatial separation through either the left or right *pars interarticularis* but the vertebra remains a single unit. Spondylolysis is seen more frequently in lumbar vertebrae relative to the vertebrae of other regions (Aufderheide and Rodriguez-Martin, 1998).

We investigate whether vertebral border shifting (VBS) offers a protective exposure against spondylolysis. This requires a systematic examination of complete vertebral columns for each skeleton. Vertebral column completeness is determined by vertebral sequence and joint congruence for each vertebral element present. Although the methodology employed to document vertebral border shifting is described elsewhere (Merbs, 1974; Barnes, 1994; Ogilvie *et al.*, 1998; Ogilvie, 2006), we provide a brief description here. A cranial border shift is defined as a superior movement of the boundary between two adjacent vertebral regions, whereas a caudal shift is defined as an inferior movement of the boundary with the skeleton retaining the modal number of twenty-four presacral vertebrae (PSV) (Merbs, 1974). Sacralization of the last lumbar vertebra is considered a cranial shift, while lumbarization of the first sacral vertebra is a caudal shift. Determining whether a shift is cranial or caudal requires complete thoracic and lumbar regions. While cervical ribs can occur, the presence of the cervical vertebrae is not as critical as this region most often retains the modal number of seven elements (Schultz, 1961). All Point Hope adult skeletons in the VBS component of this study possess seven cervical vertebrae. Vertebral element counts and the number of PSV were documented for each vertebral region and vertebrae exhibiting partial shifting were noted. Individuals with combined VBS and spondylolysis are noted. This examination utilizes a reduced Tigara sample ($n = 102$; males, $n = 49$; females, $n = 53$) composed only of individuals with completely preserved PSV regions.

Statistical procedures

The prevalence percentages for the selected postcranial lesions were calculated for each sex and age category within each cultural group. These prevalence percentages were then comparatively analyzed using chi-square following procedures outlined by Moore and McCabe (2003), with subsistence and sex as the independent factors and an $\alpha = 0.05$. In several instances, statistical analyses were not performed due to small sample size. Additionally, we have limited the number of statistical analyses in order to avoid Type I errors.

Results

Overview

Tables 7.1 and 7.2 present the frequencies for the PCSLs for the Ipiutak and Tigara samples, respectively, grouped by the sex and age categories. The "Old" adults (40+ years old) are a smaller proportion of the skeletal samples within both cultural periods (see Tables 7.1 and 7.2). The "Young" (18–39 years old) adult males and females represent the majority of the Ipiutak sample, 44% and 37%, respectively (see Table 7.1). The "Old" adult males and females represent 14% and 5%, respectively, within the Ipiutak sample. Ten Ipiutak individuals, i.e., 23% of the Ipiutak sample, exhibit the postcranial lesions examined for the current study. Nine of these ten Ipiutak individuals are males, representing 21% of the total Ipiutak sample and 36% of the Ipiutak males (see Table 7.1). For those Ipiutak males possessing a single lesion, spondylolysis is seen in four "Young" males and one "Old" male. Another "Old" male (AMNH 99.1/77) exhibits eburnation in the occipital-cervico joint (see

Table 7.1. *Frequency of Point Hope Ipiutak skeletons with lesions by lesion type*[a]

	Multiple lesions	Upper limb trauma/ fractures	Vertebral collapse/ Schmorl's node	Eburnation	Spondylolysis	Spina bifida	Periostitis
Ipiutak males ($n = 25$)							
Young ($n = 19$)	1	—	(1)	—	4	1	(1)
Old ($n = 6$)	1	(1)	(1)	1	1	—	—
Ipiutak females ($n = 18$)							
Young ($n = 16$)	1	—	(1)	(1)	—	—	—
Old ($n = 2$)	—	—	—	—	—	—	—
Total	3	(1)	(3)	1 (1)	5	1	(1)

[a] Numbers in parentheses represent lesion counts associated with individuals who exhibit multiple lesions.

Table 7.1). Three (AMNH 99.1/87, 99.1/91a, and 99.1/660) of these ten individuals exhibit multiple types of lesion (see Table 7.3 for more detail). One common morbidity characteristic seen across these three individuals is that they possess lesions that mostly reflect degeneration of vertebral column joints in combination with another lesion, e.g., upper limb trauma or eburnation. The fibular periostitis seen in the younger male (AMNH 99.1/660) may be a case of chronic sclerosing osteomyelitis (see Figure 7.1). We note also the Ipiutak male (AMNH 991.1/181; estimated age 40–44 years), previously described by Larsen and Rainey (1948), who exhibits two unhealed "green bone" perimortem wounds in the form of two embedded projectile points, one in the manubrium and another in the corpus of the sternum (see Figure 7.2).

In Tigara with estimated ages, the "Young" males and females make up the majority of the sample, 43% and 41%, respectively (see Table 7.2). The Tigara, relative to the Ipiutak, exhibit a higher prevalence of postcranial skeletal lesions as 65% (101 individuals) of the Tigara sample possess at least one lesion of interest (see Table 7.2). For Tigara males, fifty-two individuals (63% of the males) exhibit at least one skeletal lesion whereas forty-nine females (67% of the females) exhibit at least one lesion (see Table 7.2). Within the Tigara sample, twenty-five individuals exhibit multiple PCSLs of different types with fourteen of those individuals being male. Table 7.3 provides expanded descriptions for those Tigara with more than one type of PCSL.

Table 7.2. *Frequency of Point Hope Tigara skeletons with lesions by lesion type*[a]

	Multiple lesions	Lower limb trauma/fractures	Upper limb trauma/fractures	Vertebral collapse/ Schmorl's node	Spondylolysis	Spina bifida	Eburnation	TB	EA[b]
Tigara males ($n = 82$)									
Young ($n = 53$)	6	—	2	1	14	1	—	4	—
Old ($n = 6$)	2	1	1	—	5	—	2	—	—
Adult ($n = 16$)	6	2	—	—	5	—	—	—	—
Tigara females ($n = 73$)									
Young ($n = 50$)	6	—	—	1	22	—	—	1	1
Old ($n = 7$)	2	—	1	1	—	—	—	—	—
Adult ($n = 16$)	3	1	1	2	6	—	—	—	1
Total	25	4	5	5	52	1	2	5	2

[a] The numbers presented in this table are conservative frequency counts of the lesions present in the Tigara sample as individuals with multiple lesions have been counted only once.
[b] Erosive arthropathy.

Table 7.3. *Descriptions of Ipiutak and Tigara individuals with multiple types of lesions*

Cultural period:	Cat #: 99.1/	FS# and B#	Sex	Description
Ipiutak	87	36	m	Age 50–59: Fused thoracic and lumbar vertebrae (T9, T10, T11, T12 and L1), Schmorl's node L3 vertebral body (superior surface), healed fracture of the left scapular acromion and coracoid processes where the coracoid process healed as a separate floating element
Ipiutak	91		f	Age 35–44: Collapsed T8 vertebral body, left and right tibiae exhibit pronounced eburnation at the proximal ends in association with the knee joints
Ipiutak	660	JB1	m	Age 30–39: Periostitis of RT fibula, this massive gummatous lesion is seen on most of the fibular shaft with only the proximal 1/4 and distal 1/4 unaffected, Schmorl's nodes in lumbar vertebrae (L2 and L3)
Tigara	222	140	f	Age 40–44: Spondylolysis L5, eburnation LT femoral head and acetabulum
Tigara	278	197	m	Age 35–39: Dislocated left distal radio-ulnar joint (distal ulna articulates proximal to the radio-carpal joint), eburnation in both LT and RT distal humeri at the capitulum
Tigara	305	236	m	Age 35–39: Healed fractures of lower limbs, LT tibial/talar/calcaneal healed fracture with bone fusion, RT femur midshaft fracture with misalignment
Tigara	334	275	f	Age 30–34: Spondylolysis L4 and L3, healed fracture of RT distal tibia and talus
Tigara	336	282	m	Age 30–39: Spondylolysis L5, healed fracture of RT acromion process
Tigara	338	285	f	Age 30–34: Spondylolysis L5, periostitis of anterior surfaces of both LT and RT tibiae
Tigara	340	287	m	Age 50–59: Spondylolysis L4 and Baastrup's disease, healed RT radial Colles fracture
Tigara	345	294	f	Age 50–59: Eburnation of both LT and RT acetabulae and femoral heads, eburnation of left humeral trochlea
Tigara	357	308	m	Age 40–49: Fusion L4 and L5 vertebrae, eburnation LT knee (distal femur and proximal tibia)
Tigara	366	316	f	Age 25–34: Incomplete thoracic spinous process fusion, collapsed T9 anterior vertebral body
Tigara	374	325	f	Age 35–44: Collapse of ventral vertebral bodies of T6, T8, T9, T12, L1, and L3, Schmorl's node L1
Tigara	400	358	m	No Age Est.: Spondylolysis L5, T12 Schmorl's node; RT femoral head and neck lesions (mushroomed with associated acetabular expansion)

(*cont.*)

Table 7.3. (*cont.*)

Cultural period:	Cat #: 99.1/	FS# and B#	Sex	Description
Tigara	402	360	m	No Age Est.: Spondylolysis L4 and L5, fusion and vertebral body collapse of T8, T9, and T10
Tigara	417	380	f	No Age Est.: Spondylolysis L5, healed fracture of RT femoral head with associated acetabulum remodeling
Tigara	441	407	m	Age 30–34: Spondylolysis L5, T11 Schmorl's node
Tigara	465	435	f	Age 20–24: Spondylolysis L4 and L5, eburnation of C1-C2 at the articulation of the odontoid process, collapse of L1 anterior vertebral body
Tigara	468	439	m	No Age Est.: Spondylolysis L5 (unilateral through left lamina), Collapse of T11 and T12 anterior vertebral bodies
Tigara	478	453	m	Age 25–34: Spondylolysis L5, collapse of L1 anterior vertebral body, healed misaligned fracture of RT distal tibia (with expansion of talar articular region)
Tigara	488	463	m	No Age Est.: Spondylolysis L5, collapse of T7 vertebral body, healed Colles fracture of RT radius
Tigara	490	465	m	No Age Est.: Spondylolysis L4 and L5 (L4 lesion is unilateral through the LT neural arch with large callus on RT L4 pedicle, spondylolysis does not involve, normally, massive remodeling seen in the pedicle), eburnation at the capitular regions of both LT and RT humeri
Tigara	513	490	f	No Age Est.: Spondylolysis L5, TB lesion at L4
Tigara	519	496	m	No Age Est.: Healed fracture of RT femoral head and neck region (extensive mushroom expansion encroaching the femoral neck), spina bifida in sacrum
Tigara	526	503	f	Age 35–39: Fusion T9 and T10 vertebral bodies, eburnation, osteomyelitis of RT distal fibula (distal 1/3 fibula with healed misaligned fracture)
Tigara	540	516	f	No Age Est.: Spondylolysis L5; collapse of anterior vertebral bodies of T8, T11, T12, Schmorl's node T12
Tigara	541	518	m	Age 35–39: Collapse of anterior vertebral bodies, healed RT fibula distal fracture; eburnation of RT humeral capitulum and RT radial head expansion

Spondylolysis is the most pervasive PCSL in the Tigara (see Table 7.2). Spondylolysis affects fifty-two individuals as a single lesion (see Table 7.2) and sixteen more individuals with multiple lesions (see Table 7.3), with a nearly even number of males (thirty-three individuals) and females (thirty-five individuals) possessing the lesion. Although Tigara females have a higher prevalence (48%) of spondylolysis relative to males (40%), it is not a significant difference (chi-square: $\chi^2 = 0.23$, $df = 1$, $p > 0.30$). Non-spondylolytic vertebral lesions

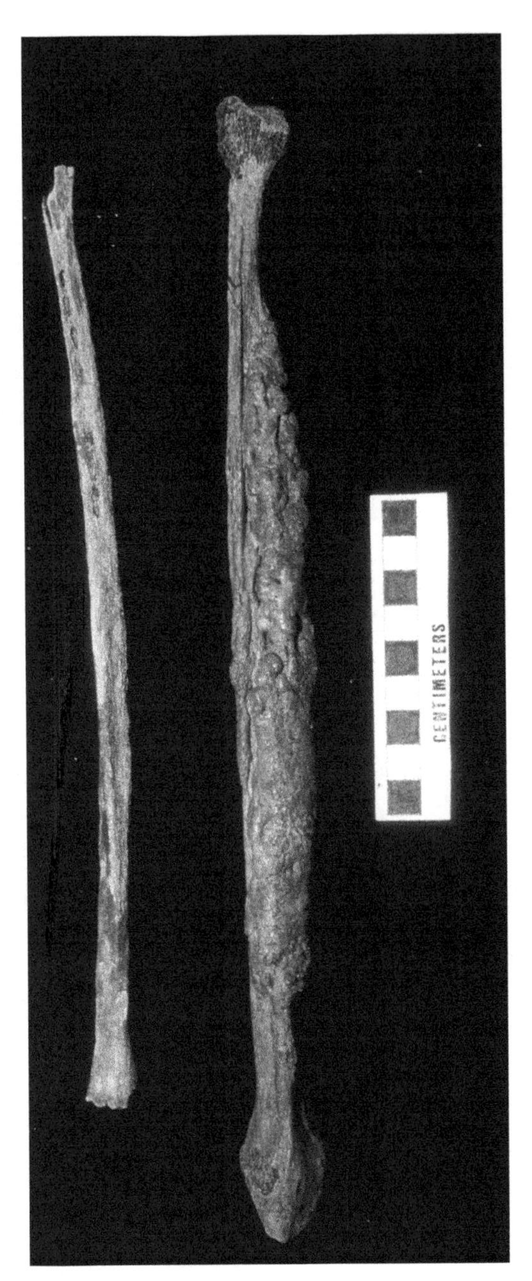

Figure 7.1. Left and right fibulae of AMNH 99.1/660, an Ipiutak adult male who exhibits chronic sclerosing osteomyelitis on the right fibula. Scale is in centimeters. Photo courtesy of C. E. Hilton.

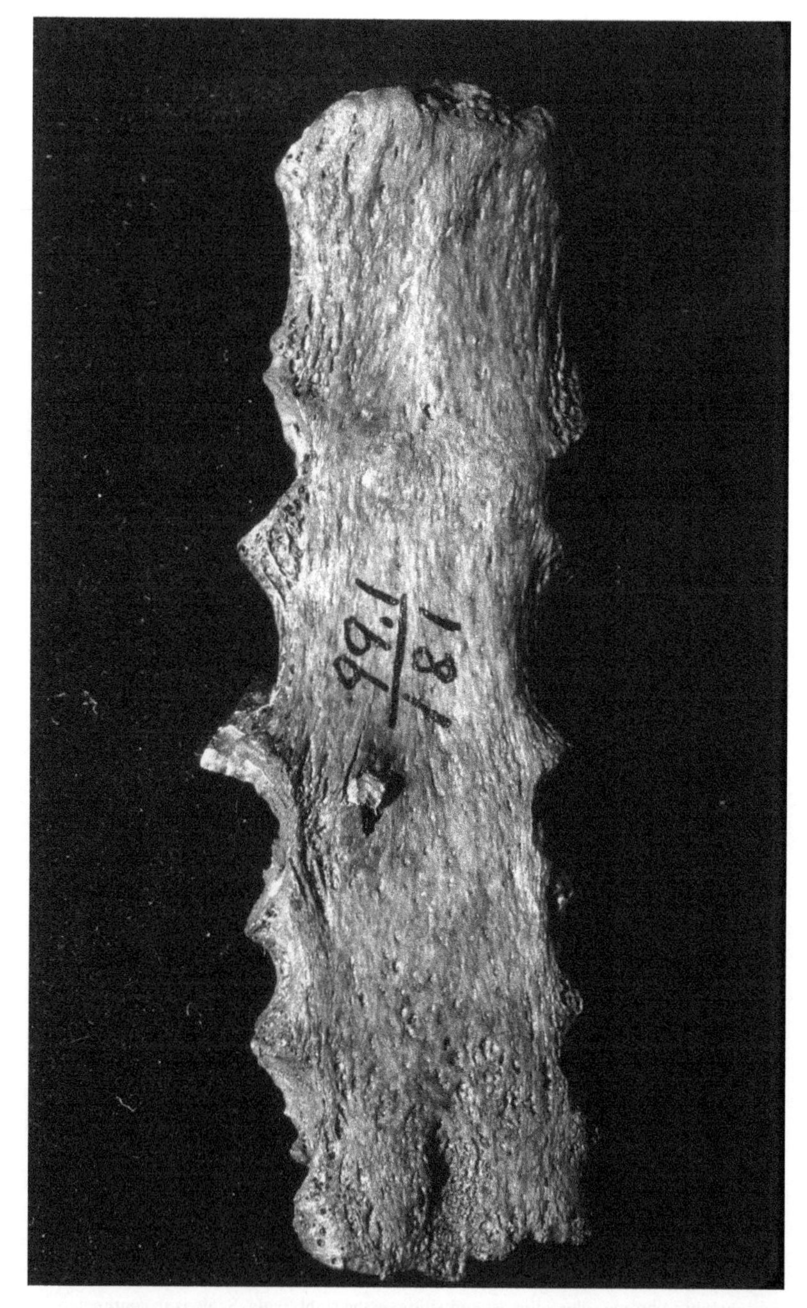

Figure 7.2. Anterior view of the sternal corpus of AMNH 99.1/181, an adult Ipiutak male with an embedded projectile point. The projectile point has entered the sternum from the right lateral aspect and has pierced the anterior sternal cortex just inferior to the accession number. Photo courtesy of C. E. Hilton.

have the next highest absolute frequency of postcranial lesions within the Tigara sample (see Tables 7.2 and 7.3). These lesions include Schmorl's nodes, vertebral body collapse, and vertebral body fusion. The non-spondylolytic lesions are seen in approximately 12% of the Tigara females and 10% of the Tigara males. Joint eburnation and trauma/fractures of the lower limbs and upper limbs have similar prevalence within the Tigara sample at 6%, 7%, and 6%, respectively.

The overall prevalence of PCSLs in the Tigara grouped by the sex and age categories are calculated from the Table 7.2 frequency counts. Within the "Old" Tigara males, 77% exhibit a PCSL whereas only 57% of the "Old" Tigara females possess a PCSL (see Table 7.2). For the "Young" females, 58% exhibit a PCSL, whereas 53% of the "Young" males have a PCSL (see Table 7.2). Tigara males as a group have a higher frequency of healed lower and upper limb fractures relative to the Tigara females (see Tables 7.2 and 7.3). Healed lower limb fractures and traumatic lesions are seen to range from a healed misaligned fracture in the femoral neck region to fusion of the distal tibia, talus, and calcaneus (see Table 7.3, Figure 7.3).

The descriptions in Table 7.3 highlight the range of lesions seen in Tigara individuals with more than one type of PCSL. Multiple PCSLs are seen in fourteen Tigara males (17% of the Tigara males) and eleven Tigara females (15% of the Tigara females; see Table 7.3). In terms of absolute frequencies, those individuals with multiple PCSL types are dominated by individuals who fall into the "Young" age category. However, as a group, 25% of the "Old" Tigara individuals have multiple types of PCSLs in comparison to only 12% of the "Young" Tigara. Additionally, almost 30% of the "Old" Tigara females possess two or more different PCSL types, a percentage nearly twice that seen in the "Old" Tigara males.

Other intriguing PCSLs are seen in the Tigara adult sample. First, four "Young" males, a "Young" female, and one "Adult" female exhibit PCSLs suggestive of tuberculosis as caused by the infectious agents associated with the *Mycobacterium tuberculosis* complex (MTBC) (see Tables 7.2 and 7.3). Second, bilateral erosive arthropathies seen in the upper limbs of at least one mature female skeleton are suggestive of an advanced systemic arthritic condition, i.e., rheumatoid arthritis, hemochromatosis, or psoriatic arthritis (see Table 7.2). The tuberculosis-like lesions and erosive arthropathies are discussed below in more detail. Finally, we note the "Young" Tigara male (AMNH 99.1/69, estimated age 30–34 years), documented previously by Larsen and Rainey (1948), who exhibits an unhealed "green bone" perimortem wound from a projectile point that penetrated the abdominopelvic cavity and became embedded in the iliac portion of the right *os coxa* (see Figure 7.4).

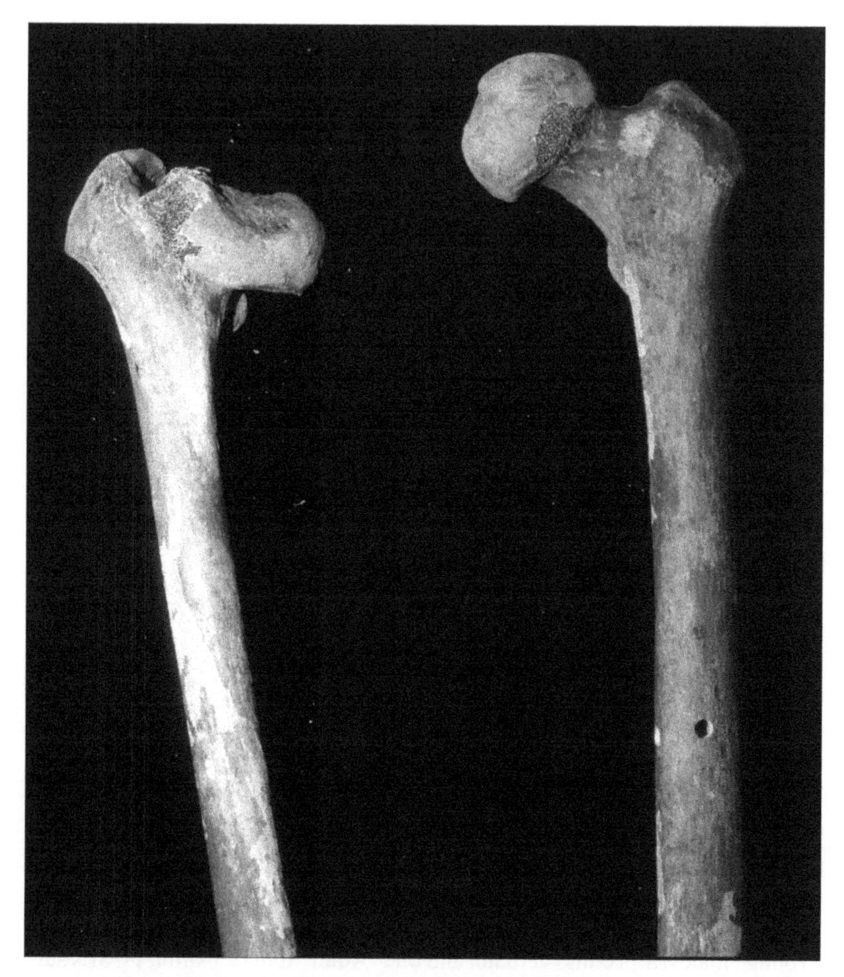

Figure 7.3. Anterior view of the left and right proximal femora of AMNH 99.1/417, an adult female Tigara. The right femur exhibits a misaligned healed fracture of the femoral head. Photo courtesy of C. E. Hilton.

Vertebral border shifting and spondylolysis

In Tigara with preserved presacral regions, we analyzed whether vertebral border shifting (VBS) provides a protective exposure against spondylolysis. Among the Tigara, VBS is seen in thirteen males (28% of the adult males) and thirteen females (25% of the adult females). As stated above, spondylolysis has the highest prevalence of any type of lesion type within both the male and female adult Tigara skeletons we examined (see Tables 7.2 and 7.3). For

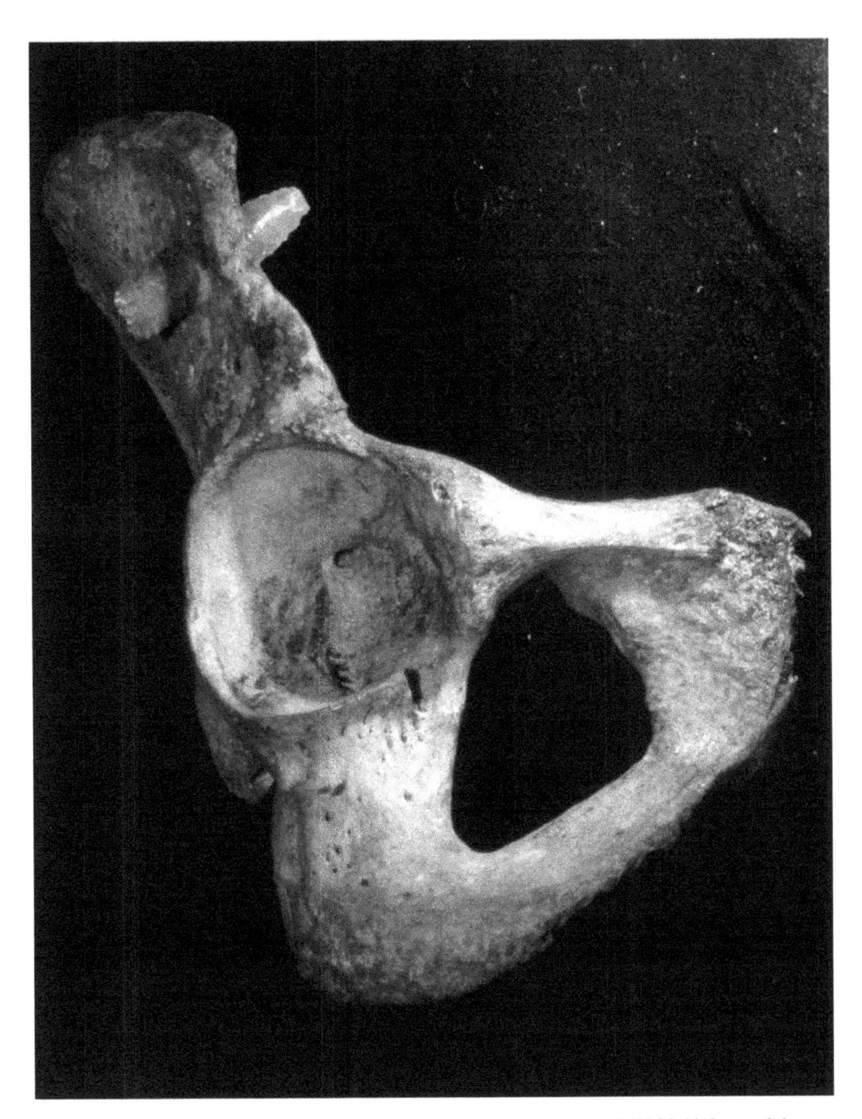

Figure 7.4. Oblique lateral view of the right *os coxa* of AMNH 99.1/69, an adult Tigara male with a projectile point embedded in the iliac blade just superior to the acetabulum. Orientation highlights the projectile point. Photo courtesy of C. E. Hilton.

Tigara females, twenty-five of them (47%) exhibit spondylolysis, with twelve possessing more than one spondylolytic lesion. For Tigara males, twenty-four of them (49%) have spondylolysis with nine possessing more than one spondylolytic lesion. Three adult Tigara females were each seen to have three lumbar vertebrae affected with spondylolysis, whereas a single adult Tigara male was seen to have four affected lumbar vertebrae.

Regarding an association between VBS and spondylolysis, nine of the twenty-six individuals with VBS possessed spondylolysis whereas thirty-eight of seventy-six individuals without VBS possessed spondylolysis. An “odds ratio” (OR) of 0.529 calculated from these frequency counts suggests that VBS has a protective effect as Tigara with VBS are 0.529 times less likely to have spondylolysis than those Tigara without VBS. However, a chi-square analysis indicates that the distribution of the frequency counts is not statistically significant ($\chi^2 = 1.845$, $df = 1$, $p = 0.174$). Small sample size for the Ipiutak precluded a similar analysis of the association between VBS and spondylolysis.

Tuberculosis

In our current study, skeletal evidence of tuberculosis-like skeletal lesions (TB), although investigated, was not identified in the Ipiutak sample. However, in the Tigara sample, four adult males (AMNH 99.1/237-170; 99.1/312-243; 99.1/330a-271; 99.1/383; 99.1/474) and two adult females (AMNH 99.1/387-342; 99.1/513-490) were identified with lesions indicative of tertiary TB (see Table 7.3; Gossett and Hilton, 2004; Latchaw and Hilton, 2006). These seven individuals represent a prevalence frequency of approximately 4.5% for tertiary TB in the Tigara sample. Five of these seven individuals fall into the “Young” age category and the remaining two are categorized simply as “Adult” (AMNH 99.1/474; 99.1/513-490). We provide descriptions for the two Tigara individuals (AMNH 99.1/237-170; 99.1/312) with the most extensive tuberculosis-like skeletal lesions.

The skeletal remains of AMNH 99.1/237-170 are those of a male with an estimated age of 30–34 years. Extensive tuberculosis-like lesions are seen in the lower thoracic vertebrae and the areas adjacent to the right acetabular region (see Figures 7.5a–d). Lytic destruction is present as smooth-edged circumferential pitting throughout the T8–L5 vertebral bodies (see Figure 7.5a and b). Porosity is seen on the ventral surfaces of several left lower ribs adjacent to the rib heads. Both the right *os coxa* and the right proximal femur exhibit extensive bone remodeling in association with the destructive manifestations of tuberculosis (see Figures 7.5c and d). The lesion-related bony remodeling in AMNH 99.1/237 includes: bone deposition encompassing the lateral margin of the right acetabulum; marked spike-shaped bone projections extending from the superior pubic ramus at the region where the *psoas major* muscle exits the abdominal cavity to insert on the lesser trochanter of the femur; a large lacuna through the cortical bone separating the acetabulum from the pelvic cavity;

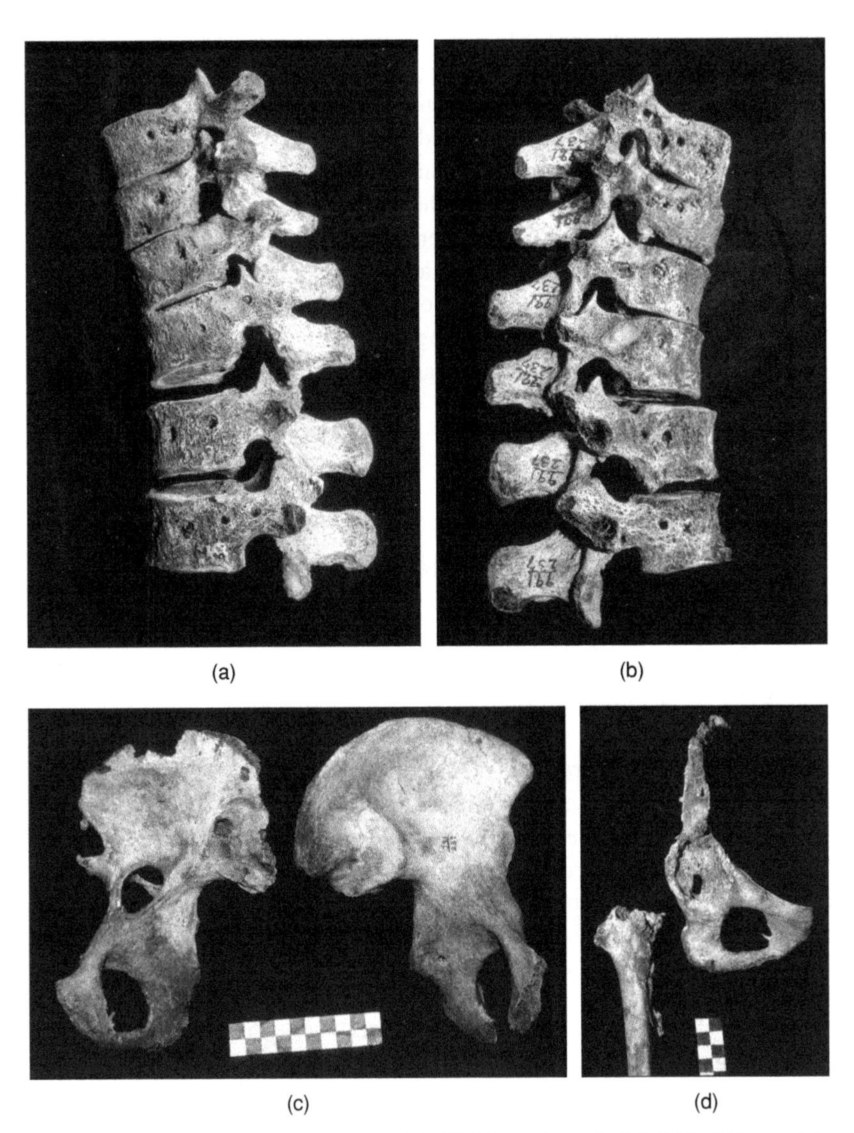

Figure 7.5. (a) Left lateral view of the T9-L2 vertebrae of AMNH 99.1/237, an adult male Tigara. The vertebral bodies exhibit smooth-edged circular pits due to tuberculosis. (b) Right lateral view of the T9-L2 vertebrae of AMNH 99.1/237. The vertebral bodies exhibit smooth-edged circular pits resulting from tuberculosis. (c) Left and right *os coxae* of AMNH 99.1/237. The right *os coxa* exhibits a lytic lesion through the acetabulum and evidence of bone modification in the area associated with the *psoas major* muscle. Postmortem damage is present along the iliac crest of the right *os coxa*. Normal left *os coxa* for comparison. Scale is in centimeters. (d) Right hip joint of AMNH 99.1/237 exhibiting antemortem destruction of the femoral head and the acetabulum due to tuberculosis. Scale is in centimeters. All photos courtesy of C. E. Hilton.

complete antemortem destruction of the right femoral head with additional lesions inferior to the lesser trochanter (see Figures 7.5c and d).

The second individual with extensive tuberculosis-like lesions is another male (AMNH 99.1/312-243) estimated to be 20–24 years at death. Smooth-edged lytic destruction is seen on the anterior surfaces of the vertebral bodies of T9–L5 (see Figures 7.6a and b). The vertebral bodies of T10–T12 exhibit nearly complete antemortem destruction that produced angular kyphosis in life (see Figures 7.6a and b). Right ribs six and seven possess porosity and the right eighth rib has an abnormal extension to the vertebra. The left *os coxa* exhibits two areas of lytic destruction, including a large smooth-walled cavitation that has destroyed a large area of the ischial tuberosity and another lesion located in the posterior acetabular region.

Bilateral arthropathy

Two Tigara females (AMNH 99.1/460-430; AMNH 99.1/522-497) exhibit bilateral erosive arthropathies (Latchaw and Hilton, 2004, 2006). One is represented by a nearly complete skeleton (AMNH 99.1/522-497) with an estimated age of 35–39 years at death. The other female is a younger adult (AMNH 99.1/460-430) lacking key diagnostic skeletal regions. Given the lack of skeletal completeness for the younger female, our primary description focuses on the 35–39-year-old Tigara female (AMNH 991/522-497).

In this older Tigara female, the bilateral erosive arthropathy predominantly affected the radiocarpal and metacarpophalangeal joints of both the right and left upper limbs (see Figures 7.7–7.10). Although the shoulder and hip joints are affected, these anatomical regions do not exhibit the same extensive bone modifications seen in the skeletal elements of the hands, wrists, and forearms. The sacroiliac joints are normal. Minimal osteophyte development is present in the lumbar vertebrae.

The carpals of both hands are ankylosed (see Figures 7.7, 7.8, and 7.10b). Carpal lesions are erosive, small, round, and appear predominantly on the subchondral region of the bone, which exhibits extensive destruction. Metacarpal and phalangeal lesions are small and round, predominantly eroding the subchondral bone. The chondro-osseous junctions of the articular surfaces are unaffected. The right second metacarpal head is severely eroded (see Figures 7.8 and 7.10b). The proximal joint surfaces of both third metacarpals are also eroded. The articular surfaces of both first metacarpals exhibit new bone deposition along their bases. Each second metacarpal in both wrists is fused to its respective trapezoid (see Figures 7.7, 7.8, and 7.10b). As noted, all the carpal bones present in each hand, except for the pisiforms, have ankylosed

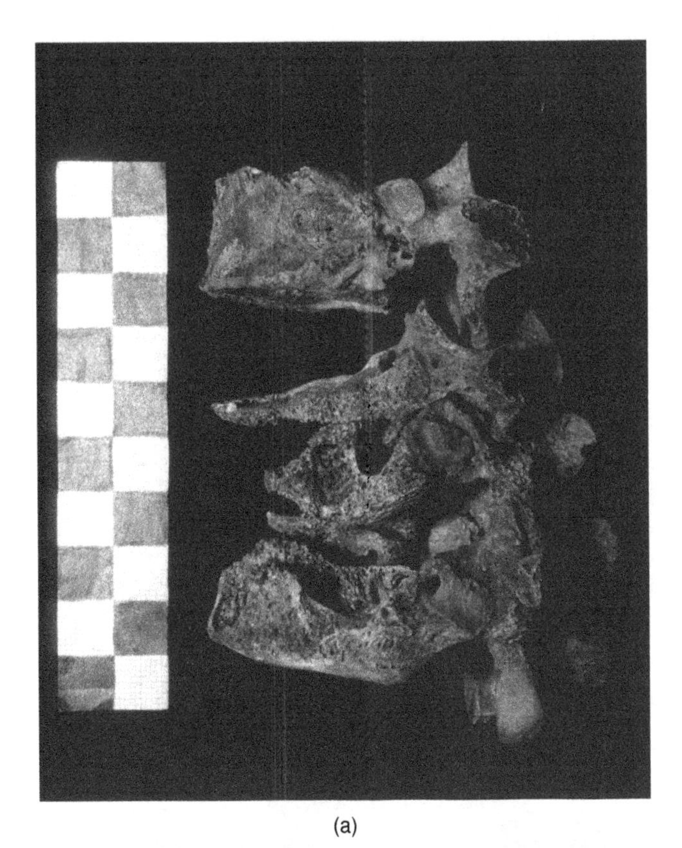

(a)

(b)

Figure 7.6. (a) Left lateral view of T8-T12 vertebrae of AMNH 99.1/312, an adult male Tigara exhibiting extensive destruction of the vertebral bodies due to tuberculosis. Scale is in centimeters. (b) Superior view of the T12-T10 vertebral bodies of AMNH 99.1/312 exhibiting smooth-edged lytic destruction due to tuberculosis. Scale is in centimeters. Both photos courtesy of C. E. Hilton.

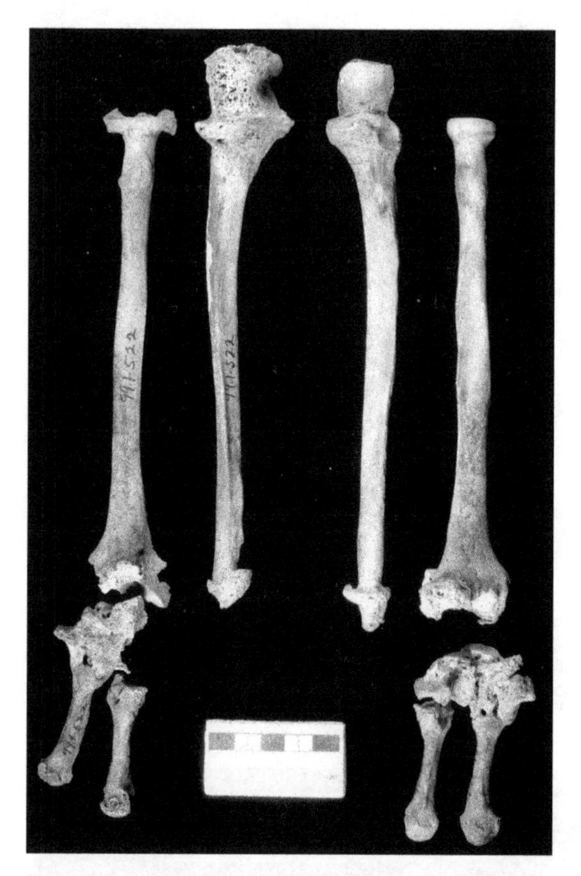

Figure 7.7. Anterior view of the skeletal elements of the right and left forearms and hands of AMNH 99.1/522, an adult female Tigara. Lesions can be seen in the right proximal and distal ulna and radius, as well as the left distal radius and ulna. Ankylosis is seen in each of the right and left carpal regions as well as fusion of each MC II to its respective trapezoid. Scale is in centimeters. See text for further discussion. Photo courtesy of C. E. Hilton.

into a carpal block (see Figures 7.8 and 7.10b). However, in the right hand, the medial side of the right carpal block is missing due to postmortem damage. The degree of fusion seen in the right carpals is so extensive that individual carpal bones are not easily distinguishable (see Figures 7.8 and 7.10b). These fused bones exhibit some porosity.

Both left and right radiocarpal joints are severely affected (see Figures 7.9a, c, and 7.10a). Joint space narrowing on both sides suggests little, if any, movement at these joints. The distal articular surface of the right radius has a rugged topography (see Figure 7.9b). The left radius has extensive growth and expansion of the distal end with a porous joint surface (see Figure 7.9a). The

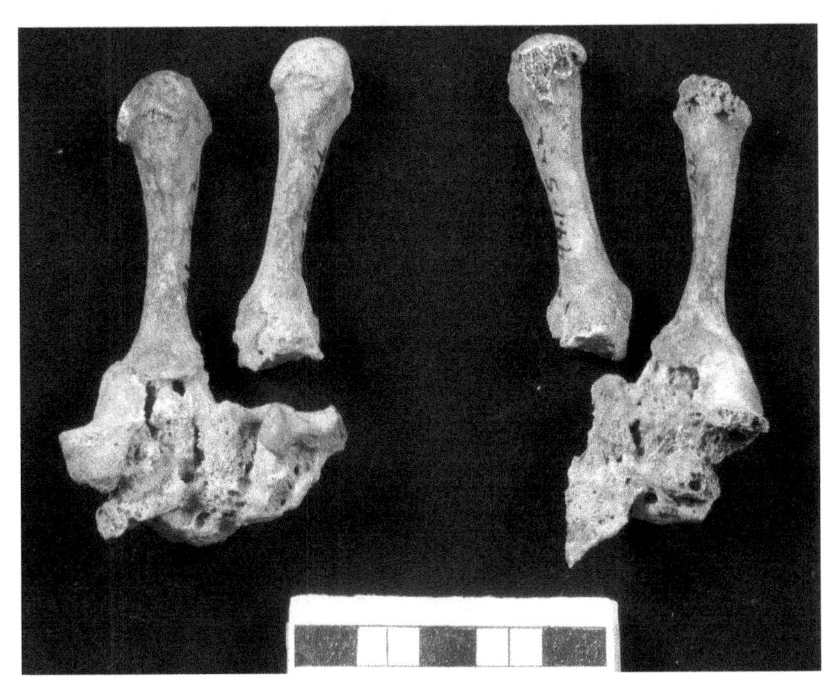

Figure 7.8. Close-up palmar view of selected left and right hand elements of AMNH 99.1/522 illustrating ankylosis of the carpal elements and fusion of each MC II. Scale is in centimeters.

joint surfaces of the corresponding articulations of the scaphoid and lunate bones are affected. The distal articular surface of the right ulna sustained modifications to its normal morphology (see Figure 7.9b). The left ulna exhibits an enlarged styloid process and new bone growth that formed an abnormal bony lip around the entire distal end (see Figure 7.9a).

The proximal ends of both the right radius and ulna exhibit lesions. In contrast, the left proximal radius and ulna do not exhibit lesions (see Figure 7.9c). The right proximal ulna has expanded abnormally in overall size. The joint surface shows smooth exposure of trabeculae with porous erosions around the periphery of the subchondral region. The proximal radius is remodeled with a flattened and expanded radial head (see Figure 7.9c). The radial head is porous and exhibits eburnation on its articular surface with the ulna (see Figure 7.9c). The subchondral surfaces of the right distal humerus exhibit changes in morphology corresponding to its articulation with the right radius and ulna. These humeral joint surfaces exhibit severe macro- and microporosity with a large area of the subchondral region destroyed. The surviving bone exhibits eburnation.

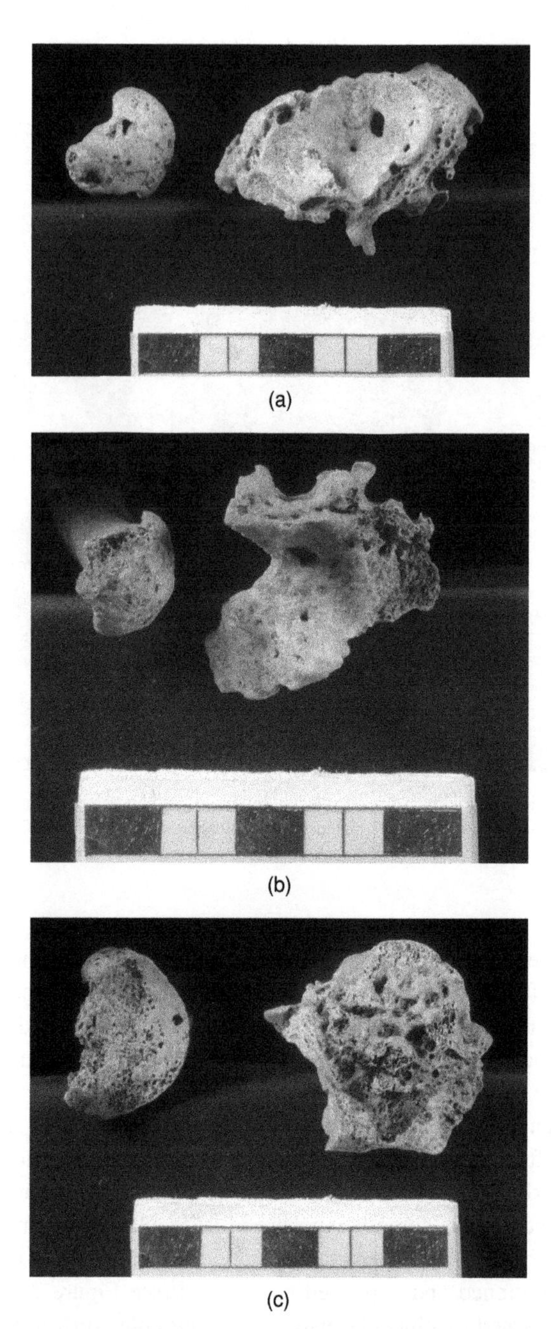

Figure 7.9. (a) Inferior aspect of the left distal ulna and radius of AMNH 99.1/522 illustrating bony alterations of the radiocarpal articular surfaces. Scale is in centimeters. (b) Inferior aspect of the right distal ulna and radius illustrating bony alterations of the radiocarpal articular surfaces. Scale is in centimeters. (c) Superior aspect of the right proximal ulna and the radial head illustrating bony destruction and modification. Scale is in centimeters. All photos courtesy of C. E. Hilton.

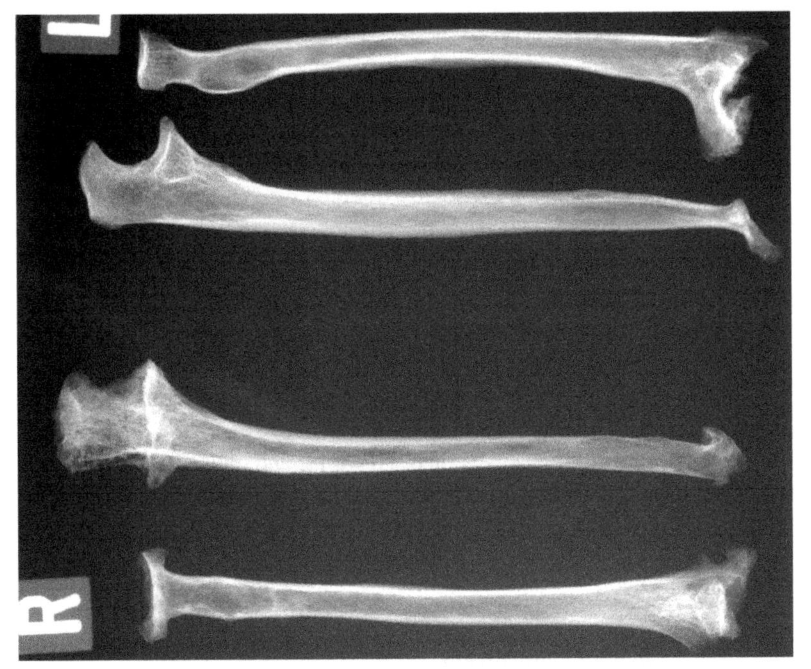

(a)

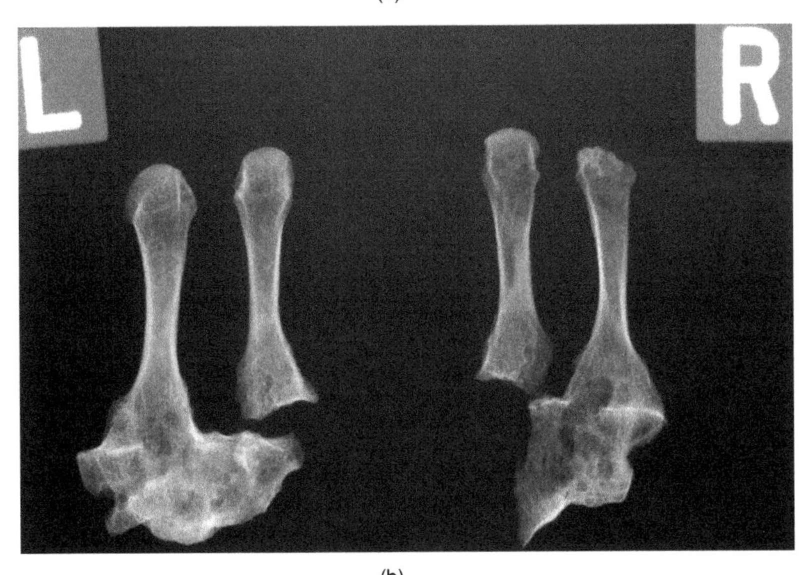

(b)

Figure 7.10. (a) Radiograph of the left and right ulnae and radii of ANMH 99.1/522. The distal ends of all four skeletal elements exhibit extensive alteration. Sclerosis can be seen along the articular surfaces of the left distal radius, right proximal radius, and the coronoid process of the right ulna. Bony modification is also seen in the coronoid process of the right ulna and the right radial head. (b) Anterior-posterior radiograph of selected skeletal elements of the left and right hands. Within each hand, ankylosis can be seen between the carpal bones and between the MC II and the trapezoid. The MC IIIs seen in the radiograph do not exhibit any signs of ankylosis to their respective capitates. Both photos courtesy of C. E. Hilton.

The axial skeleton is complete. Cervical vertebrae are unaffected but several lower thoracic vertebrae have porosity and eburnation on their articular facets. The third through fifth lumbar vertebrae exhibit evidence of osteoarthritis with lipping at the body margins, horizontally oriented osteophytes, and rough vertebral body surfaces. The right *os coxa* exhibits morphological changes at the border of the acetabular fossa. The acetabular area is porous with the exposure of smooth trabecular bone. The femoral heads correspond to the modification of the acetabulae. The *fovea capitis* on the right femoral head shows additional bone growth, corresponding with the erosion of the right acetabulum. The left acetabulum displays minor changes in morphology. The left femoral head has a sclerotic lip extending beyond the margin of the articular surface, giving the femoral head a "mushroom" appearance. The tibiae and fibulae are normal. The bones of the feet appear normal, except for a small lytic defect on the proximal articular surface of both third metatarsals. The corresponding cuneiforms also show this defect.

Discussion and conclusion

Overview

Arctic foragers and their specialized lifestyles have long been important to anthropologists interested in human cultural and biological adaptations to circumpolar environments (see Collins, 1984; Shephard and Rode, 1996 for reviews). Our study contributes additional information towards better understanding the impact of prehistoric Arctic foraging lifestyles on adult skeletal biology and morbidity through an assessment of the prevalence pattern of select adult PCSLs among the Point Hope Ipiutak and Tigara of northwest coastal Alaska. We used an epidemiological approach (Rothman, 2002; Gordis, 2009) informed by anthropological work among living hunter-gatherers, recognizing that distinctive cultural differences in resource targeting can produce significant differences in overall morbidity. The most dramatic difference between our two samples is the higher prevalence of PCSLs among the Tigara females relative to their Ipiutak counterparts, 66% versus 5.5%, respectively. Tigara males exhibit also a higher prevalence of postcranial skeletal lesions than their Ipiutak counterparts, 62% versus 36%, respectively. Although spondylolysis did not exhibit a significant association with vertebral border shifting, it is the most pervasive PCSL in both the Ipiutak and the Tigara. Spondylolysis is substantially higher in both the Tigara males and females relative to that seen in the Ipiutak. More importantly, our comparisons demonstrate that the highest PCSL prevalence percentages are seen in the later whale-centric Tigara relative to the earlier non-whale-centric Ipiutak counterparts, 64% versus 23%, respectively. Additionally, based on our observations of tuberculosis-like skeletal lesions, infectious disease in the form of tuberculosis was more prevalent in the Tigara relative to the Ipiutak. From our perspective, we view the difference in PCSL prevalence patterns as a reflection of higher levels of exposure to chronic and infectious diseases in the Tigara in association with the increased physical demands of the shift in their subsistence strategy and the close contact with their highly ranked "new" prey.

The Point Hope coastal location allowed people in both cultural periods to exploit a broad spectrum of animals as food and as sources of raw materials for the manufacture of highly specialized technology (Larsen and Rainey, 1948; Burch, 1981). Both the Ipiutak and the Tigara, clearly, possessed sophisticated cultural and social adaptations that allowed for survival in one of the world's harshest environments (Larsen and Rainey, 1948; Shephard and Rode, 1996). However, the Tigara cultural focus on targeting, pursuing, and subsequent processing of whales contributed strikingly to the increased prevalence of PCSLs in both the Tigara males and females relative to the non-whaling strategy of the Ipiutak. Based on our analyses of single and multiple PCSLs in individuals, the physical demands placed on Tigara females were more arduous than those on Ipiutak females. Given that contemporary Iñupiaq females provide substantial infrastructure in terms of physical labor to support men engaged in hunting activities (Spencer, 1977), it is not surprising to see a concomitant increase in PCSL prevalence among Tigara females in conjunction with the increase in the Tigara males. However, the high PCSL prevalence in the Tigara females compared to the Ipiutak females suggests a substantial increase in female physical labor demands within the Tigara period.

Infectious disease

As to infectious diseases, Hilton and colleagues (Gossett and Hilton, 2004; Latchaw and Hilton, 2006) provided preliminary descriptions of skeletal evidence for tuberculosis (TB) in the Tigara at Point Hope. Dabbs (2009b), subsequently, has noted not only the presence of TB in the Tigara but in the Ipiutak as well. As both cultural periods predate European contact, the suggestion of TB in precontact Point Hope may be controversial. Regardless of the controversy, evidence of precontact TB at Point Hope impacts our understanding of possible comorbidities in those Arctic foraging communities. However, for our current discussion, we focus on the strong evidence of TB we observed in the Tigara sample.

Tuberculosis in humans is caused by the *Mycobacterium tuberculosis* complex (MTBC), a group of closely related bacteria species that have strains with varying degrees of virulence to humans (Coberly and Chaisson, 2007; Nelson, 2007; Heymann, 2008; Wirth *et al.*, 2008). While for some individuals TB may take on the manifestations of a chronic disease, TB, even if treated, can become a debilitating respiratory infection that ultimately is fatal (Coberly and Chaisson, 2007; Nelson, 2007; Heymann, 2008). Tigara PCSLs related to TB are indicative of the tertiary stage of the disease (Aufderheide and Rodriguez-Martin, 1998). In the pre-antibiotic medication era, only 5–7% of individuals with TB developed the tertiary form (Steinbock, 1976; Aufderheide and Rodriguez-Martin, 1998; Roberts and Buikstra, 2003). Thus, the evidence for individuals with TB-like skeletal lesions suggests that the disease was most likely more pervasive in the Tigara community than that presented in Table 7.2, as other individuals would have had only soft-tissue TB-related lesions. Tigara individuals infected by the MTBC but asymptomatic for the disease (see Heymann, 2008, for a review) would have been infectious to others if past the latent period, i.e., the time from infection to infectiousness (Aron, 2007). Additionally, depending on the pathogenicity and virulence of the MTBC strains, other Point Hope Tigara community members may have had the non-skeletal primary and secondary forms of TB but died before it progressed to their skeletons. Such a scenario highlights at least one set of conditions that would produce an example of the "osteological paradox" (see Wood *et al.*, 1992, for a review) at Point Hope, wherein individuals with morbidity died before manifesting skeletal lesions related to illness.

Researchers have established that TB was present in the New World prior to European contact (Buikstra, 1981; Buikstra and Cook, 1981; Ortner and Putschar, 1981; Powell, 1988; Aufderheide and Rodriguez-Martin, 1998). However, it has been more difficult to determine whether the transmission of MTBC infectious agents to indigenous Iñupiaq populations of northwest coastal Alaska predate European contact (Keenleyside, 1998). The consensus is that TB was introduced to Point Hope in the 1870s by members of whaling and trading ships (Burch, 1981). However, Fortuine (2005) highlights two possible pre-contact cases of TB in Alaska, one at Barrow and the other at St. Lawrence Island.

While the evolutionary origins of *M. tuberculosis* belong within the genus *Homo*, a range of organisms do provide reservoirs for the MTBC that facilitate transmission of these bacteria to humans (Buikstra, 1981; Kelley and Lytle-Kelley, 1999; Coberly and Chaisson, 2007; Kiers *et al.*, 2008; Wirth *et al.*, 2008). Without the aid of DNA fingerprinting of preserved soft tissue samples derived from the human remains (Coberly and Chaisson, 2007; Wirth *et al.*, 2008), pinpointing the origin of TB at Point Hope during the Tigara cultural period is difficult and beyond the scope of this chapter. However, the MTBC includes *M. pinnipedii*, a bacterium known to cause TB in seals and known to infect humans (Kiers *et al.*, 2008). Additionally, *M. tuberculosis* has been documented in sea mammals, particularly cetaceans and pinnipeds (Viale, 1981; Cousins and Francis, 1990; Forshaw and Phelps, 1991; Cousins *et al.*, 1993; Thompson *et al.*, 1993; Romano *et al.*, 1995; Woods *et al.*, 1995; Bernardelli *et al.*, 1996; Bastida *et al.*, 1999; Hunter *et al.*, 1998), although these cases may in fact represent the recently recognized *M. pinnipedii*.

Researchers (Ray, 1984; Fortuine, 1986; Keenlyside, 1998) have noted that living Arctic foragers are at a higher risk for acquiring and transmitting a range of zoonotic and parasitic infections due to cultural behaviors leading to exposures to pathogens. These behaviors include close contact with animals and their waste products, the habitual ingestion of both undercooked and raw meat, and living within dwellings that generate close physical contact between individuals. These behaviors were common in recent Arctic Alaskan foragers and presumably were practiced by their prehistoric counterparts. Close contact with animals and their waste products made Alaskan foragers especially vulnerable to zoonotic diseases such as botulism, brucellosis, trichinosis, *Echinococcus* parasites, tularemia, rabies, and *Salmonella enteritidis* (Ray, 1984; Fortuine, 1986). In fact, Ray (1984) notes that botulism from fish and other meats was known to have killed entire families in the Bering Sea region. Ethnographic accounts of Iñupiat communities readily provide documentation of close contact with animals during the process of hunting, butchering, and consumption of these animals (Murdoch, 1892; Nelson, 1899; Stefansson, 1914; Spencer, 1959, 1977; Nelson, 1969; Burch, 1981, 1988; Sheehan, 1985; Chance, 1990; Worl, 1999; Sakakibara, 2007). With respect to TB, Bastida *et al.* (1999) note that cross-infections from pinnipeds to both their marine and terrestrial predators should be considered for understanding the transmission of MTBC and, thus, TB across species. In recognition of the observations of Bastida *et al.* (1999), Hilton and colleagues (Gossett and Hilton, 2004; Latchaw and Hilton, 2006) proposed that the hunting, butchering, and consumption of marine mammals provided a possible avenue of transmission for *M. tuberculosis*. Here, we also suggest that the higher prevalence of TB-like lesions among Tigara males is related to their close contact and primary access to marine mammal carcasses.

Fortuine (2005) proposes that the strain of TB in precontact Alaska may have been both sporadic and of low virulence relative to post-European contact forms. Other illnesses, if combined as comorbidities with periodic episodes of malnutrition, may have reduced an individual's immunocompetence against infection by the less pathogenic forms of the MTBC (Aron, 2007; Coberly and Chaisson, 2007; Nelson, 2007; Heymann, 2008). As documented by our descriptions of skeletal lesions, TB transforms into a long-term chronic morbidity, and, with the destruction of soft and hard tissues, TB can severely limit an individual's overall mobility and full participation in cultural activities.

Bilateral arthropathy

With respect to another morbidity that would limit individual mobility, we have expanded our previous description of a Tigara female skeleton (35–39 years of age) who exhibits a pattern of bilateral erosive arthropathy suggestive of several possible advanced arthritic conditions, i.e., rheumatoid arthritis (RA), hemochromatosis, and psoriatic arthritis (Latchaw and Hilton, 2004, 2006). Although the characteristics of the age, sex, and lesion distribution pattern warrant a RA diagnosis, a review of the manifestations of RA and other diseases on dried bone undermine such a diagnosis. Rheumatoid arthritis lesions occur on the margins of the subchondral bone adjacent to the articular surface itself. Bone erosions seen on these Point Hope skeletal elements are almost solely confined to the subchondral region. The marginal areas, with the exception of one intermediate phalanx, were completely unaffected. While RA can affect the subchondral bone in severe cases, the erosive fronts always begin in the marginal area (Leisen *et al.*, 1991). It would be unusual to see RA lesions on the articular surface without extensive damage to the adjacent area. The most striking feature of the Point Hope specimen is the complete fusion of the carpal bones in both hands.

Although ankylosis is considered a common characteristic of RA, Rothschild *et al.* (1990) argue that the presence of joint ankylosis and subchondral lesions rule out a confident RA diagnosis. Hemochromatosis and psoriatic arthritis can both appear similar to the classic RA distribution pattern and these alternative diagnoses should not be ignored. Hemochromatosis, a disorder in which the content of iron in the body is greatly increased, can also cause arthritic symptoms similar to those seen in RA. Hemochromatosis leads to organ dysfunction and, often, iron deposition in the synovial tissue of the metacarpophalangeal, interphalangeal, shoulder, elbow, hip, and knee joints (Aufderheide and Rodriguez-Martin, 1998). Inflammatory arthritis usually appears at around 40 years of age. The pattern of arthritis associated with hemochromatosis is compatible with the distribution of lesions seen in this Tigara skeleton. Hemochromatosis is a compelling diagnosis as prehistoric Point Hope foragers consumed a diet rich in iron due to an extremely high intake of meat.

Psoriatic arthritis occurs in about 5–7% of people with psoriasis. It affects men and women with the same overall frequency, and onset generally occurs between 30 and 55 years of age. Overall, 30–50% of psoriatic arthritis patients will manifest symmetric polyarthritis involving the joints of the hands, feet, wrists, ankles, knees, and elbows (Klippel, 2001). Psoriatic arthritis presents a combination of erosion and bone production, characteristics that are helpful in distinguishing it from RA (Klippel, 2001). Another line of evidence supporting a diagnosis of psoriatic arthritis is that the Tigara female's estimated age of 35–39 years at death is well within the age range for the onset of psoriatic arthritis.

Irrespective of the disease diagnosis for this Tigara female, the degree of joint ankylosis seen in her hands, wrists, and forearms would have limited her body movements and most likely impacted her ability to fully participate in all physical activities commonly seen in traditional female Iñupiaq foragers (Murdoch, 1892; Nelson, 1899; Griffin, 1930; Spencer, 1959, 1977; Burch, 1981, 1988). Additionally, traditional Iñupiaq women perform an extensive range of activities that require a tremendous degree of manual dexterity, such as manufacturing by hand many aspects of their technology (Griffin, 1930; Ammitzboll *et al.*, 1991; Aksaajuq Otak, 2005; Pedersen, 2005). Manual activities include sewing clothing for men, women, and children, dismembering animal carcasses for food, and household responsibilities such as cooking and caring for children (Griffin, 1930; Ammitzboll *et al.*, 1991; Aksaajuq Otak, 2005; Pedersen, 2005). Discerning the daily levels of pain suffered by this Tigara woman and the social support she got from other community members is difficult. Her advanced state of lesions may have developed over extended time with gradual joint mobility reduction. She may have adopted accommodative behaviors to compensate for body movement limitations and to maintain a level of self-sufficiency. Webb (1995) provides several extreme examples of individual forager self-sufficiency and resiliency in Aboriginal Australians who lived with extensive skeletal lesions, such as lower limb amputations, and yet appear to have had minimal support from others. However, among living Venezuelan foragers, social and economic support is regularly provided to individuals who exhibit limited mobility resulting from joint abnormalities (Hilton, unpublished).

Ethnographic context

Although morbidity and comorbidities are pervasive in all human societies, we utilize ethnographic research on North American Arctic foragers in order to better conceptualize the implications of the PCSLs in the precontact Point Hopers (Murdoch, 1892; Nelson, 1899; Stefansson, 1914; Griffin, 1930; Spencer, 1959; Nelson, 1969, 1977; Burch, 1981, 1988; Sheehan, 1985; Chance, 1990; Worl, 1999; Sakakibara, 2007). One limitation in these ethnohistoric and ethnographic documents is that descriptions of morbidities within Arctic foragers are most certainly biased by post-contact disease processes. Additionally, other cultural groups do differentially embody concepts of wellness that also can bias our understanding of morbidity within that group (Pike *et al.*, 2010).

Burch (1981) noted the need to recognize that the recent historic whale-hunting inhabitants of Point Hope utilized all available terrestrial and marine animal resources as integral parts of a cultural system that specialized in the resource targeting of bowhead whales. Thus, while whale hunting is important to contemporary Point Hope cultural identity (Sakakibara, 2007), a range of other key subsistence and cultural activities help to sustain this focus. We propose that whale hunting and, possibly, the underlying tasks necessary to support a whale-centric subsistence strategy appear to have contributed substantially to a dramatic overall increase in the physical demands and the risk of chronic and infectious disease burdens for precontact Tigara Point Hopers beyond that in the Ipiutak. Alvard and Nolin (2002), who have studied small-scale whaling in living indigenous Indonesians, using methods and technology similar to that seen in Iñupiaq whalers, note that whale hunting is a dangerous activity that exposes all members of a boat crew to high risk for injury and death. Appendages can be easily caught and amputated by fast moving harpoon lines as well as boats being dragged and capsized by harpooned whales (Alvard, personal communication).

Ethnographers who focused specifically on North American Arctic communities provide observations of hardship, risk, and injury to individuals (Murdoch, 1892; Ray, 1892; Nelson, 1899; Stefansson, 1914; Spencer, 1959; Nelson, 1969; Burch, 1981, 1988). Burch (1988) noted that Eskimo women engaged in extreme physical exertion especially in traveling and when retrieving and butchering the carcasses of large-bodied game animals. Murdoch (1892) noted that women had an awkward gait and frequently stood with a stooped posture. Both Murdoch (1892) and Nelson (1899) noted that "rheumatism" and complaints of "rheumatism" were frequent. Given the prevalence of eburnation seen in the prehistoric Point Hope remains, we are not surprised by the frequent complaints of rheumatism in people with similar Arctic lifestyles. Ray (1892) noted that "Eskimo" appeared rarely to reach old age and that the majority of individuals died before 40 years of age, and a person reaching the age of 60 years appeared decrepit. These qualitative ethnohistoric accounts support not only the observations of joint degeneration seen in our Ipiutak and Tigara skeletal samples but also the age distribution patterns in that "Old" adults (40+ years) represent smaller proportions of the samples relative to "Young" adults (see Tables 7.1 and 7.2).

Nelson (1969) provides one informant's account regarding the dangers of hunting on sea ice. The informant described how a man's legs and pelvis were crushed by moving sea ice as the man tried to cross an enlarging ice pile with dogs and a sled in an approaching storm. The injured hunter was subsequently rescued after the storm by other hunters but two boys who had tried to walk to shore from the sea ice from the same hunting party were subsequently found frozen to death. Murdoch (1892) mentions also a male double amputee who had lost both his feet at the ankles, although the cause of amputation was not specified. Nelson (1969) noted the dangers for both men and women of chance encounters with polar bears by recounting the story of a woman who survived a polar bear encounter. As the polar bear tried to bite her, she shoved her hand down its mouth, leaving her mitten in its throat, causing the polar bear to choke and suffocate. Sled dogs were also known to injure people by biting them (Nelson, 1899), thereby providing another possible cause of infection.

With respect to injuries sustained during conflict with other people, interpersonal violence at Point Hope was present during both cultural periods sampled, given that an adult Ipiutak male and an adult Tigara each exhibit perimortem trauma in the form of lithic projectile points embedded in a skeletal element. While ritual activities may provide some cultural buffering against interpersonal antagonism, the aggregation of people into small circumscribed physical and social spaces does produce a high level of anxiety and social tension among individuals within a forager community (Hilton, n.d.). Burch (1988) notes that, specifically among Eskimo families, personal relations could be extremely intense since the same people often spent months and even years together. Violent interactions among Eskimo living together in small settlements were not rare and these social disruptions often had ramifications beyond the principal individuals involved (Burch, 1981, 1988). Among foragers, anxiety and social tension may stem from a variety of reasons, including a succession of poor hunting returns, the lack of sharing key resources, to the discovery of marriage infidelity (Yu, 1997; Hilton, n.d.). Nelson (1899) noted that blood feuds among Eskimo around Bering Strait occurred on a regular basis and resulted in violence and death. Burch (1981, 1988) noted that assassination was the ultimate check on individuals who engaged in socially disruptive behaviors within a community and that acts of revenge might take years before being carried out. Burch (1981) describes specifically an extended period of tense social disruption at Point Hope in the 1880s brought about by a man identified as Atanauzaq who established himself as a despotic middleman between Point Hope community members and the crews of American whaling and trading ships. Atanauzaq was assassinated in 1889 by Point Hope community members after an extended period of erratic behavior that included "appropriating" other people's resources and taking other men's wives (Burch, 1981). Additionally, inter-community violence represents another important consideration as a cause for traumatic injury, especially when a community inhabits a highly desirable resource-rich location such as Point Hope (Burch, 1981). Obviously, the reasons why two prehistoric Point Hope adult males (AMNH 99.1/69 and 99.1/181) were shot with projectile weapons will remain unknown. However, the lack of evidence for parry fractures within the two samples suggests that some aspects of interpersonal violence may have lacked close physical contact associated with the postcrania and relied instead on projectile weaponry.

For foragers without antibiotics, even small lacerations and low-level infections could have produced substantial negative health consequences including death (Hilton, n.d.). While shamans were brought in for care when disease appeared mortal, a shaman's role was to determine if witchcraft was involved (Spencer, 1959, 1977). Care for illness was done mostly with "home remedies" using blubber and human urine for treatment of wounds and cuts (Spencer, 1959, 1977). Murdoch (1892) reported that in one Utiavwĩñ winter house, pots of cooked meat could be seen in the same corner as the chamber pot and the male urinal pot. Blood-letting was a common practice to relieve inflamed and aching regions of the body (Nelson, 1899), with joints often targeted (Spencer, 1959). However, shamans were unnecessary for treating broken bones and lacerations as others could set a bone fracture and wrap the injured body part using pieces of wood or whale bone in combination with caribou skins, whereas severe wounds might require a combination of caribou skin, hair, and blubber in the wrapping (Spencer, 1959, 1977).

Previous paleopathological studies

Others have investigated paleopathological lesions in samples of precontact Point Hope human skeletal remains (Lester and Shapiro, 1968; Costa 1977, 1980a, b, 1982; Schwartz *et al.*, 1995; Keenleyside, 1998, 2003, 2006; Madimenos, 2005; Dabbs, 2009a, b, 2011) but have primarily focused on craniodental lesions and their associated health issues. Several researchers (Merbs, 1983, 2002; Hart Hansen, 1989; Hart Hansen *et al.*, 1991) have also examined paleopathological lesions in other samples of North American Arctic foragers.

Lester and Shapiro (1968) investigated vertebral arch defects in Point Hope skeletal remains and reported the frequency of spondylolysis. They sampled both the Ipiutak ($n = 47$) and Tigara ($n = 248$) skeletons and included individual skeletons with at least four lumbar vertebrae present. Lester and Shapiro (1968) indicated a prevalence of spondylolysis of 21% and 45% in the Ipiutak and Tigara, respectively, but no significant differences were seen between the sexes in the distribution of spondylolysis, within either the Ipiutak or the Tigara cultural periods. Lester and Shapiro (1968) favored hereditary weakness in vertebral structure as the most likely cause for the prehistoric Point Hope spondylolysis. Surprisingly, they dismissed biomechanically stressful body postures in the etiology of spondylolysis despite noting that Iñuit people used a "stooped" standing and walking posture that placed traumatic strain on lumbar ligaments for extended periods by maintaining ninety-degree hip flexion with simultaneous knee extension. We note that such hip flexion can be seen in ethnographic photographs and films of Inuit people, and this body posture is maintained habitually by men during kayak use and especially by women during the butchering and processing of animal carcasses. Merbs (1983, 2002) has pioneered the most extensive work on spondylolysis by providing a general overview with specific attention to manifestations of spondylolysis in Canadian Inuit skeletons. Merbs (2002) examined 417 immature and mature skeletons derived from four samples associated with prehistoric Thule and historic Inuit archaeological sites within the Hudson Bay, Labrador, and Southampton Island regions. A spondylolysis prevalence of approximately 22% can be calculated from the raw frequency counts presented by Merbs (2002) as ninety individuals exhibited spondylolysis. After excluding two individuals for which sex determination was not made, males were shown to comprise 70% of the individuals with spondylolysis. The sex distribution values seen in Table 1 of Merbs (2002) allow for a calculation of the spondylolysis prevalence values for each sex. Sex designations were assigned for 209 individuals, 109 females and 100 males (Merbs, 2002). Within the sexed individuals, spondylolysis is seen in 16.5% of the females and 41% of the males. These frequency values differ from our Tigara values for both sexes. However, we agree with Merbs (2002) that spondylolysis most likely results from fatigue fracturing associated with high levels of biomechanical stresses in the lower back region in conjunction with unusual body postures, rather than a genetic predisposition for a biomechanically weak *pars interarticularis*. We suspect that an additional contributing factor for the Tigara was strenuous burden carrying, like that documented in other living foragers (Hilton and Greaves, 2008). Spencer (1977) noted the arduous and strenuous physicality needed to haul whale carcasses ashore for the butchering procedures and the transport of whale parts for storage. We presume that the lifting and transporting of heavy animal body parts contributed also to spondylolysis.

In order to address precontact health and disease, Keenleyside (1998) examined five skeletal samples of pre-European-contact Arctic foragers including four from northern coastal Alaska in order to address precontact health and disease. Of the five samples, three are from the Point Barrow region, one is from Point Hope, and the final one is an Aleut sample from the Chaluka site in the eastern Aleutian Islands. Keenleyside (1998) states that her results substantiate a perspective that these Arctic populations experienced a substantial disease burden during the precontact period. The coastal Eskimo skeletons she studied exhibited lesions indicative of iron deficiency anemia, trauma, infection, and dental health problems. Regarding postcranial trauma, Keenleyside (1998) noted that within her pooled sample of coastal Eskimo skeletons approximately 12% (13 of 112 individuals) exhibited evidence of a healed postcranial fracture. The evidence for postcranial infections was lower in the north coastal samples, as only three of the 112 individuals had lesions indicative of infection.

Using protocols outlined by Steckel and Rose (2002), Dabbs (2009a, b, 2011) assessed lesions in the Ipiutak and Tigara of Point Hope as part of an overall comparison to lower latitude prehistoric agriculturalist skeletal samples of the Western Hemisphere. Thus, Dabbs' (2010) observations minimize fine-grained differences between these two foraging populations in favor of broader-scale comparisons between Arctic foragers and agriculturalists. This difference of scale accounts for the most important differences between her findings and ours. While Dabbs noted aspects of activity levels responsible for differences in lesions between Tigara and Ipiutak, she did not deeply contextualize the data with respect to foragers *as* foragers. In examining the daily lived experiences of Arctic foragers as their own optic, we have shown through nuanced comparisons how patterns of morbidity reveal what makes these groups distinctive. At the same time, we respectfully acknowledge Dabbs' choice of a forager-agricultural comparison given the subsistence stragegy intensification we note in the more whale-centric Tigara sample. For future work, we propose a potentially more detailed comparison in disease burdens specifically between Tigara and lower latitude prehistoric forager and agricultural communities.

Summary and conclusions

In summary, our current study has sought to highlight, through the documentation and comparative analysis of adult PCSLs, the habitual lived experiences of different prehistoric subsistence strategies in order to better understand the range of variability in morbidity within the context of Arctic foraging at Point Hope, Alaska. Although a comparison with prehistoric agriculturalists was not the intention of our chapter, we note several intriguing similarities to prehistoric agriculturalists that may contribute to the manifestation of some of the lesions seen at Point Hope. One aspect of the Ipiutak and Tigara cultural horizons at Point Hope is that they are recurrent occupations that go beyond short duration encampments and represent aggregated settled communities at this location. Such long-term aggregations of multiple individuals at a single location were more frequent across foragers (see Kelly, 1995, and Binford, 2001, for reviews) than that typically characterized by anthropologists. In particular, these types of forager aggregations compel us to consider the implications on morbidity of the operational conditions in place whereby individuals within communities have reduced mobility while maintaining a complete reliance on wild food resources. First, repeated and extended occupation of Point Hope by the Tigara appears to have increased adult workloads, especially for women, relative to the earlier Ipiutak. Contributing factors to increased workloads may not only have been new task assignment and specializations related to whaling but also increased competition to limited resources, thereby requiring an intensification of the handling and processing times of resources with finite limitations. As noted previously, much labor time and effort by adults of both sexes is expended in the preparation for the whaling season (Spencer, 1977), and while successful whaling may provide a substantial payoff for residents, unsuccessful whaling may force individuals to engage in additional and necessary "catch-up" subsistence activities to make up for the loss of an expected resource. Second, decreased mobility and the establishment of a small-scale permanent settlement may also have increased the likelihood for transmitting infectious pathogens, such as MTBC, across individuals as a result of the close proximity to other households, their trash, and the close contact between the waste products of humans and their animals, i.e., dogs. However, more permanent residency may have also relaxed the physical demands on individuals with mobility-limiting diseases, such as bilateral arthropathies. Our analyses indicate also that, contrary to the "healthy forager concept" proposed by Eaton and colleagues (Eaton and Konner, 1985; Eaton *et al.*, 1988; Eaton and Eaton, 1999a, b, 2000), precontact Arctic foragers were not necessarily free of infectious diseases or chronic illnesses. The human remains from the two Point Hope precontact cultural periods, especially the Tigara, exhibit skeletal lesions indicative of chronic morbidities often thought to be primarily associated with agricultural subsistence strategies. Thus, we agree with Keenleyside (1998) that precontact Alaskan Arctic foragers had to contend with a substantial morbidity burden that could severely affect the quality of their later adult years, if they were fortunate enough to live so long.

In conclusion, the high prevalence of postcranial skeletal lesions seen in the Tigara foragers, relative to the Ipiutak, indicates that overall levels of morbidity and comorbidities were higher with the later whale-centric subsistence strategy at Point Hope. Our results are important because they imply that foraging intensity, possibly in combination with more permanent residency, increased substantially for the later Tigara relative to the earlier non-whaling Ipiutak. This idea highlights the fact that while whale hunting is a coordinated activity requiring multiple individuals engaged in a high level of sophisticated social organization and resource monitoring, it is a dangerous activity with a "boom or bust" food resource payoff that puts individuals at great risk for bodily harm and affects their later life experience. Additionally, the increased permanent residency necessary to monitor whale migrations and defend a highly desirable resource-rich location, such as Point Hope, is coupled with increased exposure to infectious pathogens and competition for food resources. Whaling appears to be an additional subsistence activity that the Tigara added to an already arduous but changing set of foraging adaptations and workloads. These results support the idea that chronic and infectious diseases were at times an endemic feature of precontact Arctic foraging societies.

Acknowledgements

Our chapter is dedicated to the memory of Professor John Martin "Jack" Campbell who captivated us with accounts of his adventures in Alaskan archaeology, his stunning photographs, and always encouraged our efforts on this project. We thank Drs. Ian Tattersall and David Hurst Thomas, and the American Museum of Natural History for permission to examine the Point Hope human remains and archaeological artifacts as well as the gracious help they extended to us during our visits. We thank Dr. Ken Mowbray and Gary Sawyer for facilitating access to the Point Hope human remains and providing logistical support during the early phases of this research. Ken and Gary always made those research visits to the AMNH very pleasurable. We also thank L. Kilgore, B. Straight, and J. Whittaker for providing comments on earlier versions of this manuscript. Several anonymous reviewers helped to greatly improve the clarity of the manuscript and we thank them for extending their time to us in this manner. Partial funding was provided by Queens College/City University of New York, Grinnell College, and from an American Museum of Natural History Collections Study Grant.

References

Aksaajuq Otak, L. (2005). Iniqsimajuq: Caribou-skin preparation in Igloolik, Nunavut. In *Arctic Clothing*. Montreal: McGill-Queen's University Press, pp. 74–9.

Alvard, M. S. and Nolin, D. A. (2002). Rousseau's whale hunt. *Current Anthropology*, 43, 533–59.

Ammitzboll, T., Bencard, J., Bodenhoff, J. *et al.* (1991). Chapter 6: Clothing. In *The Greenland Mummies*. Washington, D.C.: Smithsonian Institution Press, pp. 116–49.

Anderson, D. D. (1984). Prehistory of north Alaska. In *Handbook of North American Indians, Volume 5: Arctic*. Washington, D.C.: Smithsonian Institution Press, pp. 80–93.

Aron, J. L. (2007). Mathematical modeling: the dynamics of infection. In *Infectious Disease Epidemiology: Theory and Practice*. Second Edition. Boston: Jones and Bartlett Publishers, pp. 186–212.

Arriaza, B. T. (1995). *Beyond Death: The Chinchorro Mummies of Ancient Chile.* Washington, D.C.: Smithsonian Institution Press.

Aufderheide, A. C. and Rodriguez-Martin, C. 1998. *Human Paleopathology.* Cambridge: Cambridge University Press.

Barnes, E. (1994). *Developmental Defects of the Axial Skeleton in Paleopathology.* Niwot, CO: University Press of Colorado.

Bastida, R., Loureiro, J., Quse, V. *et al.* (1999). Tuberculosis in a wild subantarctic fur seal from Argentina. *Journal of Wildlife Diseases*, 35, 766–8.

Bernardelli, A., Bastida, R., Loureiro, J. *et al.* (1996). Tuberculosis in sea lions and fur seals from the south-western Atlantic coast. *Revue Scientifique et Technique*, 15, 985–1005.

Binford, L. R. (2001). *Constructing Frames of Reference: An Analytical Method for Archaeological Use of Hunter-Gatherer and Environmental Data Sets.* Berkeley, CA: University of California Press.

Buikstra, J. E. (1981). Introduction. In *Prehistoric Tuberculosis in the Americas.* Evanston, IL: Northwestern University Archaeological Program, pp. 1–23.

Buikstra, J. E. and Cook, D. C. (1981). Pre-Columbian tuberculosis in West-Central Illinios: Prehistoric disease in biocultural perspective. In *Prehistoric Tuberculosis in the Americas.* Evanston, IL: Northwestern University Archaeological Program, pp. 115–39.

Buikstra, J. E. and Ubelaker, D. H. (1994). *Standards for Data Collection from Human Skeletal Remains.* Fayetteville, AK: Arkansas Archaeological Survey.

Burch, E. S. (1981). *Traditional Eskimo Hunters of Point Hope, Alaska: 1800–1875.* North Slope Borough.

Burch, E. S. (1988). *The Eskimos.* Norman, OK: University of Oklahoma Press.

Chance, N. (1990). *The Iñupiat and Arctic Alaska: An Ethnography of Development.* New York, NY: Wadsworth-Thomas Learning.

Coberly, J. S. and Chaisson, R. E. (2007). Tuberculosis. In *Infectious Disease Epidemiology: Theory and Practice.* Second Edition. Boston, MA: Jones and Bartlett Publishers, pp. 653–97.

Collins, H. B. (1984). History of research before 1945. In *Handbook of North American Indians, Volume 5: Arctic.* Washington, D.C.: Smithsonian Institution Press, pp. 8–16.

Costa, R. L. (1977). *Dental Pathology and Related Factors in Archaeological Eskimo Samples from Point Hope and Kodiak Island, Alaska.* Ph.D. University of Pennsylvania.

Costa, R. L. (1980a). Incidence of caries and abscesses in archaeological Eskimo skeletal samples from Point Hope and Kodiak Island, Alaska. *American Journal of Physical Anthropology*, 52, 501–14.

Costa, R. L. (1980b). Age, sex and antemortem loss of teeth in prehistoric Eskimo samples from Point Hope and Kodiak Island, Alaska. *American Journal of Physical Anthropology*, 53, 579–87.

Costa, R. L. (1982). Periodontal disease in the prehistoric Ipiutak and Tigara skeletal remains from Point Hope, Alaska. *American Journal of Physical Anthropology*, 59, 97–110.

Cousins, D. V. and Francis, B. R. (1990). Tuberculosis in captive seals: Bacteriological studies on an isolate belonging to the Mycobacterium tuberculosis complex. *Research in Veterinary Science*, 48, 196–200.

Cousins, D. V., Williams, S. N., Reuter, R. *et al.* (1993). Tuberculosis in wild seals and characterization of the seal bacillus. *Australian Veterinary Journal*, 70, 92–7.

Dabbs, G. (2009a). *Health and Nutrition at Prehistoric Point Hope, Alaska: Application and Critique of the Western Hemisphere Health Index*. Ph.D. University of Arkansas.

Dabbs, G. (2009b). Resuscitating the epidemiological model of differential diagnosis: Tuberculosis at prehistoric Point Hope, Alaska. *Paleopathology Newsletter*, 148, 11–24.

Dabbs, G. R. (2011). Health status among prehistoric Eskimos from Point Hope, Alaska. *American Journal of Physical Anthropology*, 146, 94–103.

Dumond, D. E. (1984). Prehistory: Summary. In *Handbook of North American Indians, Volume 5: Arctic*. Washington, D.C.: Smithsonian Institution Press, pp. 72–9.

Eaton, S. B. and Eaton, B. B. (1999a). Hunter-gatherers and human health. In *The Cambridge Encyclopedia of Hunters and Gatherers*. Cambridge: Cambridge University Press, pp. 449–56.

Eaton, S. B. and Eaton, B. B. (1999b). The evolutionary context of chronic degenerative diseases. In *Evolution in Health and Disease*. Oxford: Oxford University Press, pp. 251–9.

Eaton, S. B. and Eaton, S. B. (2000). Paleolithic vs. modern diets: Selected pathophysiological implications. *European Journal of Nutrition*, 39, 67–70.

Eaton, S. B. and Konner, M. (1985). Paleolithic nutrition: A consideration of its nature and current implications. *The New England Journal of Medicine*, 312 (5), 283–9.

Eaton, S. B., Shostak, M. and Konner, M. (1988). *The Paleolithic Prescription: A Program of Diet & Exercise and a Design for Living*. New York, NY: Harper & Row.

Ferguson, R. B. (1997). Violence and warfare in prehistory. In *Troubled Times: Violence and Warfare in the Past*. Amsterdam: Gordon and Breach, pp. 321–55.

Forshaw, D. and Phelps, G. R. (1991). Tuberculosis in a captive colony of pinnipeds. *Journal of Wildlife Diseases*, 27, 288–95.

Fortuine, R. (1986). Early evidence of infections among Alaskan natives. *Alaska History*, 2 (1), 39–56.

Fortuine, R. (1989). *Chills and Fever: Health and Disease in the Early History of Alaska*. Fairbanks, AK: University of Alaska Press.

Fortuine, R. (2005). *"Must We All Die?": Alaska's Enduring Struggle with Tuberculosis*. Fairbanks, AK: University of Alaska Press.

Giddings, J. L. (1968). *Ancient Men of the Arctic*. London: Secker & Warburg.

Gordis, L. (2009). *Epidemiology*. Philadelphia, PA: Saunders Elsevier.

Gossett, S. and Hilton, C. E. (2004). Tuberculosis in Iñupiats from Point Hope, Alaska: A possible maritime resource connection. *American Journal of Physical Anthropology*, S30, 133.

Griffin, N. M. (1930). *The Roles of Men and Women in Eskimo Culture*. Chicago, IL: The University of Chicago.

Hart Hansen, J. P. (1989). The mummies from Qilakitsoq: Paleopathological aspects. *Meddelelser om Grønland (Man & Society)*, 12, 69–82.
Hart Hansen, J. P., Meldgaard, J. and Nordqvist, J. (1991). *The Greenland Mummies*. Washington, D.C.: Smithsonian Institution Press.
Heymann, D. L. (2008). *Control of Communicable Diseases Manual, 19th Edition: An Official Report of the American Public Health Association*. American Public Health Association.
Hill, K. and Hurtado, M. (1996). *Aché Life History*. New York, NY: Aldine de Gruyter Publishers.
Hilton, C. E. and Greaves, R. D. (2008). Seasonality and sex differences in travel distance and resource transport in Venezuelan foragers. *Current Anthropology*, 49, 144–53.
Holliday, T. W. and Hilton, C. E. (2010). Body proportions of circumpolar peoples as evidenced from skeletal data: Ipiutak and Tigara (Point Hope) versus Kodiak Island Inuit. *American Journal of Physical Anthropology*, 142, 287–302.
Howell, N. (1979). *Demography of the Dobe !Kung*. New York, NY: Academic Press.
Hunter, J. E. B., Duignan, P. J., Dupont, C. *et al.* (1998). First report of potentially zoonotic tuberculosis in fur seals in New Zealand. *New Zealand Medical Journal*, 111, 130–2.
Keenleyside, A. (1998). Skeletal evidence of health and disease in pre-contact Alaskan Eskimos and Aleuts. *American Journal of Physical Anthropology*, 107, 51–70.
Keenleyside, A. (2003). An unreduced dislocated mandible in an Alaskan Eskimo: A case of altruism or adaptation. *International Journal of Osteoarchaeology*, 13, 384–89.
Keenleyside, A. (2006). Skeletal biology: Arctic and subarctic. In *Handbook of North American Indians, Volume 3: Environment, Origins, and Population*. Washington, D.C.: Smithsonian Institution Press, pp. 524–31.
Kelley, M. A. and Lytle-Kelley, K. (1999). Considerations on past and present non-human sources of atypical and typical mycobacteria. In *Tuberculosis: Past and Present*. Szeged: Golden Book-TB Foundation, pp.
Kelly, R. L. (1995). *The Foraging Spectrum: Diversity In Hunter-Gatherer Lifeways*. Washington, D.C.: Smithsonian Institution Press.
Kiers, A., Klarenbeek, A., Mendelts, B., van Soolingen, D. and Koeter, G. (2008). Transmission of *Mycobacterium pinnipedii* to humans in a zoo with marine mammals. *International Journal of Tuberculosis and Lung Disease*, 12, 1469–73.
Klippel, J. H. (2001). *Primer on the Rheumatic Diseases*. Twelfth Edition. Atlanta, GA: Athritis Foundation.
Larsen, C. S. (1997). *Bioarchaeology: Interpreting Behavior from the Human Skeleton*. Cambridge: Cambridge University Press.
Larsen, C. S. and Kelly, R. L. (1995). *Bioarchaeology of the Stillwater Marsh: Prehistoric Human Adaptation in the Western Great Basin*. Anthropological Papers of the American Museum of Natural History 77. New York, NY: American Museum of Natural History.
Larsen, H. and Rainey, F. (1948). *Ipiutak and the Arctic Whale Hunting Culture*. Anthropological Papers of the American Museum of Natural History 42. New York, NY: American Museum of Natural History.

Latchaw, M. R. and Hilton, C. E. (2004). Bilateral erosive arthropathy in the upper limbs: An Inuit case from Pt. Hope, Alaska. *American Journal of Physical Anthropology*, S38, 132.
Latchaw, M. R. and Hilton, C. E. (2006). Health and disease in the Ipiutak and Tigara of Point Hope, Alaska. *American Journal of Physical Anthropology*, S42, 118–119.
Leisen, J. C. C., Duncan, H., Riddle, J. M. (1991). Rheumatoid erosive arthropathy as seen in macerated (dry) bone specimens. In *Human Paleopathology: Current Syntheses and Future Options*. Washington, D.C.: Smithsonian Institution Press, pp. 211–15.
Lester, C. W. and Shapiro, H. L. (1968). Vertebral arch defects in the lumbar vertebrae of pre-historic American Eskimos. *American Journal of Physical Anthropology*, 28, 43–8.
Madimenos, F. (2005). *Dental Evidence for Division of Labor Among the Prehistoric Ipiutak and Tigara of Point Hope, Alaska*. M.A. Louisiana State University.
Merbs, C. F. (1974). The effects of cranial and caudal shift in the vertebral column of Northern populations. *Arctic Anthropology*, 6, 12–9.
Merbs, C. F. (1983). *Patterns of Activity-induced Pathology in a Canadian Inuit Population*. National Museum of Man Mercury Series, Archaeological Survey of Canada No. 119. Ottawa: National Museum of Man.
Merbs, C. F. (2002). Spondylolysis in Inuit skeletons from Arctic Canada. *International Journal of Osteoarchaeology*, 12, 279–90.
Moore, D. S. and McCabe, G. P. (2003). *Introduction to the Practice of Statistics*. Fourth Edition. New York, NY: W. H. Freeman and Company.
Murdoch, J. (1892). Ethnological results of the Point Barrow expedition. *Annual Report of the Bureau of American Ethnology*, 9, 19–441.
Nelson, K. E. (2007). Epidemiology of infectious disease: General principles. In *Infectious Disease Epidemiology: Theory and Practice*. Second Edition. Boston, MA: Jones and Bartlett, pp. 25–62.
Nelson, R. K. (1969). *Hunters of the Northern Ice*. Chicago, IL: The University of Chicago Press.
Nelson, W. E. (1899). The Eskimo about the Bering Strait. *Annual Report of the Bureau of American Ethnology*, 18 (Part I), 3–518.
Ogilvie, M. D. (2006). Vertebral border shifts in two pre-Iñupiat groups from Pt. Hope, Alaska. *American Journal of Physical Anthropology*, S38, 187.
Ogilvie, M. D., Hilton, C. E. and Ogilvie, C. D. (1998). Lumbar anomalies in the Shanidar 3 Neandertal. *Journal of Human Evolution*, 35, 597–610.
Ortner, D. J. and Putschar, W. G. J. (1981). *Identification of Pathological Conditions in Human Skeletal Remains*. Washington, D.C.: Smithsonian Institution Press.
Pedersen, K. (2005). Eskimo sewing techniques in relation to contemporary sewing techniques: See through a copy of a Qilakitsoq costume. In *Arctic Clothing*. Montreal: McGill-Queen's University Press, pp. 70–3.
Pike, I., Straight, B., Hilton, C., Osterle, M. and Lanyasunya, A. (2010). Documenting the health of consequences of endemic warfare in three pastoralist communities in Northern Kenya: A conceptual framework. *Social Science and Medicine*, 70, 45–52.

Powell, M. L. (1988). *Status and Health in Prehistory: A Case Study of the Moundville Chiefdom*. Washington, D.C.: Smithsonian Institution Press.
Rainey, F. (1941). The Ipiutak Culture at Point Hope, Alaska. *American Anthropologist*, 43, 364–75.
Rainey, F. G. (1947). *The Whale Hunters of Tigara*. Anthropological Papers of the American Museum of Natural History 41(2). New York, NY: American Museum of Natural History.
Rainey, F. (1971). *The Ipiutak Culture: Excavations at Point Hope, Alaska*. Reading, MA: Addison-Wesley.
Ray, D. J. (1984). Bering Strait Eskimo. In *Handbook of North American Indians, Volume 5: Arctic*. Washington, D.C.: Smithsonian Institution Press, pp. 285–302.
Ray, P. H. (1892). Appendix 6: Ethnographic sketch of the natives of Point Barrow. Ethnological results of the Point Barrow expedition. *Annual Report of the Bureau of American Ethnology*, 9, 19–441.
Roberts, C. A. and Buikstra, J. E. (2003). *The Bioarchaeology of Tuberculosis: A Global View on a Reemerging Disease*. Gainesville, FL: University Press of Florida.
Romano, M. I., Alito, A., Bigi, F., Fisanotti, J. C. and Cataldi, A. (1995). Genetic characterization of mycobacteria from South American wild seals. *Veterinary Microbiology*, 47, 89–98.
Rothman, K. J. (2002). *Epidemiology: An Introduction*. New York, NY: Oxford University Press.
Rothschild, B. M., Woods, R. J. and Ortel, W. (1990). Rheumatoid arthritis in the buff: Erosive arthritis in representative defleshed bones. *American Journal of Physical Anthropology*, 82, 441–9.
Sakakibara, C. (2007). *"Cetaceousness" and Global Warming among the Iñupiat of Arctic Alaska*. Ph.D. University of Oklahoma.
Schultz, A. (1961). Vertebral column and thorax. *Primatologia*, 4, 1–66.
Schwartz, J. H., Brauer, J. and Gordon-Larsen, P. (1995). Brief communication: Tigara (Point Hope, Alaska) tooth drilling. *American Journal of Physical Anthropology*, 97, 77–82.
Sheehan, G. W. (1985). Whaling as an organizing focus in Northwestern Alaskan Eskimo Societies. In *Prehistoric Hunter-Gatherers: The Emergence of Cultural Complexity*. New York, NY: Academic Press, pp. 123–53.
Shephard, R. J. and Rode, A. (1996). *The Health Consequence of Modernization: Evidence from Circumpolar Regions*. Cambridge: Cambridge University Press.
Spencer, R. F. (1959). *The North Alaskan Eskimo: A Study in Ecology and Society*. Bureau of American Ethnology, Bulletin 171. Washington, D.C.: Smithsonian Institution Press.
Spencer, R. F. (1977). Arctic and Sub-Arctic. In *The Native Americans*. Second Edition. New York, NY: Harper & Row, pp. 56–113.
Steckel, R. H. and Rose, J. C. (2002). *The Backbone of History: Health and Nutrition in the Western Hemisphere*. Cambridge: Cambridge University Press.
Stefansson, V. (1914). *The Stefansson–Anderson Arctic Expedition of the American Museum: Preliminary Ethnological Report, Part I*. Anthropological Papers of the

American Museum of Natural History 14. New York, NY: American Museum of Natural History.

Steinbock, R. T. (1976). *Paleopathological Diagnosis and Interpretation: Bone Diseases in Ancient Human Populations*. Springfield, IL: Charles C. Thomas.

Thompson, P. J., Cousins, D. V., Gow, B. L. *et al.* (1993). Seals, seal trainers, and mycobacterial infection. *American Review of Respiratory Diseases*, 147, 164–7.

Viale, D. (1981). Lung pathology in stranded cetaceans on the Mediterranean coasts. *Aquatic Mammals*, 8, 96–100.

Webb, S. (1995). *Palaeopathology of Aboriginal Australians: Health and Disease across a Hunter-Gatherer Continent*. Cambridge: Cambridge University Press.

Wirth, T., Hildebrand, F., Allix-Béguec, C. *et al.* (2008). Origin, spread and demography of the *Mycobacterium tuberculosis* complex. *PLoS Pathogens*, 4 (9), 1–10.

Wood, J. W., Milner, G. R., Harpending H. C. and Weiss, K. M. (1992). The osteological paradox: Problems in inferring prehistoric health from skeletal samples. *Current Anthropology*, 33, 343–58.

Woods, R., Cousins, D. V., Kirkwood, R. and Obendorf, D. L. (1995). Tuberculosis in a wild Australian fur seal (*Arctocephalus pusillus doriferus*) from Tasmania. *Journal of Wildlife Diseases*, 31: 83–6.

Worl, R. (1999). Inupiat Arctic whalers. In *The Cambridge Encyclopedia of Hunters and Gatherers*. Cambridge: Cambridge University Press, pp. 61–5.

Yu, P. (1997). *Hungry Lightning: Notes of a Woman Anthropologist in Venezuela*. Albuquerque, NM: University of New Mexico Press.

8 *Bone strength and subsistence activities at Point Hope*

LAURA L. SHACKELFORD

Introduction

Long bone robusticity, which is associated with high activity and mobility levels in the Pleistocene (Trinkaus *et al.*, 1994; Churchill *et al.*, 1996, 2000; Holt, 2003; Shackelford, 2007; Holt and Formicola, 2008), is assumed to generally decrease in Holocene and recent humans (Frayer, 1980; Jacobs, 1985; Smith, 1985; Marchi, 2008; Holt *et al.*, 2012). The archaeological site of Point Hope, Alaska, however, appears to represent an exceptionally robust population relative to other samples of similar antiquity. Excavations at Point Hope, Alaska, conducted from 1939 to 1941 by a team led by Helge Larsen and Froelich Rainey, were the source of thousands of artifacts and hundreds of human skeletons mostly attributed to two precontact archaeological cultures, the Ipiutak, dating from *c.* 1,600 to 1,100 years BP, and the later Tigara, dating from *c.* 800 to 300 years BP (Larsen and Rainey, 1948; Jensen, Chapter 2, this volume). Archaeological and artifactual evidence indicates both cultural groups exploited fish and sea mammals, but the Ipiutak relied heavily on caribou hunting (Larsen and Rainey, 1948; Giddings, 1967). The Tigara population was similar to modern Inuit populations occupying the same region, with whaling as the primary subsistence activity (Larsen and Rainey, 1948). Although not tested directly at the time, the presumably high level of appendicular robusticity associated with the Point Hope skeletons was attributed to the intense work effort involved in these subsistence activities in an arctic environment (Larsen and Rainey, 1948; Giddings, 1967).

The present study attempts to determine whether the effects of an active, Arctic subsistence pattern are discernible in the long bones of the Point Hope Inuit postcranial skeletons. In order to assess the work effort related to subsistence activities in the Point Hope sample, it is necessary to address two simple, interrelated questions: (1) Are the individuals from Point Hope robust

The Foragers of Point Hope: The Biology and Archaeology of Humans on the Edge of the Alaskan Arctic, eds. C. E. Hilton, B. M. Auerbach, L. W. Cowgill. Published by Cambridge University Press.

relative to other samples from the same general time period; and (2) if so, is this robusticity related to their presumed subsistence activities? Robusticity of the upper and lower limbs is analyzed in Holocene samples representing different subsistence activities. If there is a relationship between robusticity and mechanical loading, then these activities are expected to result in localized patterns of diaphyseal strength congruent with their behavior.

Cross-sectional properties of long bones and mechanical loading

Experimental and clinical research has demonstrated the relationship between cross-sectional properties of cortical bone and mechanical loading, as well as the plasticity of long bone diaphyses. While decades of experimental research have correlated mechanical loading with an osteogenic response, it has also demonstrated that bone is most strongly influenced by a subset of factors that include strain magnitude, strain rate, strain frequency, strain gradient, acceleration and rest intervals (Lanyon *et al.*, 1982; Lanyon and Rubin, 1984, 1985; Biewener and Taylor, 1986; Biewener and Bertram, 1993; Judex and Zernicke, 2000; Goodship *et al.*, 2009; Judex and Carlson, 2009). These various factors are undoubtedly linked, making it difficult to parse out the specific contribution or signal of any one influence on the skeleton. Exercise research and anthropological studies have demonstrated similar results in humans, with high-impact exercises such as squash, tennis, volleyball, or gymnastics leading to greater bone remodeling than high-force, low-impact activities such as swimming, cycling, or even weight-lifting, which require a large muscular force but lack a gravitational component (King *et al.*, 1969; Jones *et al.*, 1977; Fehling *et al.*, 1995; Robinson *et al.*, 1995; Burr *et al.*, 1996; Haapasalo *et al.*, 1996; Frost, 1997; Shaw and Stock, 2009; Warden, 2009; Bogenschutz *et al.*, 2011).

Studies of the effects of populations involved in vigorous, repetitive activities such as rowing or intense, long-distance mobility further illustrate the complexity of identifying the relationship between skeletal robusticity and specific activities. In an analysis of the effects of rowing on humeral strength, Weiss (2003) found that male ocean-rowers had more robust humeri than did male river-rowers or non-rowers. Females from the same populations, however, also had relatively more robust humeri despite the fact that they did not row. While this research suggests that bone remodeling has some relationship with activity, it is not directly attributable to a specific behavior. The behavioral repertoire, particularly for the upper limbs, is complex and there is not a single, clear environmental factor associated with cross-sectional size or shape.

Since the lower limb is used primarily for locomotion, differences in lower limb robusticity are most commonly explained as differences in the intensity or repetitiveness of terrestrial mobility, with the most relevant assessments of mobility interpreted from the midshaft femur to the midshaft tibia (Shackelford, 2005; Stock, 2006; Shaw and Stock, 2009). Multiple studies, however, have identified variable effects of terrain on lower limb strength, particularly the shape of the femoral midshaft (Ruff, 1999; Marchi *et al.*, 2006; Marchi, 2008). In contrast, body size and body proportions are strongly correlated with climate, and the proximal femur in particular is influenced by body laterality associated with ecogeographic patterning (Ruff, 1995; Ruff, 2000; Holliday and Ruff, 2001). However, a recent analysis suggests that body proportions may also influence femoral and tibial midshaft shape (Shaw and Stock, 2011).

Given the multiple influences on skeletal robusticity, can we distinguish patterns of behavior through localized bone strength? If so, is this relationship strong enough to differentiate groups that vary in mobility and subsistence activities? When looking at the groups from Point Hope and the comparative samples specifically, does robusticity in the upper limbs correlate with differences in activities presumably related to resource exploitation, and are differences in lower limb robusticity most commonly explained as differences in the intensity or repetitiveness of their inferred mobility?

Materials

In addition to the Point Hope sample, data were collected from five Holocene populations to maximize variations in environment, subsistence activities and body size. Given that age affects cortical bone remodeling (Ruff and Hayes, 1981, 1983; Ruff *et al.*, 1994), these samples are restricted to prime-age (20–50 years old) individuals. Data were collected on fifty individuals in each sample or the maximum number of available specimens when fewer than fifty skeletons were present. An equal number of males and females were included when possible. A brief overview of each comparative sample is provided below and in Table 8.1.

The Point Hope sample comprises fifty-three adults (twenty-seven males and twenty-six females) from collections of the American Museum of Natural History. Twenty-nine Ipiutak skeletons and twenty-four Tigara skeletons were included. The genetic relationship or continuity between the Ipiutak and Tigara was unknown at the time this study was undertaken (but see Maley, Chapter 4, this volume). Analyses of cranial variation between the Ipiutak and Tigara (Debetz, 1959; Keenleyside, 2006) as well as the recognition of a significant cultural shift after the Ipiutak period (Anderson, 1984) suggest that there is not a direct ancestral relationship between them. This archaeological evidence has been further substantiated by analyses of molar microwear that indicate diverse

Table 8.1. *Samples used in analyses*

Sample	Geographic origin	Date	*N* (M; F)	Subsistence
Point Hope Inuit	Ipiutak and Tigara cultures, Alaska	*c*. 1,600–1,100 years BP; *c*. 750–200 years BP	53 (27; 26)	Whaling, caribou hunting
Jomon	Eastern Honshu, Japan	Late/Final Jomon 4,000–2,500 years BP	50 (25; 25)	Hunting, fishing, gathering
Andaman Islanders[a]	Bay of Bengal, SE Asia	Protohistoric *c*. 1860–1900 CE	31 (16; 15)	Canoe for transport and food; swimming; limited terrestrial mobility; sexual division of labor
Libya	Wadi Tanezzuft, southwestern Fezzan, Libya	Middle Pastoral (6,100–5,000 years BP)	37 (23; 14)	Semi-nomadic herders; sexual division of labor
Egypt	Nile Valley, Egypt	Predynastic-Middle Kingdom (5,050–3,736 years BP)	42 (19; 23)	Agriculture and animal husbandry
East Africa[b]	Kenya, Uganda	Early twentieth century	40 (21; 19)	Non-urban; engaged in manual labor, no mechanized transport

[a] Data provided by J. Stock.
[b] Data provided by C. Ruff.

dietary habits between the Ipiutak and Tigara, with the Tigara demonstrating a higher density of microwear features suggestive of a very abrasive diet similar to prehistoric coastal Inuit who consumed high amounts of frozen meat (El Zaatari, 2008, and Chapter 6, this volume).

Despite the distinct archaeology and diets, the Ipiutak and Tigara were similar in their subsistence behavior, with both engaging in high-intensity hunting of maritime and/or terrestrial mammals (Larsen and Rainey, 1948; Giddings, 1967). Given the needs of this study, these samples were evaluated using ANOVA and determined to be statistically indistinguishable from one another with respect to their cross-sectional features. Thus, they are considered together as a single sample for these analyses.

Two comparative samples are from populations known to have engaged in high-intensity subsistence behaviors. The first is a sample of prehistoric Jomon hunter-gatherers from the Late to Final Jomon period (4,000 to 2,500 years BP) who occupied temperate, coastal regions of eastern Honshu, Japan ($N = 50$; twenty-five males, twenty-five females). The Jomon hunter-gatherers occupied

Japan *c.* 13,000–2,000 years BP and were likely descended from Pleistocene nomads whose origins were in Northeastern or Central Asia (Imamura, 1996; Omoto and Saitou, 1997; Hammer *et al.*, 2006). The Jomon show a great deal of continuity as a hunting-fishing-gathering society, although there is evidence of changes in subsistence and ways of life through time (Mizoguchi, 2002). The later Jomon phases demonstrated a radiating mobility pattern, and during this period the population's characteristic pattern of robusticity appeared (Mizoguchi, 2002; Ohtani *et al.*, 2005).

The second high-intensity sample is indigenous Andaman Islanders (AI) from the Bay of Bengal ($N = 31$; sixteen males, fifteen females). The skeletal collection represents the period immediately following the first permanent settlement of the islands by Europeans in 1858 and comes from the Great and Little Andaman Islands (Man, 1878, 1883, 1885; Brander, 1880). The AI exploited both marine and terrestrial resources (Myka, 1993; Stock and Pfeiffer, 2001), but terrestrial mobility was restricted within the islands due to tribal conflict, defense of territories, and high population density. Marine mobility, however, was very high. The AI regularly used canoes for transportation and food procurement, although only the males paddled canoes. All individuals, regardless of age or sex, habitually spent hours in the water, swimming (Man, 1883; Stock and Pfeiffer, 2001). J. Stock generously provided these data.

Three additional comparative samples come from African populations with varying levels of mobility. A sample of semi-nomadic herders from the Libyan Sahara is dated to the Middle Pastoral period, *c.* 6,100–5,000 years BP ($N = 37$; twenty-three males, fourteen females). This population resided in the Tanezzuft valley, west of the Acacus scarp. The central Sahara experienced fluctuating wet and dry periods during the Holocene, with a wet period from 13,000 to 5,000 years BP followed by a dramatic desertification around 5,000 years BP (Cremaschi and di Lernia, 2001). Being fed by a large river and occupied by a large oasis, the Tanezzuft valley escaped this aridification, and consequently became a focus for animal and human activity (Cremaschi and di Lernia, 2001). Archaeological evidence indicates that Middle Pastoral populations of Wadi Tanezzuft were associated with sheep/goat herding and engaged in seasonal movements from the fluvial valleys in summer seasons to western mountainous zones in winter seasons (Cremaschi and di Lernia, 1999; Tafuri *et al.*, 2006). Stress-related skeletal markers suggest that there was a sexual division of labor, with males involved in subsistence activities involving prolonged and strenuous walking (Arrighetti *et al.*, 2002).

An Egyptian sample ($N = 42$; nineteen males, twenty-three females) is archaeologically derived and comes from three time periods: (1) Predynastic (>5,050 years BP) from the sites of Mesaeed and Keneh; (2) Old Kingdom (5,050–4,131 years BP) from the site of Giza; and (3) Middle Kingdom

Table 8.2. *Summary statistics of body size variables, male samples*

		Pt. Hope Inuit	Jomon	Libya	AI	Egypt	East Africa
Body mass (kg)	*Mean*	64.9	59.3	62.9	41.8	61.5	56.7
	S.E.	0.998	0.812	1.57	0.906	0.751	1.80
	N	31	24	19	15	21	21
Stature (cm)	*Mean*	162.5	157.2	171.2	151.4	166.7	166.4
	S.E.	1.17	0.853	1.98	1.16	0.912	2.02
	N	31	23	14	15	21	21
Bi-iliac breadth (mm)	*Mean*	275.0	261.7	260.8	201.2	260.9	242.9
	S.E.	2.83	3.00	8.57	1.18	2.06	3.40
	N	25	24	9	15	21	21
Brachial index	*Mean*	73.7	82.2	77.3		78.3	79.3
	S.E.	0.377	0.422	1.16	—	0.501	0.300
	N	28	21	8		23	21
Crural index	*Mean*	80.8	83.8	86.1	85.0	82.7	85.5
	S.E.	0.495	0.571	1.12	0.491	0.582	0.531
	N	30	16	5	14	9	21

(3,941–3,736 years BP) from the site of Sheik Farag. The period over which these sites span represents the introduction of full-time sedentary agriculture and animal husbandry throughout Egypt, which first appeared in the region *c*. 5,000 BCE. (Wetterstrom, 1993). The shift from foraging to agriculture is associated with new settlement patterns, including year-round occupation sites and consequent reduced mobility, although these early phases likely maintained some semi-nomadic movement between the Nile and eastern Sahara (Wetterstrom, 1993; Bard, 1994; Wilkinson, 2003).

Finally, an East African sample ($N=40$; twenty-one males, nineteen females) was acquired from cadavers collected in the 1940s and 1950s in Kenya and Uganda. No information was available for any of the individuals regarding occupation. Given that both samples were collected prior to 1960, it is assumed that they were not highly urbanized, i.e., that these individuals most likely engaged in some manual labor and did not rely on mechanized transport (Goldthorpe and Wilson, 1960, as cited in Ruff, 1995). C. Ruff generously provided these data.

Methods

Standard linear metrics were taken on each individual to quantify body size, body shape, and long bone length, as available (Tables 8.2 and 8.3). Given their circumpolar location, the Point Hope Inuit sample is of considerable interest

Table 8.3. *Summary statistics of body size variables, female samples*

		Pt. Hope Inuit	Jomon	Libya	AI	Egypt	East Africa
Body mass (kg)	*Mean*	56.2	48.6	54.9	40.6	53.5	51.9
	S.E.	1.16	1.630	1.33	1.091	1.11	1.24
	N	27	23	13	15	25	19
Stature (cm)	*Mean*	150.5	146.5	157.6	145.7	158.7	158.7
	S.E.	1.04	0.624	1.04	0.813	1.05	1.66
	N	27	23	8	15	26	19
Bi-iliac breadth (mm)	*Mean*	266.1	259.6	265.0	198.1	254.4	236.1
	S.E.	3.75	3.40	13.7	0.860	2.52	2.59
	N	25	24	8	15	25	19
Brachial index	*Mean*	71.4	78.08	77.6		78.5	78.5
	S.E.	0.591	0.445	1.392	—	0.744	0.618
	N	23	19	5		16	19
Crural index	*Mean*	80.1	83.5	87.4	84.1	84.3	85.4
	S.E.	0.475	0.371	1.133	0.510	1.15	0.393
	N	25	16	2	13	9	19

with respect to skeletal features that have long been associated with climatic variables (Ruff, 1991; Holliday and Ruff, 2001). In particular, body breadth and relative limb lengths of Point Hope were compared to African and Asian skeletons to assess the extent of their cold-adapted features.

Variation in the rigidity and shape of long bone diaphyseal cross-sectional properties were evaluated for Point Hope and other Holocene samples. Robusticity was quantified using cross-sectional properties of the distal humerus (35% of humeral length with 0% identified as the distal end of the bone) and the midshaft and proximal femur (50% and 80% of femoral length, respectively, with 0% identified as the distal end of the bone). Humans demonstrate a high frequency of bilateral asymmetry in their upper limbs but relatively little asymmetry in their lower limbs (Auerbach and Ruff, 2006). For this reason, humeral data were collected bilaterally and right and left humeri were analyzed separately. Femoral measurements were taken unilaterally with an equal number of right and left femora included.

Two-dimensional cross-sectional data were collected on all bones by combining external contour molds with biplanar radiography or by computed tomographic (CT) scanning (for detailed description of methods, bone orientations, etc., see Shackelford [2005]). These two non-invasive methods produce comparable data (O'Neill and Ruff, 2004; Stock and Shaw, 2007). Cross-sectional properties were calculated using a PC version of SLICE (Eschman, 1992).

Here, robusticity refers to the strengthening of a skeletal element through the addition of bone tissue (Ruff *et al.*, 1993). In order to evaluate robusticity, long bones were modeled as hollow beams and mathematical formulae for predicting the bone's integrity under specific types of loading were applied to diaphyseal cross-sections (Ruff and Hayes, 1983). Long bone strength was evaluated using the amount and distribution of cortical bone in a cross-section and the polar moment of area, which is considered to be the most relevant indicator of a bone's performance under a variety of loads (Lieberman *et al.*, 2004). The amount of cortical area (CA) in a cross-section approximates the resistance of the bone to loading in tension or compression. The second moment of area, or area moment of inertia (I), is used to determine the bending rigidity of a bone in a given plane. As such, a ratio of the anteroposterior and mediolateral second moments of area (I_x/I_y) was used to estimate the shape of a cross-section. A bone with a ratio of second moments of area close to one will be roughly circular because approximately equal amounts of bone are distributed along the anteroposterior and mediolateral axes; the bone will become increasingly elliptical as this ratio diverges from one. The sum of any two second moments of area calculated about orthogonal axes is the polar moment of area (J), a measure of torsional strength and bending rigidity (Nordin and Frankel, 2000; Daegling, 2002).

To control for body size variation between samples, cortical properties were standardized by body mass (CA) or by the product of body mass and bone length squared (J) except for the proximal femoral section, which was standardized by body mass multiplied by bi-iliac breadth (Polk *et al.*, 2000; Ruff, 2000). Body mass was estimated using stature and bi-iliac breadth, given that this is a non-mechanical assessment of body mass (Ruff, 1991, 1994, 2000). If stature and body mass were unavailable, femoral head diameter was used to estimate body mass, given the mechanical significance of the femoral head for weight bearing in bipedal locomotion (Ruff *et al.*, 1991; McHenry, 1992; Grine *et al.*, 1995). Body mass estimates from these two methods give similar results when applied to fossil humans (Auerbach and Ruff, 2004). Values for standardized cross-sectional measurements for each group are provided in Tables 8.4 through 8.7.

Statistical analysis

Data for males and females were evaluated separately. For cross-sectional properties, samples were evaluated for differences in robusticity. Standardized cortical properties were evaluated for normality using graphical methods and verified by a Shapiro–Wilks W test (Sokal and Rohlf, 1995). Differences between

Table 8.4. *Summary statistics of humeral cross-sectional properties, male samples*

		Inuit	Jomon	Libyan	AI	Egyptian	East African
Right 35% CA[a]	*Mean*	340.1	392.7	302.4	457.7	331.1	340.4
	S.E.	11.3	11.9	18.5	17.6	13.1	12.8
	N	27	24	10	11	20	21
Right 35% J^b	*Mean*	225.2	350.9	173.6	249.7	180.3	186.5
	S.E.	10.4	11.5	18.7	15.9	12.1	11.5
	N	26	21	8	11	19	21
Right 35% I_x/I_y	*Mean*	1.10	0.91	1.04	1.13	0.98	1.19
	S.E.	0.03	0.03	0.05	0.05	0.04	0.04
	N	27	24	11	11	20	21
Left 35% CA[a]	*Mean*	324.3	387.5	305.0	429.3	310.4	
	S.E.	11.5	12.0	15.1	16.9	13.1	—
	N	26	24	15	12	20	
Left 35% J^b	*Mean*	218.2	349.1	155.6	235.9	168.6	
	S.E.	10.5	11.5	14.6	15.2	12.1	—
	N	25	21	13	12	19	
Left 35% I_x/I_y	*Mean*	1.06	0.91	1.05	1.09	1.02	
	S.E.	0.03	0.03	0.04	0.05	0.04	—
	N	26	24	17	12	20	

[a] CA is standardized by body mass.
[b] J is standardized by body mass × humeral length^2.

Table 8.5. *Summary statistics of humeral cross-sectional properties, female samples*

		Inuit	Jomon	Libyan	AI	Egyptian	East African
Right 35% CA[a]	*Mean*	304.6	324.2	260.1	320.0	276.1	263.0
	S.E.	11.6	11.9	20.2	15.8	13.1	13.1
	N	24	23	8	13	19	19
Right 35% J^b	*Mean*	180.2	242.2	139.1	154.1	151.8	146.7
	S.E.	10.4	11.4	22.8	14.1	13.6	11.7
	N	24	20	5	13	14	19
Right 35% I_x/I_y	*Mean*	1.24	0.96	1.20	1.08	1.04	1.19
	S.E.	0.04	0.04	0.07	0.05	0.04	0.04
	N	24	24	8	13	19	19
Left 35% CA[a]	*Mean*	286.6	307.5	265.6	318.9	269.3	
	S.E.	11.3	12.0	21.3	16.2	12.6	—
	N	25	22	7	12	20	
Left 35% J^b	*Mean*	176.2	225.6	140.6	158.2	165.1	
	S.E.	10.9	13.3	20.2	16.1	12.9	—
	N	24	16	7	11	17	
Left 35% I_x/I_y	*Mean*	1.23	1.00	1.13	1.14	1.14	
	S.E.	0.04	0.04	0.08	0.06	0.05	—
	N	25	23	7	12	20	

[a] CA is standardized by body mass.
[b] J is standardized by body mass × humeral length^2.

Table 8.6. *Summary statistics of femoral cross-sectional properties, male samples*

		Inuit	Jomon	Libyan	AI	Egyptian	East African
50% CA[a]	*Mean*	736.7	763.4	684.1	840.6	643.3	636.7
	S.E.	16.7	17.8	22.5	22.5	20.0	19.0
	N	27	24	15	15	19	21
50% J^b	*Mean*	566.3	623.3	436.1	450.0	402.7	363.8
	S.E.	18.7	21.2	27.0	25.1	22.3	21.2
	N	27	21	13	15	19	21
50% I_x/I_y	*Mean*	1.22	1.35	1.38	1.28	1.20	1.18
	S.E.	0.05	0.05	0.06	0.06	0.04	0.05
	N	27	24	15	15	19	21
80% CA[a]	*Mean*	805.4	734.7	684.9		655.1	637.4
	S.E.	16.1	17.1	21.7	—	18.8	18.3
	N	27	24	15		20	21
80% J^c	*Mean*	399.7	391.1	308.4		300.0	307.7
	S.E.	14.2	14.2	26.4	—	15.6	15.2
	N	24	24	7		20	21
80% I_{max}/I_{min}	*Mean*	1.70	1.63	1.67		1.70	1.65
	S.E.	0.07	0.08	0.10	—	0.08	0.08
	N	27	24	14		20	21

[a] CA is standardized by body mass.
[b] J is standardized by body mass × femoral length^2.
[c] J is standardized by body mass × bi-iliac breadth.

Table 8.7. *Summary statistics of femoral cross-sectional properties, female samples*

		Inuit	Jomon	Libyan	AI	Egyptian	East African
50% CA[a]	*Mean*	744.0	675.6	592.9	686.5	567.1	528.4
	S.E.	16.0	17.0	23.5	21.0	18.2	18.7
	N	26	23	12	15	20	19
50% J^b	*Mean*	563.8	492.3	350.1	332.6	338.7	314.0
	S.E.	16.1	17.5	25.9	21.1	19.3	18.8
	N	26	22	10	15	18	19
50% I_x/I_y	*Mean*	1.21	1.03	1.20	1.21	1.03	1.21
	S.E.	0.04	0.04	0.06	0.05	0.05	0.05
	N	26	23	12	15	20	19
80% CA[a]	*Mean*	814.5	684.3	613.9		581.8	548.4
	S.E.	17.9	19.0	28.9	—	20.4	20.9
	N	26	23	10		20	19
80% J^c	*Mean*	370.9	308.4	228.5		242.9	244.5
	S.E.	12.6	12.9	23.4	—	14.2	14.2
	N	24	23	7		19	19
80% I_{max}/I_{min}	*Mean*	1.81	1.90	1.58		2.03	1.69
	S.E.	0.06	0.07	0.10	—	0.07	0.07
	N	26	23	10		20	19

[a] CA is standardized by body mass.
[b] J is standardized by body mass × femoral length^2.
[c] J is standardized by body mass × bi-iliac breadth.

samples were evaluated using model II ANOVA and post-hoc Tukey HSD tests for multiple comparisons. Bilateral asymmetry of cortical properties of the humeri was quantified using the following formula: % asymmetry = 100 × (maximum − minimum)/minimum (Churchill and Formicola, 1997; Stock and Pfeiffer, 2004). Asymmetry between groups was tested using model II ANOVA.

Behavioral expectations

Previous analyses have attempted to categorically quantify mobility or terrain in order to elucidate the relationship between robusticity and behavioral factors (Ruff, 1999; Stock and Pfeiffer, 2001; Stock, 2002). These quantifications are difficult to make and inherently imprecise, however, given that mobility can be defined in multiple ways (Kelly, 1992; Stock, 2002). In order to make testable predictions about robusticity, expectations for each sample were categorized as low, moderate, or high for the upper and lower limbs based on the group's primary mode of mobility and subsistence, following Stock (2002). Primary mode of mobility and subsistence for each group was determined based on archaeological or ethnographic information. When possible, this categorical grouping also factored in information regarding relevant material technology for a sample (for example, bows and arrows or projectiles).

The Inuit and AI samples are expected to have the highest upper limb robusticity because of the large loads on the upper limb generated by marine mobility. The AI relied on the canoe for transport, and the Inuit used kayaks for whaling (Larsen and Rainey, 1948; Giddings, 1967; Stock and Pfeiffer, 2001). This would be true especially in the AI males, since only the males paddled canoes (Man, 1883, 1885; Stock and Pfeiffer, 2001). The Jomon sample is also expected to have generated large loads on the upper limb since they habitually hunted with bows and due to their generally high levels of activity as a hunting-fishing-gathering population (Kimura and Takahashi, 1982; Sakaue, 1998; Ohtani *et al.*, 2005).

The expectation of robusticity in the upper limbs of the Egyptian sample is unclear, given that this is a Predynastic agricultural society. There have been mixed results with respect to how the transition to agriculture affected upper limb robusticity in Native American populations (Bridges, 1989; Larsen, 1995; Bridges *et al.*, 2000). Early agricultural populations in the New World have been identified as having variably robust upper limbs, so following this model the Egyptian sample was expected to have moderately robust upper limbs.

The highest levels of lower limb robusticity are expected in the Jomon, Libyan, and Inuit samples given their generally high terrestrial mobility, as the

Table 8.8. *Robusticity expectations for samples*

Sample	Activity/Subsistence	Upper limb mobility category	Lower limb mobility category
Point Hope Inuit	Kayaks; harpoons, bows	High	Moderate
Jomon	Bow hunting, fishing	High	High
Andaman Islanders	Canoes, daily swimming	High	Low
Libyan	Nomadic herders	Low	High
Egyptian	Agriculture, animal husbandry	Moderate	Moderate
East African	1940–50s non-urbanized	Low	Low

Jomon were active hunter-gatherers, the Libyan nomadic herders performed seasonal migrations into mountainous terrain surrounding the Tanezzuft valley (Arrighetti *et al.*, 2002; di Lernia, 2002), and at least the Ipiutak practiced caribou hunting (Larsen and Rainey, 1948; Giddings, 1967). In contrast, the AI are expected to present the most gracile lower limbs, as their limited island territory restricted their terrestrial mobility (Stock and Pfeiffer, 2001; Stock, 2002). A summary of terrestrial and marine mobility and material foraging technologies for each sample is provided in Table 8.8.

This is a very simplified picture of behavioral variation between samples. While mobility – be it terrestrial or marine – may be a fundamental behavioral trait and the most repetitive load applied to the skeleton, it ignores other aspects of mechanical loading of the limbs (Stock and Pfeiffer, 2001; Weiss, 2003). Additionally, terrestrial mobility can be variably defined, and some studies have suggested that terrain may have a greater effect on lower limb robusticity than mobility (Ruff, 1999; Marchi *et al.*, 2006; Marchi, 2008). In the upper limb, these simple categorizations neglect the many activities and tools each of these groups most likely performed and utilized. At the same time, these samples are designated based on activities of the males of their group, and this will be considered when interpreting robusticity. However, these broad generalities about behavioral categorizations allow for testable predictions about robusticity in the upper and lower limbs.

Results

Humeri

There is a consistent pattern among male samples in measures of strength and rigidity in the upper limbs: the Jomon males have the highest average axial (CA) and torsional (*J*) strength, the AI and Point Hope Inuit are intermediate, and

Table 8.9. *Percentage differences between male samples in humeral cortical properties*[a]

	Right 35% CA[b]	Right 35% J^b	Right 35% $I_x/I_y{}^b$	Left 35% CA[c]	Left 35% J^c	Left 35% $I_x/I_y{}^c$
Point Hope–Jomon	15.4*	11.4*	−17.3*	19.5*	60.0*	−14.2*
Point Hope–Libya	−11.1	−22.9	−5.5	−6.0	−28.7*	−0.9
Point Hope–Andaman	34.6*	10.9	2.7	32.4*	8.1	2.8
Point Hope–Egypt	−2.6	−19.9	−10.9	−4.3	−22.7*	−3.8
Point Hope–East Africa	0.1	−17.2	8.2	—	—	—
Jomon–Libya	23.0*	−30.8*	14.3	−21.3*	−55.4*	15.4
Jomon–Andaman	16.6*	−0.5*	24.2*	10.8	−32.4*	19.8*
Jomon–Egypt	−15.7*	−28.1*	7.7	−19.9*	−51.7*	12.1
Jomon–East Africa	−13.3*	−25.7*	30.8*	—	—	—
Libya–Andaman	51.4*	43.8*	8.7	40.8*	51.6*	3.8
Libya–Egypt	9.5	3.9	−5.8	1.8	8.4	2.9
Libya–East Africa	12.6	7.4	14.4	—	—	—
Andaman–Egypt	−27.7*	−27.8*	−13.3	−27.7*	−28.5*	−6.4
Andaman–East Africa	−25.6*	−25.3*	5.3	—	—	—
Egypt–East Africa	2.8	3.4	21.4*	—	—	—

[a] Percentage difference calculated as [(group 2 – group 1)/group 1] × 100.
[b] Each pairwise comparison evaluated at $\alpha = 0.003$ to maintain a family-wide 0.05 α level.
[c] Each pairwise comparison evaluated at $\alpha = 0.005$ to maintain a family-wide 0.05 α level.
* Significant differences between groups.

the three African samples have the lowest average values (Tables 8.4 and 8.9; Figure 8.1). Although this does not coincide with strict behavioral expectations, this is consistent with other studies of the Jomon, which have found that this group is extremely robust relative to other recent humans (Kimura and Takahashi, 1982; Sakaue, 1998; Ohtani *et al.*, 2005). Table 8.9 summarizes these values and the differences between samples. For the right and left humeri, the Jomon males have significantly greater axial (CA) and torsional (J) strength than all other samples. The AI have significantly greater axial and torsional strength than all groups except for Point Hope. The AI sample also has a relatively large amount of cortical bone in the humeral diaphyses, and this is also observed in other long bones of the sample. This is due to greater percentages of cortical area that result from low sub-periosteal areas, thick cortical bone and small medullary cavities in the AI diaphyseal sections (Stock and Pfeiffer, 2001).

The same general pattern is maintained in the female samples (Tables 8.5 and 8.10, Figure 8.2). The Jomon females are consistently the most robust sample, although significant differences in strength occur between the Jomon and all other groups only in right and left torsional strength (J). The remaining samples

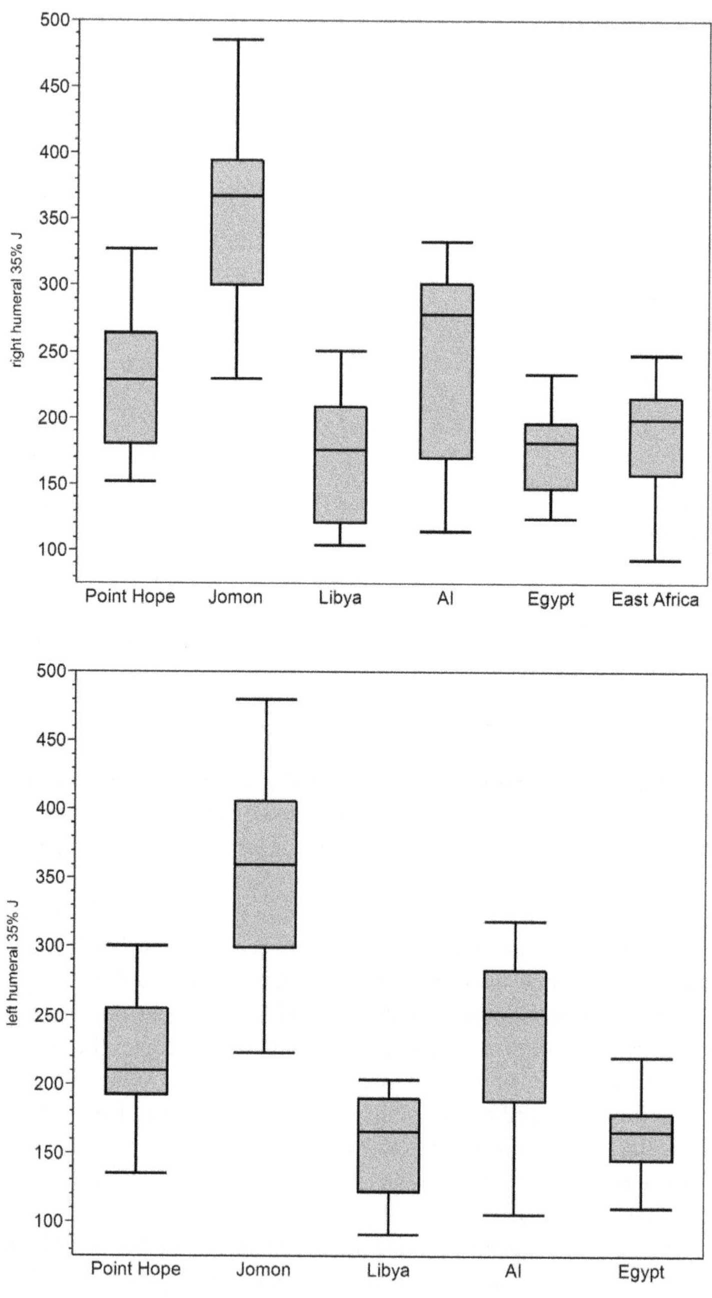

Figure 8.1. Box plots of relative bending strength (polar moment of area, *J*) for the right (top) and left (bottom) humeri of males from each sample.

Table 8.10. *Percentage differences between female samples in humeral cortical properties*[a]

	Right 35% CA[b]	Right 35% J^b	Right 35% $I_x/I_y{}^b$	Left 35% CA[c]	Left 35% J^c	Left 35% $I_x/I_y{}^c$
Point Hope–Jomon	6.4	34.4*	−22.9*	7.3	28.0*	−18.7*
Point Hope–Libya	−14.6	−22.8	−3.2	−7.3	−20.2	−8.1
Point Hope–Andaman	5.1	−14.5	−12.9	11.3	−10.2	−7.3
Point Hope–Egypt	−9.4	−15.8	−16.1*	−6.0	−6.3	−7.3
Point Hope–East Africa	−13.7	−18.6	−4.0	—	—	—
Jomon–Libya	−19.8	−42.6*	25.0*	−13.6	−37.7*	13.0
Jomon–Andaman	−1.3	−36.4*	12.5	3.7	−29.9*	14.0
Jomon–Egypt	−14.8	−37.3*	8.3	−12.4	−26.8*	14.0
Jomon–East Africa	−18.9*	−39.4*	24.0*	—	—	—
Libya–Andaman	23.0	10.8	−10.0	20.1	12.5	0.9
Libya–Egypt	6.2	9.1	−13.3	1.4	17.4	0.9
Libya–East Africa	1.1	5.5	−0.8	—	—	—
Andaman–Egypt	−13.7	−1.5	−3.7	−15.6	4.4	0.0
Andaman–East Africa	−17.8	−4.8	10.2	—	—	—
Egypt–East Africa	−4.7	−3.4	14.4	—	—	—

[a] Percentage difference calculated as [(group 2 – group 1)/group 1] × 100.
[b] Each pairwise comparison evaluated at $\alpha = 0.003$ to maintain a family-wide 0.05 α level.
[c] Each pairwise comparison evaluated at $\alpha = 0.005$ to maintain a family-wide 0.05 α level.
* Significant differences between groups.

are not significantly different from one another. The Jomon and Point Hope females demonstrate the greatest humeral rigidity, the AI and Egyptian females are intermediate, and the Libyan and East African females are the most gracile.

Point Hope and Jomon groups, as well as the Libyan females, demonstrate the lowest levels of asymmetry in humeral strength and rigidity (Table 8.11). The AI and Egyptian males have the highest percentages of asymmetry in cortical area (29.2% and 28%, respectively) and polar moments of area (45.6% and 27.6%, respectively). The AI and Egyptian females also have the highest levels of humeral asymmetry among these samples, particularly for torsional strength (55% and 33.2%, respectively).

The ratio of second moments of area (I_x/I_y) indicates the distribution of bone in a cross-section. For both sexes, the groups fall on a continuum, with the Jomon having the lowest and Point Hope having the highest average ratios of second moments of area (Tables 8.4 and 8.5; Figure 8.3). Humeral diaphyses of the Jomon males (0.91 and 0.91 for right and left humeri, respectively) and females (0.96 and 1.00 for right and left humeri, respectively) are roughly circular. In contrast, the rowing groups – AI and Point Hope – have among the highest ratios of second moments of area (AI: 1.13 and 1.09; Point Hope:

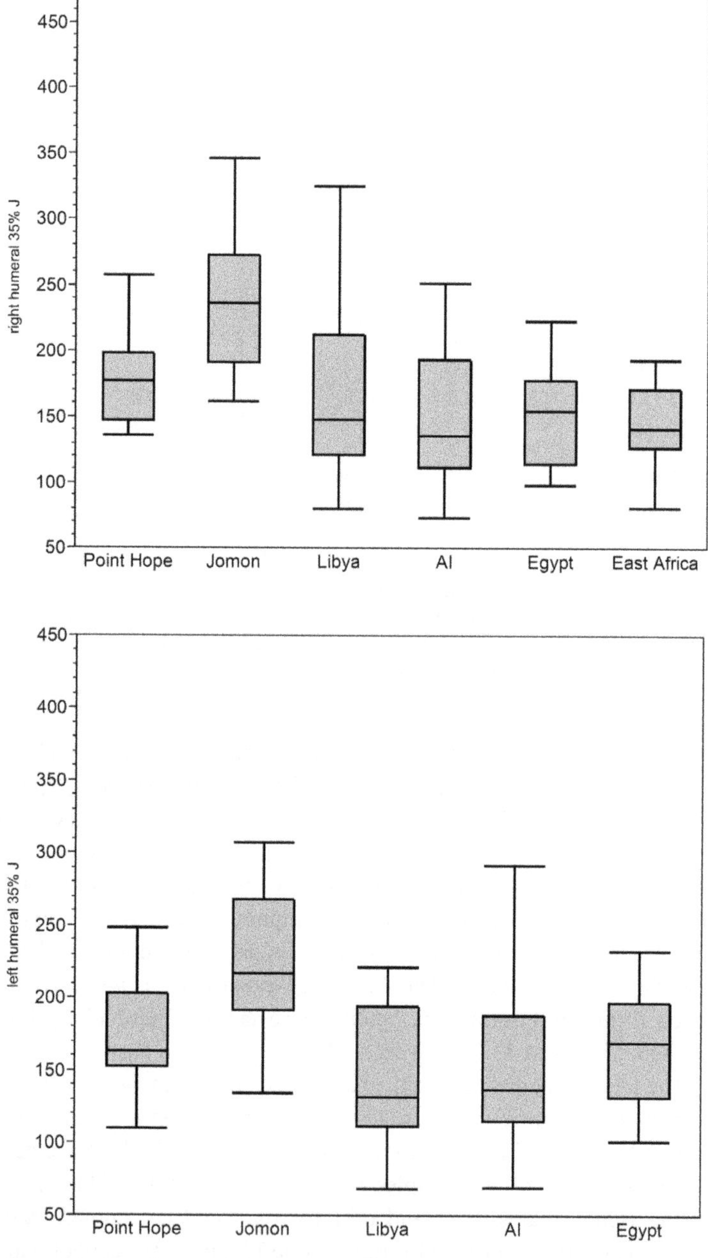

Figure 8.2. Box plots of relative bending strength (polar moment of area, *J*) for the right (top) and left (bottom) humeri of females from each sample.

Table 8.11. *Percentage humeral asymmetry in measures of cortical area (CA) and polar moments of area (J) for male and female samples*

Sample		Males CA	Males *J*	Females CA	Females *J*
Andamanese	*Mean*	29.2	45.6	37.5	55.0
	S.E.	7.02	10.5	5.48	10.3
	N	8	8	12	11
Egyptian	*Mean*	28.1	27.6	12.7	33.2
	S.E.	4.48	5.44	6.73	9.20
	N	19	18	13	8
Point Hope	*Mean*	9.28	13.4	9.89	11.0
	S.E.	1.71	2.42	1.73	2.28
	N	23	22	22	21
Jomon	*Mean*	5.73	7.53	5.70	10.4
	S.E.	1.02	1.82	1.5	3.52
	N	23	18	21	14
Libyan	*Mean*	8.37	24.5	1.07	9.53
	S.E.	1.55	5.87	3.69	3.43
	N	6	6	3	3
p-value		*0.3290*	*0.7733*	*0.8186*	*0.9643*

1.10 and 1.06, for right and left male humeri, respectively), indicating humeri that are relatively strong in an anteroposterior plane.

Femora

Of the levels of the femur that were evaluated, the midshaft has the greatest correlation with mobility, and we therefore expect the most mobile samples to have the largest average values for rigidity. For both the males and females, the Point Hope and Jomon groups have significantly greater bending and torsional strength than all other samples (Tables 8.12 and 8.13; Figure 8.4). The remaining samples are statistically indistinguishable from one another. The Point Hope females statistically also have greater CA and *J* than the Jomon females (though the Point Hope and Jomon males are not statistically different). There are no differences in diaphyseal shape between the male samples at the femoral midshaft. Among the females, only the Point Hope and Jomon samples are statistically different in diaphyseal shape, with the Jomon females significantly more circular than the Point Hope females. These shape indices at the femoral midshaft have been correlated with terrestrial mobility and terrain (Marchi *et al.*, 2006; Marchi, 2008).

At the proximal femur, the Point Hope and Jomon males and females again demonstrate the greatest average axial and torsional strength (Table 8.6).

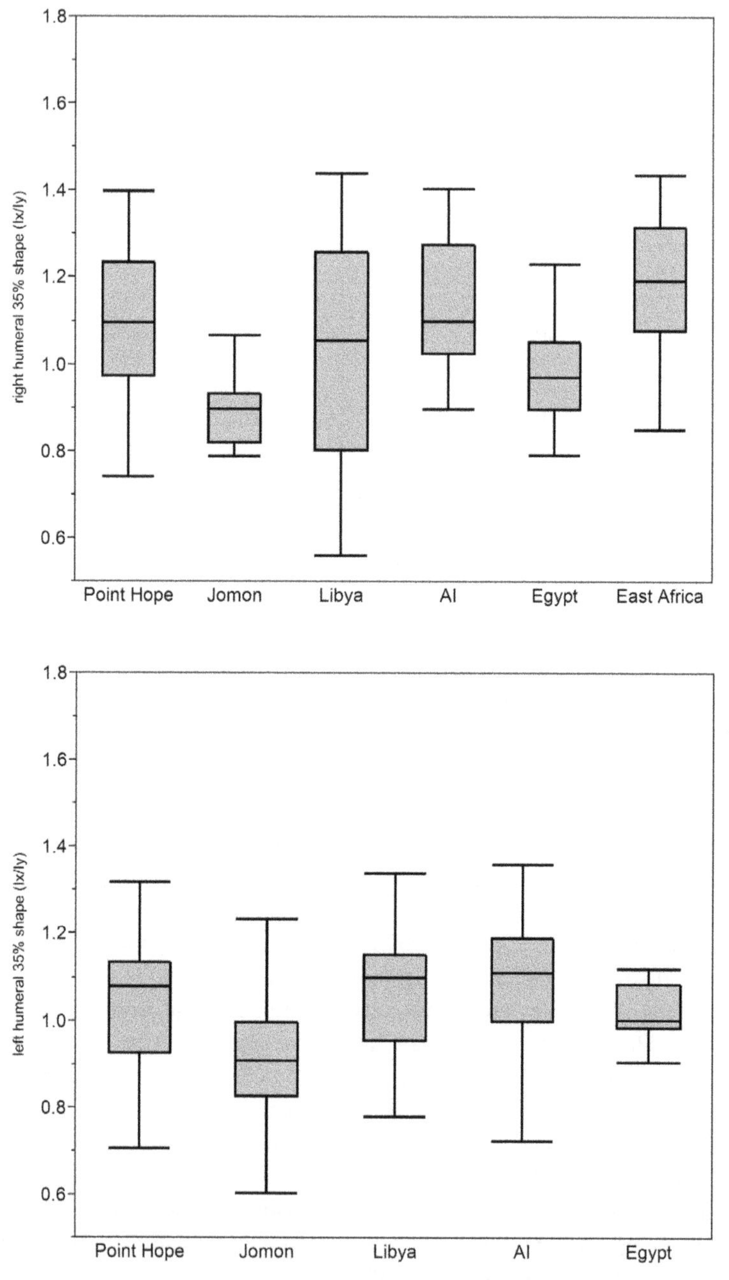

Figure 8.3. Box plots of distal humeral cross-sectional shape (ratio of second moments of area) for males from each sample. Top: right humerus; bottom: left humerus.

Table 8.12. *Percentage differences between male samples in femoral cortical properties*[a]

	Femoral 50% CA^b	Femoral 50% J^b	Femoral 50% I_x/I_y^{b}	Femoral 80% CA^c	Femoral 80% J^c	Femoral 80% I_{max}/I_{min}^{3c}
Point Hope–Jomon	3.6	10.1	10.7	−8.8*	−2.2	−4.1
Point Hope–Libya	−7.1	−23.0*	13.1	−15.0*	−22.8*	−1.8
Point Hope–Andaman	14.1*	−20.5*	4.9	—	—	—
Point Hope–Egypt	−12.7*	−28.9*	−1.6	−18.7*	−24.9*	0.0
Point Hope–East Africa	−13.6*	−35.8*	−3.3	−20.9*	−23.0*	−2.9
Jomon–Libya	−10.4	−30.0*	2.2	−6.8	−21.1	2.5
Jomon–Andaman	10.1	−27.8*	−5.2	—	—	—
Jomon–Egypt	−15.7*	−35.4*	−11.1	−10.8*	−23.3*	4.3
Jomon–East Africa	−16.6*	−41.6*	−12.6	−13.2*	−21.3*	1.2
Libya–Andaman	22.9*	3.2	−7.2	—	—	—
Libya–Egypt	−6.0	−7.7	−13.0	−4.4	−2.7	1.8
Libya–East Africa	−6.9	−16.6	−14.5	−6.9	−0.2	−1.2
Andaman–Egypt	−23.5*	−10.5	−6.3	—	—	—
Andaman–East Africa	−24.3*	−19.2	−7.8	—	—	—
Egypt–East Africa	−1.0	−9.7	−1.7	−2.7	2.6	−2.9

[a] Percentage difference calculated as [(group 2 – group 1)/group 1] × 100.
[b] Each pairwise comparison evaluated at $\alpha = 0.003$ to maintain a family-wide 0.05 α level.
[c] Each pairwise comparison evaluated at $\alpha = 0.005$ to maintain a family-wide 0.05 α level.
* Significant differences between groups.

Table 8.13. *Percentage differences between female samples in femoral cortical properties*[a]

	Femoral 50% CA^b	Femoral 50% J^b	Femoral 50% I_x/I_y^{b}	Femoral 80% CA^c	Femoral 80% J^c	Femoral 80% I_{max}/I_{min}^{3c}
Point Hope–Jomon	−9.2*	−12.7*	−14.9*	−16.0*	−16.9*	5.1
Point Hope–Libya	−20.3*	−37.9*	−0.8	−24.6*	−38.4*	−12.6
Point Hope–Andaman	−7.7	−41.0*	0.0	—	—	—
Point Hope–Egypt	−23.8*	−39.9*	−14.9	−28.6*	−34.5*	12.5
Point Hope–East Africa	−29.0*	−44.3*	0.0	−32.7*	−34.1*	−6.6
Jomon–Libya	−12.2	−28.9*	16.5	−10.3	−25.9*	−16.8
Jomon–Andaman	1.6	−32.4*	17.5	—	—	—
Jomon–Egypt	−16.1*	−31.2*	0.0	−15.0*	−21.2*	7.0
Jomon–East Africa	−21.8*	−36.2*	17.5	−19.9*	−20.7*	11.1
Libya–Andaman	15.8*	−5.0	0.8	—	—	—
Libya–Egypt	−4.4	−3.3	14.2	−5.2	6.3	28.6*
Libya–East Africa	−10.9	−10.3	0.8	−10.7	7.0	6.8
Andaman–Egypt	−17.4*	1.8	−14.9	—	—	—
Andaman–East Africa	−23.0*	−5.6	0.0	—	—	—
Egypt–East Africa	−6.8	−7.3	−14.9	−5.7	0.66	−16.9*

[a] Percentage difference calculated as [(group 2 – group 1)/group 1] × 100.
[b] Each pairwise comparison evaluated at $\alpha = 0.003$ to maintain a family-wide 0.05 α level.
[c] Each pairwise comparison evaluated at $\alpha = 0.005$ to maintain a family-wide 0.05 α level.
* Significant differences between groups.

Point Hope males have significantly greater axial strength than all other males and significantly greater torsional strength than all but the Jomon males. In both measures, the Jomon males also have significantly greater values than all other groups, with the notable exception of Point Hope (Table 8.12). The Point Hope females have significantly greater CA and *J* values than all other groups, and the Jomon females are distinguished from all of the African samples (Table 8.13). The Libyan and Egyptian samples are intermediate in strength and the East African sample is the most gracile in all femoral cross-sectional properties.

Although the shape of the proximal femoral diaphysis has been associated with body breadth and climate rather than mobility (Ruff, 2000), and despite the view of Point Hope as an extremely polar-adapted population, there are no significant differences in the distribution of cortical bone at the proximal femur among male samples (Table 8.12, Figure 8.5). Among the females, there are significant differences in the average ratio of second moments of area only between the Egyptian and Libyan samples and the Egyptian and East African samples. The Libyan female sample has the lowest average ratio of second moment of area, representing a mediolaterally broad proximal section (Table 8.13, Figure 8.5). The Egyptian females have the highest ratio, representing the most anteroposteriorly reinforced proximal femoral diaphysis.

Discussion and conclusions

With respect to the groups that lived at Point Hope, the results of these analyses demonstrate that relative to samples of similar antiquity, both males and females were generally robust. The humeri of the Point Hope males are similar to those of other populations with significant marine mobility. Despite the intense workloads that are associated with this population on land and sea, however, the upper limbs do not reach the levels of robusticity seen in at least one other active hunting-fishing-gathering group. Interestingly, the humeri of the females from Point Hope had average robusticity with respect to the full comparative sample. Both sexes do, however, have a unique pattern of anteroposterior reinforcement of the distal humerus that may be related to inferred whaling and hunting activities. In the lower limbs, the Point Hope males and females demonstrate relatively high levels of femoral rigidity at the midshaft that distinguish them from other Holocene samples and may be attributable to increased mobility.

More generally, results of these analyses suggest that there are generalized patterns of robusticity among groups that parallel loading based on activities, although they do not follow strict behavioral expectations. Levels of robusticity

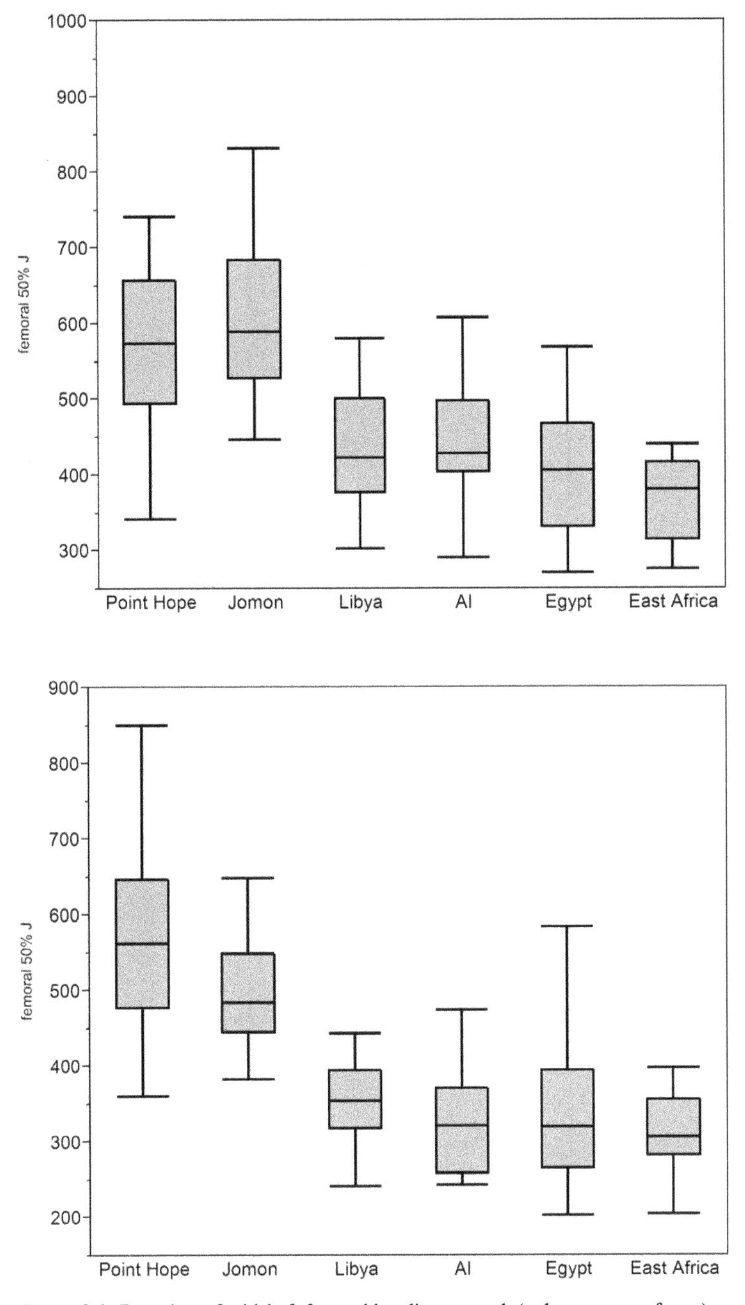

Figure 8.4. Box plots of midshaft femoral bending strength (polar moment of area) for males (top) and females (bottom) of each sample.

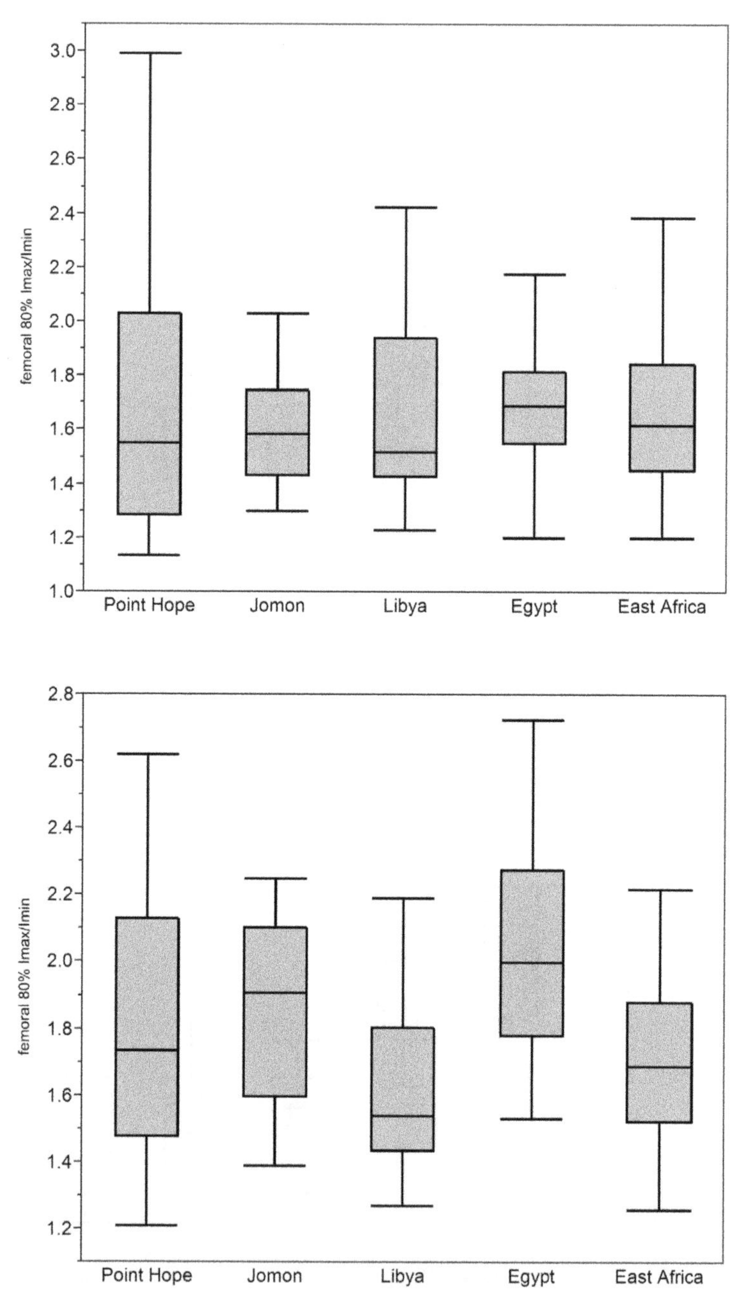

Figure 8.5. Box plots of proximal femoral cross-sectional shape (ratio of second moments of area) for males from each comparative sample. Top: right humerus; bottom: left humerus.

in the upper limb generally followed behavioral expectations. Given their high marine mobility and upper limb loading, the Point Hope, Andamanese, and Jomon groups were expected to have the highest levels of humeral robusticity. The Jomon, however, was the only sample to consistently display significant differences in robusticity from the non-marine groups. The Jomon demonstrated generalized robusticity throughout the skeleton, which is not unexpected based on several previous analyses of these hunting-gathering-fishing populations. The Point Hope Inuit and Andamanese had high levels of axial and torsional strength relative to the East African, Libyan, and Egyptian samples, although there were few significant differences. The presence of relatively robust humeri in the Egyptian females is notable, especially given the degree of asymmetry that they demonstrate and the unremarkable robusticity of the Egyptian males. Studies of robusticity associated with the transition to agriculture in Native American populations identified an increase in upper limb robusticity in females when compared to males, which suggests a greater change in activity patterns for females with this subsistence shift (Bridges, 1989; Ruff and Larsen, 1990). Further research may show a similar trend in the Egyptian Predynastic.

With respect to humeral asymmetry, Point Hope, Jomon, and the Libyan females (Table 8.11) are comparable to modern Euroamericans, Amerindians, and "non-specialized" control subjects in percentage of asymmetry in cortical areas (4.9–7.6%) and polar moments of area (8.7–11.8%) at the mid-distal humerus (Trinkaus *et al.*, 1994; Shaw and Stock, 2009). In contrast, the degree of asymmetry found in the AI and Egyptian samples is similar to that found in individuals performing specialized, unilateral tasks (Table 8.11). Analyses of professional tennis players and cricketers reveal asymmetries of up to 34% in cortical area and 57% in polar moment of area (Trinkaus *et al.*, 1994; Shaw and Stock, 2009). Similarly, non-pathological Neandertals have pronounced bilateral asymmetry in measures of diaphyseal rigidity (24% for CA and 57% for *J*) which have been associated with habitual lateralized tasks incurred with subsistence activities (Trinkaus *et al.*, 1994; Churchill *et al.*, 1996).

While the amount of cortical bone in long bone diaphyses may be informative about levels of activity, distribution of cortical bone in the diaphysis has been used to infer patterns of activity (Trinkaus *et al.*, 1991). The rigidity of a bone is dependent on both the amount and distribution of cortical bone, with bone distributed farther from the centroid of a cross-section better able to withstand bending deformation in that plane (Bertram and Swartz, 1991). The humeral diaphyses of the Jomon are roughly circular while those of the AI and Point Hope are relatively strong in an anteroposterior direction. A circular diaphyseal cross-section has been associated with adaptations to torsional loading encountered during throwing, such as in cricket or baseball (Ogawa and Yoshida, 1998; Shaw and Stock, 2009), and inferred for activities such as spear throwing in

Late Pleistocene modern humans (Churchill *et al.*, 1996, 2000). These activities create shear stresses that increase towards the edges of the cross-section in all directions rather than preferentially in a single plane, making a circular diaphysis the most efficient to resist the mechanical forces (Nordin and Frankel, 2000; Shaw and Stock, 2009). In contrast, anteroposteriorly reinforced humeri have been associated with forces induced by spear-thrusting in Neandertals (Churchill *et al.*, 1996; Schmitt *et al.*, 2003). Differences in both amount and distribution of cortical bone in the humeri of the comparative samples are likely related to the multiple activities being performed by individuals within these populations.

There was less consistency in the patterns observed in the lower limbs. There are several explanations for this. There may be relatively little difference in mobility or lower limb loading between the groups sampled here, especially given the levels of loading that may be required to affect remodeling of cortical bone. It is also likely that many of the differences between samples are not obvious due to methodology. Subtle differences in populations may not be identifiable with small sample sizes. There are also problems associated with coarsely grouping samples by a general mode of mobility or subsistence pattern.

Most interesting among the results from the femora, however, was the difference in diaphyseal shape between the Point Hope and Jomon females. Differences in the femoral midshaft shape have been used to infer differences in levels of mobility (Holt, 2003; Holt *et al.*, 2012) as well as mobility in different types of terrain (Ruff, 1999; Marchi *et al.*, 2006; Marchi, 2008). The more elliptical shape of the Point Hope femoral midshaft, therefore, could be interpreted as higher levels of mobility in these groups or mobility on rugged terrain.

Also of note was the pattern of cortical bone distribution at the proximal femur. This cross-sectional level is generally associated with body breadth, making it more closely correlated with body shape than habitual activity (Ruff, 1991, 1999). However, the lack of shape differences between male samples and the pattern of shape differences between female samples does not support this interpretation. The Point Hope sample has a relatively large average body mass, and the brachial and crural indices indicate body proportions that are typical of what is expected for an arctic environment (Tables 8.2 and 8.3). In contrast, the Libyan and Egyptian samples are most similar in longitude, climate, and body proportions (Tables 8.2 and 8.3), yet they are significantly different in proximal femoral shape (females only, Table 8.13).

Given the complexities of interpreting cross-sectional geometry, evaluating the populations from Point Hope relative to other Holocene groups achieves several stated goals. While researchers have generally assumed robusticity is reduced in post-Pleistocene populations, Point Hope individuals maintained high measures of skeletal robusticity in the upper and lower limbs. They are not extreme in their levels of robusticity, however, relative to other active populations of the same time period. Based on analyses of localized robusticity in the humeri and femora of these groups, it is likely that habitual activity may explain some of the robusticity of these Holocene populations. However, as with other recent studies, interpreting this relationship is complicated by the many other factors involved in bone adaptation. In this case, climate and environmental adaptation likely play significant roles in determining how well the skeletons of the Point Hope Ipiutak and Tigara peoples reflected their specialized subsistence strategies.

Acknowledgements

I would like to thank Benjamin Auerbach, Libby Cowgill, and Charles Hilton for allowing me to participate in the analysis of the Point Hope sample. I am grateful for their patience and support. Thank you to Jay Stock and Chris Ruff for providing data. Thank you also to three anonymous reviewers who greatly improved the quality of this contribution.

References

Anderson, D. D. (1984). Prehistory of north Alaska. In *Handbook of North American Indians, Volume 5: Arctic*. Washington, D.C.: Smithsonian Institution Press, pp. 80–93.

Arrighetti, B., Reale, B., Ricci, F. and Borgognini Tarli, S. M. (2002). Skeletal markers of stress at Site 96/129. In *Sand, Stones and Bones: The Archaeology of Death in the Wadi Tanezzuft Valley (5000–2000 BP)*. Rome: Universita degli Studi di Roma "La Sapienza".

Auerbach, B. M. and Ruff, C. B. (2004). Human body mass estimation: A comparison of "morphometric" and "mechanical" methods. *American Journal of Physical Anthropology*, 125, 331–42.

Auerbach, B. M. and Ruff, C. B. (2006). Limb bone bilateral asymmetry: Variability and commonality among modern humans. *Journal of Human Evolution*, 50, 203–18.

Bard, K. A. (1994). The Egyptian predynastic: A review of the evidence. *Journal of Field Archaeology*, 21, 265–88.

Bertram, J. E. A. and Swartz, S. (1991). The "law of bone transformation": A case for crying Wolff? *Biological Reviews of the Cambridge Philosophical Society*, 66, 245–73.

Biewener, A. A. and Bertram, J. E. A. (1993). Skeletal strain patterns in relation to exercise training during growth. *Journal of Experimental Biology*, 185, 51–69.

Biewener, A. A. and Taylor, C. R. (1986). Bone strain: a determinant of gait and speed. *Journal of Experimental Biology*, 123, 383–400.

Bogenschutz, E. D., Smith, H. D. and Warden, S. J. (2011). Midhumerus adaptation in fast-pitch softballers and the effect of throwing mechanics. *Medicine and Science in Sports and Exercise*, 43, 1698–1706.

Brander, E. S. (1880). Remarks on the aborigines of the Andaman Islands. *Proceedings of the Royal Society of Edinburgh*, 10, 415–24.

Bridges, P. S. (1989). Changes in activities with the shift to agriculture in the Southeastern United States. *Current Anthropology*, 30, 385–94.

Bridges, P. S., Blitz, J. H. and Solano, M. C. (2000). Changes in long bone diaphyseal strength with horticultural intensification in west-central Illinois. *American Journal of Physical Anthropology*, 112, 217–38.

Burr, D. B., Milgrom, C., Fyhrie, D. *et al.* (1996). In vivo measurement of human tibial strains during vigorous activity. *Bone*, 18, 405–10.

Churchill, S. E. and Formicola, V. (1997). A case of marked bilateral asymmetry in the Upper Limbs of an Upper Palaeolithic male from Barma Grande (Liguria), Italy. *International Journal of Osteoarchaeology*, 7, 18–38.

Churchill, S. E., Formicola, V., Holliday, T. W., Holt, B. M. and Schumann, B. A. (2000). The Upper Palaeolithic population of Europe in an evolutionary perspective. In *Hunters of the Golden Age: The Mid Upper Paleolithic of Eurasia 30,000–20,000 BP*. Leiden: University of Leiden, pp. 31–57.

Churchill, S., Weaver, A. H. and Niewoehner, W. A. (1996). Late Pleistocene human technological and subsistence behavior: Functional interpretations of upper limb morphology. *Quaternaria Nova*, 6, 413–47.

Cremaschi, M. and di Lernia, S. (1999). Holocene climatic changes and cultural dynamics in the Libyan Sahara. *African Archaeological Review*, 16, 211–38.

Cremaschi, M. and di Lernia, S. (2001). Environment and settlements in the Mid-Holocene palaeo-oasis of Wadi Tanezzuft (Libyan Sahara). *Antiquity*, 75, 815–25.

Daegling, D. J. (2002). Estimation of torsional rigidity in primate long bones. *Journal of Human Evolution*, 43, 229–39.

Debetz, G. (1959). The skeletal remains of the Ipiutak cemetery. *Proceedings of the 33rd International Congress of Americanists*, 31, 57–64.

di Lernia, S. (2002). Dry climatic events and cultural trajectories: Adjusting middle Holocene pastoral economy of the Libyan Sahara. In *Droughts, Food and Culture: Ecological Change and Food Security in Africa's Later Prehistory*. New York, NY: Plenum, pp. 225–50.

El Zaatari, S. (2008). Occlusal molar microwear and the diets of the Ipiutak and Tigara populations (Point Hope) with comparisons to the Aleut and Arikara. *Journal of Archaeological Science*, 35, 2517–22.

Eschman, P. N. (1992). *SLCOMM Version 1.6*. Albuquerque, NM: Eschman Archeological Services.

Fehling, P. C., Alekel, L., Clasey, J., Rector, A. and Stillman, R. J. (1995). A comparison of bone mineral densities among female athletes in impact loading and active loading sports. *Bone*, 17, 205–10.

Frost, H. M. (1997). Why do marathon runners have less bone than weight lifters? A vital biomechanical view and explanation. *Bone*, 20, 183–9.
Frayer, D. W. (1980). Sexual dimorphism and cultural-evolution in the late Pleistocene and Holocene of Europe. *Journal of Human Evolution*, 9, 399–415.
Giddings, J. L. (1967). *Ancient Men of the Arctic*. New York, NY: Knopf.
Goodship, A. E., Lawes, T. J. and Rubin, C. T. (2009). Low-magnitude high-frequency mechanical signals accelerate and augment endochondral bone repair: Preliminary evidence of efficacy. *Journal of Orthopedic Research*, 27 (7), 922–30.
Grine, F. E., Jungers, W. L., Tobias, P. V. and Pearson, O. M. (1995). Fossil *Homo* femur from Berg-Aukas, northern Namibia. *American Journal of Physical Anthropology*, 97, 151–85.
Haapasalo, H., Sievanen, H., Kannus, P., Oja, P. and Vuori, I. (1996). Humeral dimensions after long term unilateral loading: A DXA study of Finnish competitive tennis players. *Bone*, 18, 106S.
Hammer, M. F., Karafet, T. M., Hwayong, P. *et al.* (2006). Dual origins of the Japanese: Common ground for hunter-gatherer and farmer Y-chromosomes. *Journal of Human Genetics*, 51, 47–58.
Holliday, T. W. and Ruff, C. B. (2001). Relative variation in human proximal and distal limb segment lengths. *American Journal of Physical Anthropology*, 116, 26–33.
Holt, B. M. (2003). Mobility in Upper Paleolithic and Mesolithic Europe: Evidence from the lower limb. *American Journal of Physical Anthropology*, 122, 200–15.
Holt, B. M. and Formicola, V. (2008). Hunters of the Ice Age: The biology of Upper Paleolithic people. *Yearbook of Physical Anthropology*, 51, 70–99.
Holt, B., Ruff, C., Niskanen, M. *et al.* (2012). Postcranial robusticity trends in Europe across the last 30,000 years. *American Journal of Physical Anthropology*, 147, 167.
Imamura, K. (1996). *Prehistoric Japan: New Perspectives on Insular East Asia*. Honolulu, HI: University of Hawaii Press.
Jacobs, K. H. (1985). Evolution in the postcranial skeleton of Late Glacial and early Postglacial European hominids. *Zeitschrift fur Morphologie und Anthropologie*, 75, 307–26.
Jones, H. N., Priest, J. D., Hayes, W. C., Tichenor, C. C. and Nagel, D. A. (1977). Humeral hypertrophy in response to exercise. *Journal of Bone and Joint Surgery*, 59A, 204–8.
Judex, S. and Carlson, K. (2009). Is bone's response to mechanical signals dominated by gravitational loading? *Medicine and Science in Sports and Exercise*, 41 (11), 2037–43.
Judex, S. and Zernicke, R. F. (2000). High-impact exercise and growing bone: relation between high strain rates and enhanced bone formation. *Journal of Applied Physiology*, 88 (6), 2183–91.
Keenleyside, A. (2006). Skeletal biology: Arctic and subarctic. In *Handbook of North American Indians, Volume 3: Environment, Origins, and Population*. New York, NY, American Museum of Natural History, pp. 524–531.
Kelly, R. L. (1992). Mobility/sedentism: Concepts, archaeological measures and effects. *Annual Review of Anthropology*, 21, 43–66.

Kimura, T. and Takahashi, H. (1982). Mechanical properties of cross section of lower limb long bones in Jomon man. *Journal of the Anthropological Society of Nippon*, 90, 105–18.

King, J. W., Brelsford, H. J. and Tullow, H. S. (1969). Analysis of the pitching arm of the professional baseball pitcher. *Clinical Orthopaedics and Related Research*, 67, 116–23.

Lanyon, L. E. and Rubin, C. T. (1984). Static vs. dynamic loads as an influence on bone remodeling. *Journal of Biomechanics*, 17 (12), 897–905.

Lanyon, L. E. and Rubin, C. T. (1985). The effect on bone remodeling of static and graded dynamic loads. *Journal of Bone and Joint Surgery*, 67 (2), 318.

Lanyon, L. E., Goodship, A. E., Pye, C. J. and Macfie, J. H. (1982). Mechanically adaptive bone remodeling. *Journal of Biomechanics*, 15 (3), 141–54.

Larsen, C. S. (1995). Biological changes in human populations with agriculture. *Annual Review of Anthropology*, 24, 185–213.

Larsen, H. and Rainey, F. (1948). *Ipiutak and the Arctic Whale Hunting Culture.* Anthropological Papers of the American Museum of Natural History 42. New York, NY: American Museum of Natural History.

Lieberman, D. E., Polk, J. D. and Demes, B. (2004). Predicting long bone loading from cross-sectional geometry. *American Journal of Physical Anthropology*, 123, 156–71.

Man, E. H. (1878). The Andaman Islands. *Journal of the Royal Anthropological Institute*, 7, 105–9.

Man, E. H. (1883). On the aboriginal inhabitants of the Andaman Islands. *Journal of the Royal Anthropological Institute*, 12, 69–434.

Man, E. H. (1885). On the Andaman Islands and their inhabitants. *Journal of the Royal Anthropological Institute*, 14, 253–72.

Marchi, D. 2008. Relationships between lower limb cross-sectional geometry and mobility: The case of a Neolithic sample from Italy. *American Journal of Physical Anthropology*, 137, 188–200.

Marchi, D., Sparacello, V. S., Holt, B. M. and Formicola, V. (2006). Biomechanical approach to the reconstruction of activity patterns in Neolithic western Liguria, Italy. *American Journal of Physical Anthropology*, 131, 447–55.

McHenry, H. M. (1992). Body size and proportions in early hominids. *American Journal of Physical Anthropology*, 87, 407–31.

Mizoguchi, K. (2002). *An Archaeological History of Japan 30,000 BC to AD 700.* Philadelphia, PA: University of Pennsylvania Press.

Myka, F. P. (1993). *Decline of Indigenous Populations: The Case of the Andaman Islanders.* Jaipur: Rawat Publications.

Nordin, M. and Frankel, V. H. (2000). *Basic Biomechanics of the Musculoskeletal System.* Third Edition. New York, NY: Lippincott Williams & Wilkins.

Ogawa, K. and Yoshida, A. (1998). Throwing fracture of the humeral shaft: An analysis of 90 patients. *American Journal of Sports Medicine*, 26, 242–6.

Ohtani, E., Baba, H. and Kohara, Y. (2005). Morphological characteristics of earliest Jomon human remains from Tochibara rock shelter, Kita-Aiki, Nagano, Central Japan. *American Journal of Physical Anthropology*, S40, 160.

Omoto, K. and Saitou, N. (1997). Genetic origins of the Japanese: Partial support for the dual structure hypothesis. *American Journal of Physical Anthropology*, 102, 437–46.
O'Neill, M. C. and Ruff, C. B. (2004). Estimating human long bone cross-sectional geometric properties: A comparison of noninvasive methods. *Journal of Human Evolution*, 47, 221–35.
Polk, J. D., Demes, B., Jungers, W. L. *et al.* (2000). A comparison of primate, carnivoran and roden limb bone cross-sectional properties: Are primates really unique? *Journal of Human Evolution*, 39, 297–325.
Robinson, T. L., Snow-Harter, C., Taaffe, D. R. *et al.* (1995). Gymnasts exhibit higher bone mass than runners despite similar prevalence of amenorrhea and oligomenorrhea. *Journal of Bone and Mineral Research*, 10, 26–35.
Ruff, C. B. (1991). Climate and body shape in hominid evolution. *Journal of Human Evolution*, 21, 81–105.
Ruff, C. B. (1994). Morphological adaptation to climate in modern and fossil hominids. *Yearbook of Physical Anthropology*, 37, 65–107.
Ruff, C. B. (1995). Biomechanics of the hip and birth in early *Homo*. *American Journal of Physical Anthropology*, 98, 527–74.
Ruff, C. B. (1999). Skeletal structure and behavioral patterns of prehistoric Great Basin populations. In *Prehistoric Lifeways in the Great Basin Wetlands: Bioarchaeological Reconstruction and Interpretation*. Salt Lake City, UT: University of Utah Press, pp. 290–320.
Ruff, C. B. (2000). Body size, body shape, and long bone strength in modern humans. *Journal of Human Evolution*, 38, 269–90.
Ruff, C. B. and Hayes, W. C. (1981). Age and sex-differences in geometrical properties of the human femur and tibia. *Journal of Biomechanics*, 14, 488.
Ruff, C. B. and Hayes, W. C. (1983). Cross-sectional geometry of pecos pueblo femora and tibiae: A biomechanical investigation. 2: Sex, age, and side differences. *American Journal of Physical Anthropology*, 60, 383–400.
Ruff, C. B. and Larsen, C. S. (1990). Postcranial biomechanical adaptations to subsistence strategy changes on the Georgia coast. In *The Archaeology of Mission Santa Catalina de Guale 2: Biocultural Interpretations of a Population in Transition*. New York, NY: American Museum of Natural History, pp. 94–120.
Ruff, C. B., Scott, W. W. and Liu, A. Y. C. (1991). Articular and diaphyseal remodeling of the proximal femur with changes in body-mass in adults. *American Journal of Physical Anthropology*, 86, 397–413.
Ruff, C. B., Trinkaus, E., Walker, A. and Larsen, C. S. (1993). Postcranial robusticity in *Homo*. I: Temporal trends and mechanical interpretation. *American Journal of Physical Anthropology*, 91, 21–53.
Ruff, C. B., Walker, A. and Trinkaus, E. (1994). Postcranial robusticity in *Homo*. III: Ontogeny. *American Journal of Physical Anthropology*, 93, 35–54.
Sakaue, K. (1998). Bilateral asymmetry of the humerus in Jomon people and modern Japanese. *Anthropological Science*, 105, 231–46.
Schmitt, D., Churchill, S. E. and Hylander, W. L. (2003). Experimental evidence concerning spear use in Neandertals and modern humans. *Journal of Archaeological Science*, 30, 103–14.

Shackelford, L. L. (2005). *Regional Variation in the Postcranial Robusticity of Late Upper Paleolithic Humans*. Ph.D. Washington University.

Shackelford, L. L. (2007). Regional variation in the postcranial robusticity of Late Upper Paleolithic humans. *American Journal of Physical Anthropology*, 133, 655–68.

Shaw, C. and Stock, J. (2009). Habitual throwing and swimming correspond with upper limb diaphyseal strength and shape in modern human athletes. *American Journal of Physical Anthropology*, 140, 160–72.

Shaw, C. and Stock, J. (2011). The influence of body proportions on femoral and tibial midshaft shape in hunter-gatherers. *American Journal of Physical Anthropology*, 144, 22–9.

Smith, F. H. (1985). Continuity and change in the origin of modern Homo sapiens. *Zeitschrift fur Morphologie und Anthropologie*, 75, 197–222.

Sokal, R. R. and Rohlf, F. J. (1995). *Biometry*. Third Edition. New York, NY: W. H. Freeman and Company.

Stock, J. (2002). Climate, terrestrial mobility and the patterning of lower limb robusticity among Holocene foragers. *American Journal of Physical Anthropology*, S34, 148–149.

Stock, J. T. (2006). Hunter-gatherer postcranial robusticity relative to patterns of mobility, climatic adaptation, and selection for tissue economy. *American Journal of Physical Anthropology*, 131, 194–204.

Stock, J. and Pfeiffer, S. (2001). Linking structural variability in long bone diaphyses to habitual behaviors: Foragers from the southern African Later Stone Age and the Andaman Islands. *American Journal of Physical Anthropology*, 115, 337–48.

Stock, J. and Pfeiffer, S. (2004). Long bone robusticity and subsistence behaviour among Later Stone Age foragers of the forest and fynbos biomes of South Africa. *Journal of Archaeological Science*, 31, 999–1013.

Stock, J. and Shaw, C. (2007). Which measures of diaphyseal robusticity are robust? A comparison of external methods of quantifying the strength of long bone diaphyses to cross-sectional geometric properties. *American Journal of Physical Anthropology*, 134, 412–23.

Tafuri, M. A., Bentley, R. A., Manzi, G. and di Lernia, S. (2006). Mobility and kinship in the prehistoric Sahara: Strontium isotope analysis of Holocene human skeletons from the Acacus Mts. (southwestern Libya). *Journal of Anthropological Archaeology*, 25, 390–402.

Trinkaus, E., Churchill, S. E., Villemeur, I. *et al.* (1991). Robusticity versus shape: The functional interpretation of Neandertal appendicular morphology. *Journal of the Anthropological Society of Nippon*, 99, 257–78.

Trinkaus, E., Churchill, S. E. and Ruff, C. B. (1994). Postcranial robusticity in Homo. II: Humeral bilateral asymmetry and bone plasticity. *American Journal of Physical Anthropology*, 93, 1–34.

Warden, S. J. (2009). Breaking the rules for bone adaptation to mechanical loading. *Journal of Applied Physiology*, 100, 1441–2.

Weiss, E. (2003). Effects of rowing on humeral strength. *American Journal of Physical Anthropology*, 121, 293–302.

Wetterstrom, W. (1993). Foraging and farming in Egypt: The transition from hunting and gathering to horticulture in the Nile Valley. In *The Archaeology of Africa: Food, Metal and Towns*. London: Routledge, pp. 165–226.
Wilkinson, T. A. H. (2003). *Genesis of the Pharoahs: Dramatic New Discoveries Rewrite the Origins of Ancient Egypt*. New York, NY: Thames & Hudson.

9 *Postcranial growth and development of immature skeletons from Point Hope, Alaska*

LIBBY W. COWGILL

Introduction

The Point Hope osteological collection represents an ideal opportunity to study growth and development in a precontact Arctic community. Among the nearly 400 human skeletons recovered from the Ipiutak site and the nearby Tigara site, a large and well-preserved sample of immature individuals was recovered. The Point Hope immature sample therefore remains one of the largest samples of northern latitude immature skeletal material in the world. In general, immature remains are highly valued because they can provide a glimpse of health, fertility, demography, disease, and activity patterns in an understudied and poorly preserved segment of the bioarchaeological record. The Point Hope immature collection in particular permits not only generalized analyses of human growth and development, but also a more specific window into Arctic subsistence and biology to cold climates in particular.

The immature sample, for example, has the potential to shed light on developmental patterns in one of the most challenging climates inhabited by humans. The site of Point Hope is located 200 kilometers inside the Arctic Circle at sixty-eight degrees north latitude. Average January high temperatures are approximately −15.7 °C (3.7 °F) (Holliday and Hilton, 2010). This type of cold stress is likely to produce a variety of behavioral and biological patterns in the people at Point Hope, particularly prior to the advent of adequate cultural and technological buffering. One way we can explore biological patterns related to temperature stress is through the analysis of variation in ecogeographic body proportions. A wide body of research has successfully applied Bergmann's rule (1847) and Allen's rule (1877) to ecogeographic variation in adult human body proportions (Heirnaux and Froment, 1976; Trinkaus, 1981;

The Foragers of Point Hope: The Biology and Archaeology of Humans on the Edge of the Alaskan Arctic, eds. C. E. Hilton, B. M. Auerbach, L. W. Cowgill. Published by Cambridge University Press.

Holliday and Trinkaus, 1991; Ruff, 1991, 1994, 2002; Holliday and Falsetti, 1995; Holliday 1997a, b, 1999, 2002; Pearson, 2000; Holliday and Ruff, 2001; Weinstein, 2005; Auerbach, 2007, 2010, 2012; Temple *et al.*, 2008; Temple and Matsumura, 2011). In general, populations in high latitudes frequently display relatively wide bodies, high body masses for stature, short limbs relative to trunk length and foreshortened distal extremities, whereas populations at low latitudes have relatively narrow bodies, low body masses for stature, long limbs, and long distal limb extremities. This documented variation in adult human form is likely produced by selection for minimized surface area relative to volume for heat retention in cold climates and maximized surface area to volume for heat dispersal in warm climates. More recent studies have investigated variation in ecogeographic body proportions during ontogeny (Temple *et al.*, 2011; Cowgill *et al.*, 2012). The immature individuals from Point Hope, however, provide a view of developmental biology at the limits of human climatic tolerance.

In addition, research on the effects of subsistence shifts on immature individuals can be undertaken with the Point Hope material. As discussed elsewhere in this volume (Jensen, Chapter 2; Mason, Chapter 3), the Point Hope sites contain several cultural periods. Archaeological studies of the Ipiutak sequence indicate that these individuals relied seasonally on caribou found in more inland areas of Alaska, with deposits containing high frequencies of caribou skeletal remains, caribou-derived artifacts such as antler flint flakers, and the presence of bows and arrows that were traditionally used to hunt caribou (Nelson, 1899; Larsen and Rainey, 1948; Rainey, 1971). In contrast, Tigara archaeological artifacts point to a heavy dependence on maritime resources including walruses, seals, and whales. The Tigara component displays a high frequency of whale-hunting implements, such as large harpoons and floatation devices, and whale bones that were frequently utilized for house construction (Nelson, 1899; Larsen and Rainey, 1948; Rainey, 1971).

These two differing subsistence strategies are likely to engender different mechanical strains that can be detected via biomechanical analyses such as cross-sectional geometry. A wide variety of evidence has accrued in support of the idea that variation in mechanical loading influences diaphyseal morphology (Chamay and Tschantz, 1972; Jones *et al.*, 1977; Woo *et al.*, 1981; Carter and Beaupré, 2001; Martin, 2003), and the application of engineering principles to bone has been used extensively over the last thirty years to analyze behavioral patterns and mobility levels in past populations (Lovejoy and Trinkaus, 1980; Bridges, 1989; Ruff *et al.*, 1993; Larsen and Kelly, 1995; Churchill and Formicola, 1997; Trinkaus, 1997). Furthermore, previous research suggests differing patterns of postcranial robusticity exist among populations engaged in terrestrial versus marine mobility (Stock and Pfeiffer, 2001; Weiss, 2003), and these contrasts may also be detectable in different time periods at Point Hope.

In addition, subsistence-related activities are learned skills that develop over time during childhood and adolescence (Blurton-Jones *et al.*, 1989, 1994; Hawkes *et al.*, 1995; de Boer *et al.*, 2000; Bird and Bliege Bird, 2000, 2002, 2005; Bird-David, 2005; Tucker and Young, 2005). Recent literature has explored the complex interchange between genetic and environmental influences on bone functional adaptation during growth (Ruff *et al.*, 1994; Ruff, 2003; Cowgill, 2010; Cowgill *et al.*, 2010; Holmes and Ruff, 2011; Temple *et al.*, n.d.). While strong evidence exists that bone robusticity is linked to genetic factors (Cowgill, 2010), mechanical loading during growth also influences diaphyseal strength and shape (Ruff, 2003; Cowgill *et al.*, 2010). Therefore, it may be possible to detect differences in immature postcranial strength between Point Hope and other samples, and between different time periods at Point Hope.

In light of these considerations, the goals of this chapter will be pursued in two phases. First, patterns of ecogeographic body proportions and upper and lower limb activity levels will be evaluated in immature individuals from Point Hope relative to a large, geographically diverse set of samples from across the New and Old World. While immature individuals from Point Hope will likely display more cold-adapted body proportions than other individuals in the comparative sample, few differences should be detectable between cultural horizons. Second, variation within the Point Hope sample related to subsistence shifts between changing cultural periods will be explored. While samples are considerably smaller when considering individual cultural horizons, it may be possible to identify patterns of change in body proportions and activity patterns between the Ipiutak and Tigara periods.

Materials and methods

The sample

The primary data for this analysis consist of femoral and humeral cross-sectional properties and long bone measurements from seven Holocene human skeletal samples (Cowgill, 2010). Measurements were collected from a total of 570 immature individuals under the age of eighteen, although actual sample size may vary by analysis. The seven samples were selected to represent the broadest possible range of time periods, geographic locations, and subsistence strategies. Individuals displaying indicators of obvious developmental pathology were excluded, although observations of non-specific developmental

Table 9.1. *Sample descriptions, sizes, antiquity, latitudes, and locations*

Sample	Original location	Approx. time period	*N*	Average latitude	Sample location
California Amerindian	Northern California	500–4,600 years BP	91	~39° N	Phoebe Hearst Museum at the University of California, Berkeley (Berkeley, CA)
Dart	Johannesburg, South Africa	20th century	73	26° S	School of Medicine, University of Witwatersrand (Johannesburg, South Africa)
Indian Knoll	Green River, Kentucky	4,143–6,415 years BP	95	37° N	University of Kentucky, Lexington (Lexington, KY)
Kulubnarti	Batn el Hajar, Upper Nubia	Medieval	99	21° N	University of Colorado, Boulder (Boulder, CO)
Luis Lopes	Lisbon, Portugal	20th century	47	39° N	Bocage Museum (Lisbon, Portugal)
Mistihalj	Bosnia-Herzegovina	Medieval (15th century)	52	~43° N	Peabody Museum at Harvard University (Cambridge, MA)
Point Hope	Point Hope, Alaska	200–1,600 years BP	65	68° N	American Museum of Natural History (New York, NY)

stress (Harris lines, cribra orbitalia, porotic hyperostosis) were not considered grounds for exclusion.

While details of the comparative sample have been published elsewhere (Cowgill, 2010), and are summarized in Table 9.1, they are discussed at greater length here for additional clarity. The California Amerindian sample used in this analysis is derived from twenty-eight sites in the Alameda, Sacramento, and San Joaquin counties of north-central California, primarily clustered along the San Francisco Bay and the Sacramento and San Joaquin River valleys. California Amerindians of this area are best characterized as precontact, semi-sedentary foraging populations, reliant on deer, elk, antelope, fishing, and extensive exploitation of acorns. Indian Knoll is an Archaic Period shell-midden site located on the Green River in Kentucky (Webb, 1946). Individuals from Indian Knoll were likely semi-sedentary with prolonged residences at seasonally occupied sites, relatively high population densities, and relied heavily on a narrow spectrum of essential resources, such as deer, turkey, mussels, nuts, and a variety of locally collected plant materials (Winters, 1974). The site of Kulubnarti is located in Upper Nubia in the Batn el Hajar region, approximately 130 kilometers south of Wadi Halfa, where two medieval Christian cemeteries containing 406 burials were excavated at Kulubnarti in 1979 (Van Gerven *et al.*, 1995). With marginal subsistence, individuals traditionally lived in small villages, participated in small-scale agriculture, and likely suffered from chronic nutritional difficulty combined with bouts of infectious disease during growth (Van Gerven *et al.*, 1990). Mistihalj is a medieval burial site located on the border between Bosnia-Herzegovina and Montenegro. The remains at Mistihalj are culturally associated with the Vlakhs, an indigenous Balkan ethnic group, who primarily engaged in breeding sheep, horses, mules, and cattle, and migrated seasonally over varied terrain (Alexeeva *et al.*, 2003). The Dart Collection is an ethnically mixed, native African cadaver sample derived from hospitals in the Transvaal region in South Africa (Saunders and Devito, 1991). Approximately 74% of all individuals died prior to 1950, and approximately 92% of the individuals within this sample are Bantu-speaking South African Blacks. Due to the diversity of this region, it is difficult to classify this area as exclusively rural or urban. Finally, the Luis Lopes skeletal collection consists of twentieth-century Portuguese from several cemeteries in Lisbon. In general, the sample is best categorized as an urban population of low to middle socioeconomic status (Cardoso, 2005).

Within the Point Hope sample, variation was explored relative to two cultural periods within the site. In order to maximize sample size, the Point Hope sample was divided into early and late time periods. The early time period includes the Ipiutak and Norton cultural horizons. The late period consists of exclusively Tigara individuals. The immature sample includes twelve individuals from the early period and fifty-two individuals from the Tigara period.

Aging and sex

Age was undocumented for six of the seven samples used in this study, and crown and root formation evaluated from lateral mandibular radiographs was used whenever dental and postcranial remains were reliably associated. Crown and root formation was assessed following the developmental standards set by Smith (1991) for permanent dentition and Liversidge and Molleson (2004) for deciduous dentition. Each set of dentition was scored twice on two consecutive days, and individual teeth that produced different formation stage scores were evaluated a third time to resolve inconsistencies. When no dentition was directly associated with the postcranial remains, chronological age was predicted from within-sample least squares regression of femoral, tibial, or humeral length on age for each of the comparative samples in order to maximize sample size (Cowgill, 2010). By developing age-prediction equations specific to each sample, difficulties arising from the application of a formula developed on individuals differing in body size or proportions to an archaeological target sample are partially mitigated. Given the well-known difficulties of determining sex in immature samples (Scheuer and Black, 2000), this was not attempted here and both males and females were analyzed together. As adult males and females differ in both body proportions and postcranial robusticity, this may possibly bias results.

Cross-sectional geometry

Biomechanical data for this analysis consist of the midshaft total areas, cortical areas, and polar second moment of areas of immature femora and right humeri. In addition, levels of humeral asymmetry at midshaft are also explored. Biomechanical length for unfused humeri and femora were measured following Trinkaus and colleagues (2002a, b). Cross-sectional levels were chosen to best approximate the 50% section level in fused elements. For humeri, 50% of intermetaphyseal biomechanical length was used, as the proximal and distal epiphyses contribute approximately equally in length to the measurement of biomechanical length in fused elements (Ruff, personal communication). In immature femora, however, 50% of diaphyseal length was calculated as 45.5% of femoral intermetaphyseal length, as this measurement best corresponds to the location of the 50% level in individuals with fused distal femoral epiphyses due to the relatively larger contribution of the distal epiphysis to biomechanical length in fused femora (Ruff, 2003). Levels of bilateral asymmetry in humeral total area, cortical area, and polar second moments of area were determined as the difference between the larger and smaller side, expressed as a percentage ([(max – min)/min] × 100) (Trinkaus *et al.*, 1994).

All cross-sectional properties were collected using a method similar to O'Neill and Ruff's (2004) "latex cast method" (LCM) and the method used by Sakaue (1998), which rely on anteroposterior and mediolateral radiographs and silicone molding putty. In order to reconstruct the femoral and humeral cross-sectional properties, the external surface of the diaphysis was molded with Cuttersil Putty Plus™ silicone molding putty. Anterior, posterior, medial, and lateral cortical bone widths were measured with digital calipers, and measurements were corrected for parallax distortion by comparing external breadths measured on the radiograph with external breaths measured on the element. Once corrected for parallax, the four cortical bone measurements were plotted onto the two-dimensional copy of the original mold, and the endosteal contours were interpolated by using the subperiosteal outline as a guide. The resultant sections were enlarged on a digitizing tablet, and the endosteal and periosteal contours digitized. Cross-sectional properties were computed from the sections in a PC-DOS version of SLICE (Nagurka and Hayes, 1980; Eschman, 1992).

Body proportions and metrics

To provide a comprehensive exploration of changes in body proportions over the course of growth, ontogenetic changes in brachial and crural indices and body mass relative to femoral length were evaluated. Stature was not predicted because of potential mismatching between immature stature prediction formulae and the target samples. Body mass was evaluated relative to femoral length in order to evaluate body weight relative to a rough approximation of stature. Maximum diaphyseal lengths were recorded from all immature humeri, radii, femora, and tibiae to the nearest millimeter. Brachial indices were calculated as radial length/humeral length × 100. Crural indices were calculated as tibial length/femoral length × 100.

Body mass

Body mass was calculated both to standardize cross-sectional properties and to evaluate relative to long bone length for interpretation of variation in body proportions. Body mass was predicted based on formulae developed specifically for immature individuals, which predict body mass from femoral distal metaphyseal M-L breadth and femoral head size (Ruff, 2007). To remove the effect of body mass on humeral and femoral cross-sectional properties, logged cross-sectional properties were regressed on logged body mass (total and cortical area) or logged body mass × beam length^2 (polar second moment of area) using least squares regression. Standardized residuals, which are the raw residuals divided by the standard deviation of residuals, were then used in comparisons of population differences.

Statistical analysis

Samples were evaluated as a whole as opposed to being divided into discrete age categories in order to increase statistical power. When examining Point Hope relative to the other comparative samples, standardized residuals were compared among samples using one-way analysis of variance (ANOVA) to establish population differences and post-hoc tests (Tukey's HSD) to identify which populations specifically differed. When the cultural periods within Point

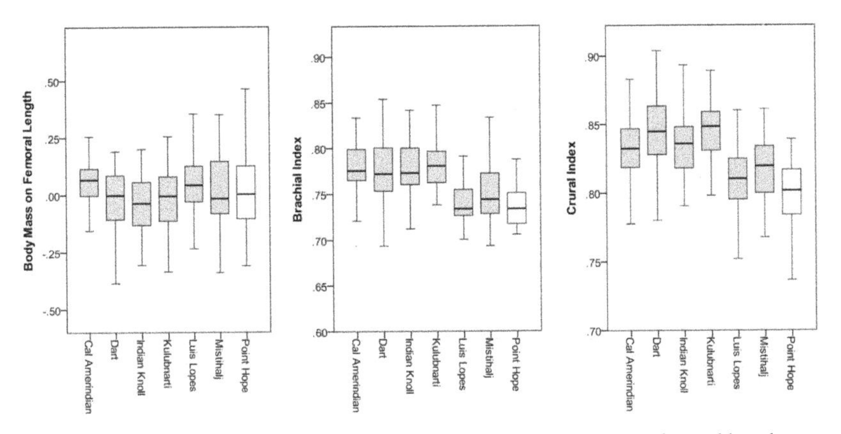

Figure 9.1. Box plots of standardized residuals for body mass on femoral length, brachial index, and crural index for Point Hope and the comparative samples.

Hope were analyzed, significantly smaller samples were employed. Therefore, Mann–Whitney *U* tests were used to identify any differences between the early and late periods.

Results

Ecogeographic body proportions across samples

Logged body mass was regressed on logged femoral length using the best fit equation and standardized residuals were evaluated via ANOVA ($p < 0.001$). Standardized residuals are shown in Figure 9.1. While significant differences do exist among the comparative samples (primarily highlighting high body mass for femoral length among California Amerindians), no significant differences exist between Point Hope and the other samples. In general, relative body mass among individuals at Point Hope is intermediate compared to the other samples.

Previous research on brachial and crural indices has shown that correlations between limb indices and age are low (Cowgill *et al.*, 2012), therefore, unaltered brachial and crural indices are analyzed here (Figure 9.1). Significant differences among populations were detected in both the upper and lower limbs ($p < 0.001$). Immature individuals from Point Hope display significantly lower brachial indices than all samples except Mistihalj and Luis Lopes, and lower crural indices than all samples except Luis Lopes ($p < 0.001$). Mean brachial and crural indices are 74 and 80, respectively.

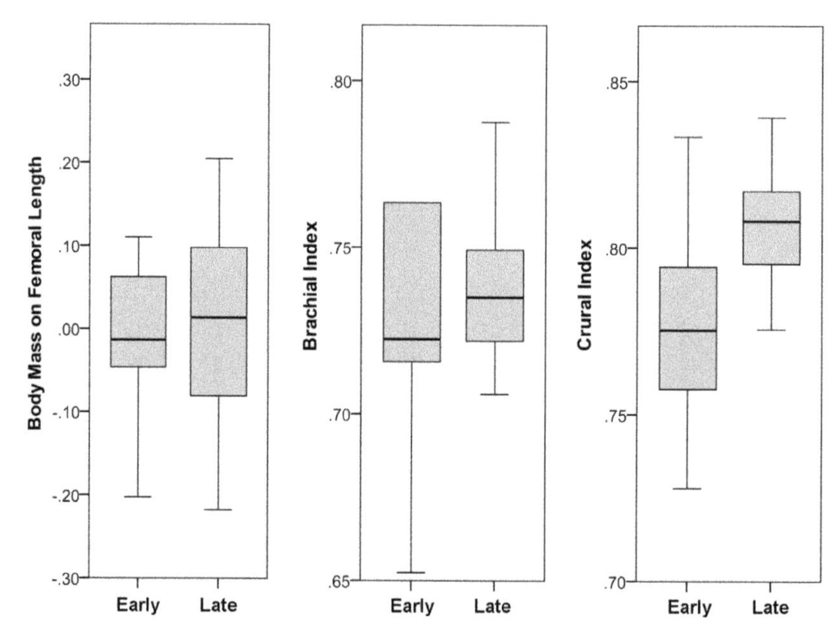

Figure 9.2. Box plots of standardized residuals for body mass on femoral length, brachial index, and crural index for the early and late temporal periods at Point Hope.

Ecogeographic body proportions within Point Hope

Logged body mass was regressed on logged femoral length using the best-fit equation and standardized residuals were evaluated with Mann–Whitney *U* tests. Box plots of standardized residuals for early and late time periods are shown in Figure 9.2. No significant differences in relative body mass exist between the two time periods. In addition, while there are no differences between time periods in brachial index, crural indices do differ significantly between early and late time periods ($p = 0.042$, early $\bar{x}$: 78, late $\bar{x}$: 80).

Cross-sectional geometry across samples

Logged cross-sectional properties were regressed on logged body mass and body mass × beam length^2, and standardized residuals were compared using ANOVA (Figures 9.3 and 9.4). Significant differences among samples were found for all cross-sectional properties of the humerus and the femur ($p < 0.001$). Post-hoc comparisons between sample pairs show that Point Hope

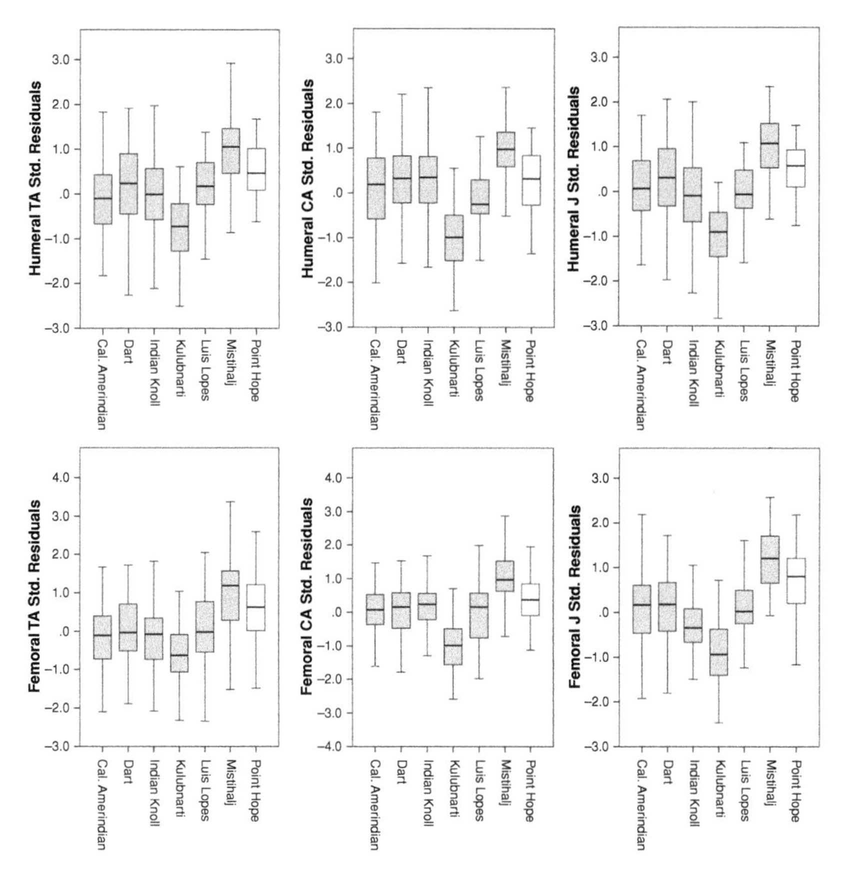

Figure 9.3. Box plots of standardized residuals for humeral and femoral total area (TA), cortical area (CA), and polar second moment of area (*J*) for Point Hope and the comparative samples.

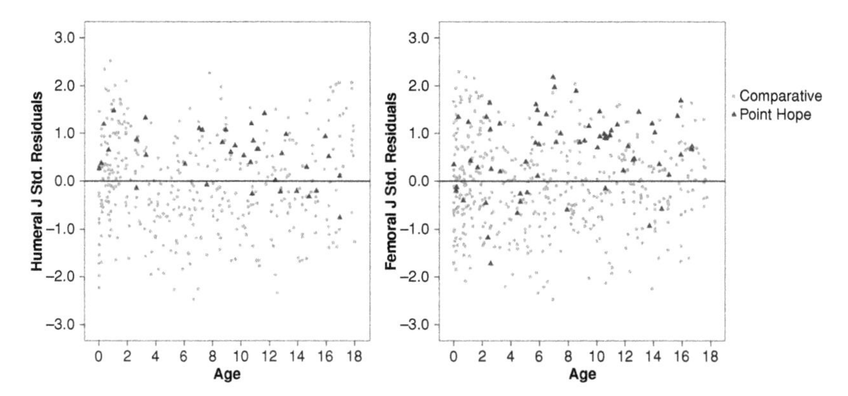

Figure 9.4. Scatter plots of standardized residuals for humeral and femoral polar second moment of area (*J*) for Point Hope and the comparative samples.

Table 9.2. *Comparison of cross-sectional properties between Point Hope and comparative samples*

	Point Hope					
	Femoral TA	Femoral CA	Femoral *J*	Humeral TA	Humeral CA	Humeral *J*
Cal. Amerindian	<0.001*	0.212	0.001*	0.001*	<0.001*	0.124
Dart	0.007*	0.184	0.004*	0.004*	<0.001*	0.783
Indian Knoll	<0.001*	0.757	<0.001*	<0.001*	<0.001*	0.010*
Kulubnarti	<0.001*	<0.001*	<0.001*	<0.001*	<0.001*	<0.001*
Luis Lopes	0.014*	0.133	0.005*	0.005*	<0.001*	0.078
Mistihalj	0.268	0.004*	0.011	0.011	0.268	0.113

* Indicate significant comparisons

humeri and femora are stronger than average, but not as robust as immature individuals from the Mistihalj sample (Table 9.2).

Cross-sectional geometry within Point Hope

Logged cross-sectional properties were regressed on logged body mass and body mass $\times$ beam length^2, and standardized residuals were compared using Mann–Whitney *U* tests (Figure 9.5). None of the comparisons of cross-sectional properties between early and late temporal periods at Point Hope significantly differed.

Humeral asymmetry across samples

Percentage differences in TA, CA, and *J* vary significantly across all groups. However, none of these differences are between individuals at Point Hope and the immature individuals in other samples (Figure 9.6). Most of the significant differences highlight the very high levels of asymmetry at Indian Knoll and the low levels at Kulubnarti. In general, levels of humeral asymmetry among individuals at Point Hope are moderate to high, but these differences do not reach significance.

Humeral asymmetry within Point Hope

Figure 9.7 shows differences in humeral asymmetry between early and late cultural horizons at Point Hope. More humeral asymmetry exists in the early

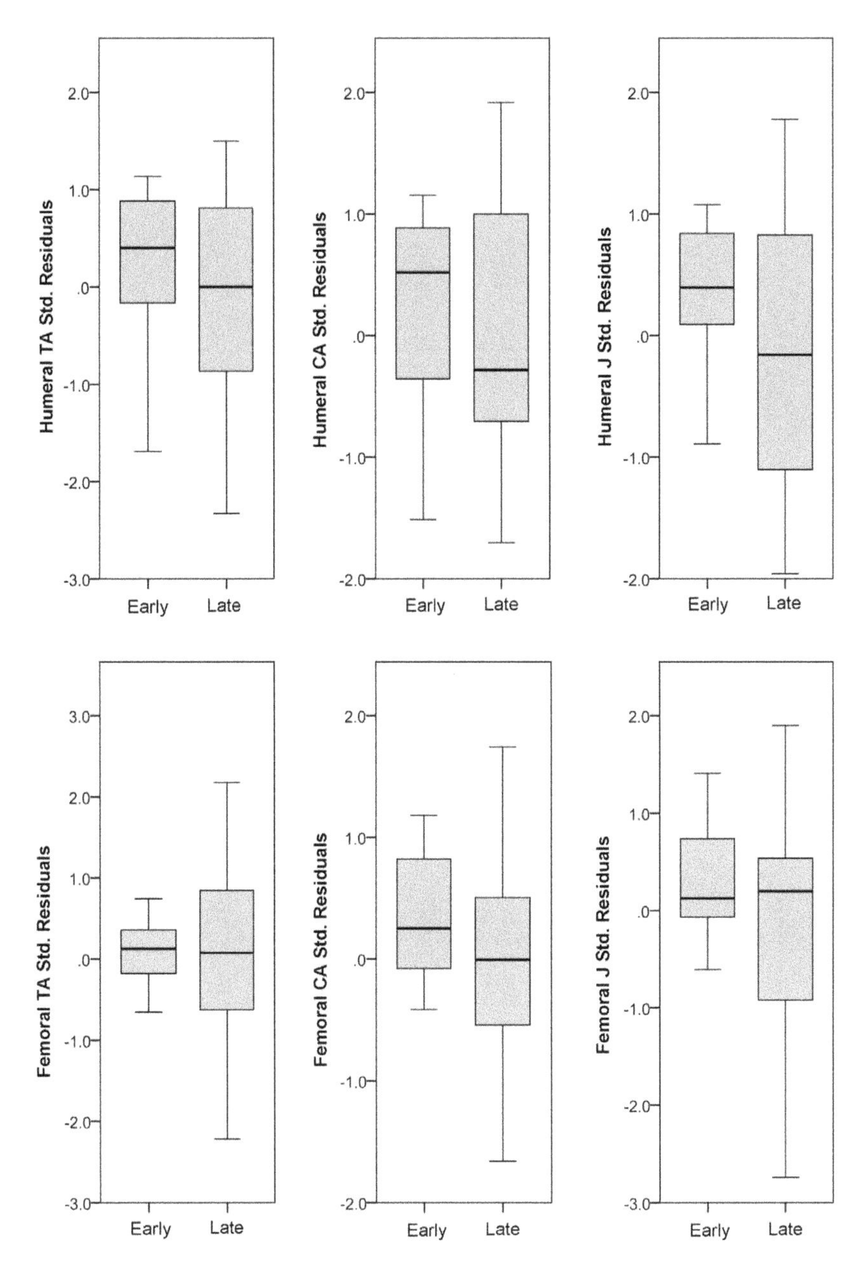

Figure 9.5. Box plots of standardized residuals for humeral and femoral total area (TA), cortical area (CA), and polar second moment of area (*J*) for the early and late temporal periods at Point Hope.

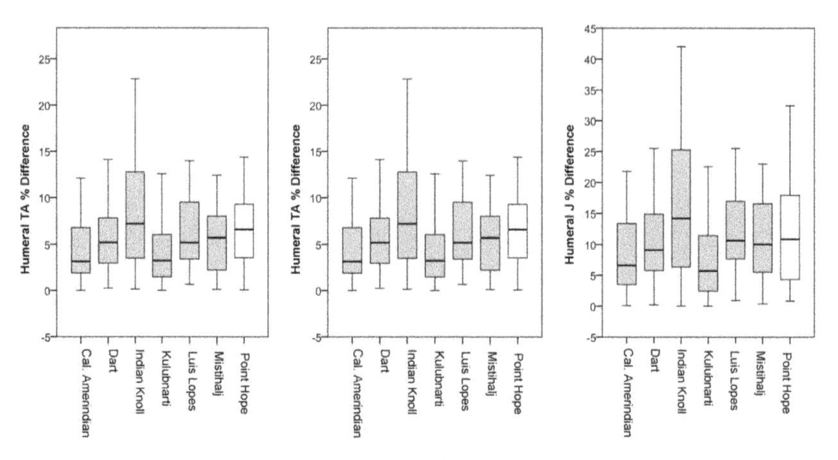

Figure 9.6. Box plots for percentage difference between sides for the humerus in total area (TA), cortical area (CA), and polar second moment of area (*J*) for Point Hope and the comparative samples.

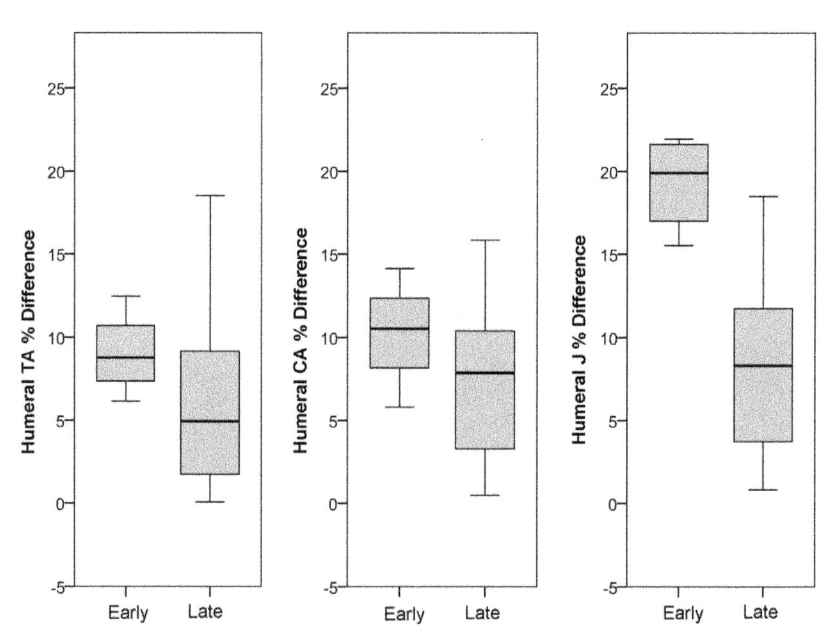

Figure 9.7. Box plots for percentage difference between sides for the humerus in total area (TA), cortical area (CA), and polar second moment of area (*J*) for the early and late temporal periods at Point Hope.

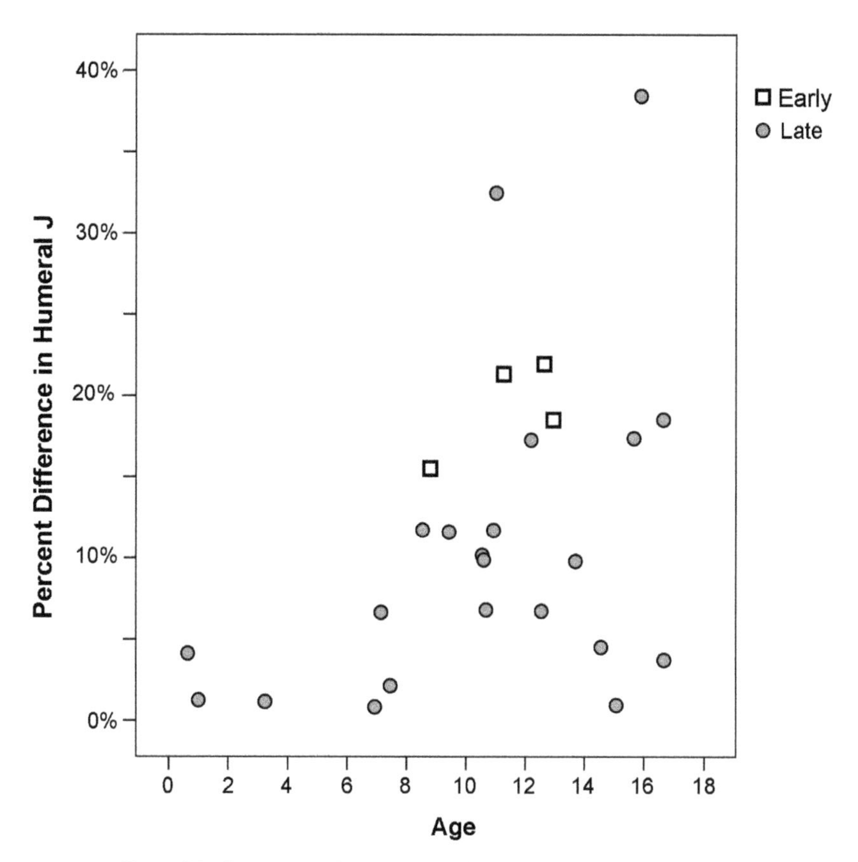

Figure 9.8. Scatter plot of percentage difference between sides for the humerus in the polar second moment of area (J) for Point Hope and the comparative samples.

time periods than the later; these differences are only significant, however, in the analysis of the polar second moment of area ($p = 0.023$, Early $\bar{x}$: 19.3%, Late $\bar{x}$: 10.4%) (Figure 9.8).

Discussion

Body proportions

Body mass relative to femoral length among immature individuals at Point Hope is intermediate compared to other individuals in the global sample. Intralimb proportions at Point Hope, however, differ strongly from the African and Native American samples, but not necessarily all European groups. While individuals from Point Hope always exhibit the lowest brachial and crural indices, mean values do not always differ from those among the Luis Lopes (Portugal) and Mistihalj (Bosnia-Herzegovina) samples. This is similar to the results reported by Holliday and Hilton (2010) in adults: Ipiutak and Tigara adults are indeed "cold adapted," but, in several cases, they were statistically indistinguishable from cold-adapted European populations, and other Arctic groups exposed to less intense cold stress.

These results also support previous research suggesting that body proportions are stable across ontogeny and that populations maintain their positions relative to other groups during growth (Temple *et al.*, 2011; Cowgill *et al.*, 2012). This implies that strong genetic correlations exist between adult body proportions and body proportions during ontogeny. While the possibility that brachial and crural indices are developmentally plastic cannot be completely excluded (plasticity in utero still remains a possibility), the stability and consistency of intralimb indices hints at genetic underpinnings. If variation in limb indices is primarily heritable, population continuity or discontinuity should be detectable.

When examining differences between the Ipiutak and Tigara individuals, all body proportions are similar between the two groups, except crural index. This is similar to Ipiutak and Tigara adults, where few differences could be detected (Holliday and Hilton, 2010; Auerbach, Chapter 10, this volume). Given lack of correction for multiple comparisons (see Cowgill, 2010, for a discussion), and the marginally significant p-value ($p = 0.042$), it is difficult to determine whether the difference in crural index between the two periods is biologically meaningful or a statistical artifact. Given the similarity in overall body proportions, continuity between the two temporal horizons cannot be ruled out. However, in light of the correlation between limb indices and climate, convergence on cold-adapted morphology is equally plausible if the two populations represent discrete migratory groups.

Cross-sectional geometry

Immature individuals at Point Hope are relatively robust compared to other immature individuals in the comparative sample. In addition, there is a wide range of variation among comparative samples in postcranial strength, although patterns are basically similar upper and lower limb. Mistihalj shows very high levels of upper and lower limb strength; very low levels of both are found in individuals from Kulubnarti. Point Hope is second to only Mistihalj in overall levels of postcranial strength. These results are consistent with Shackelford (Chapter 8, this volume), who interprets the high levels of postcranial strength as the result of mechanical loading from active lifestyles in an Arctic foraging context. However, levels of robusticity, like body proportions, maintain the same relative level compared to the other samples from birth to age eighteen (Figure 9.4). This pattern paints a somewhat more complex picture. If robust populations are robust prior to the onset of subsistence-related activity, the level of postcranial strength documented in adult populations must be at least partially the product of maternal environment and/or genetic propensity for increased bone deposition.

Although the samples are small, the lack of differences between early and late periods at Point Hope highlight this complexity. In a biomechanical sense, there are large differences between caribou hunting and whaling. However, in both adults and juveniles, this difference is not detectable via cross-sectional geometry. This suggests several possible explanations. First, sample sizes in this analysis may be too small. But second, it may be possible that, while cross-sectional geometry is adequate for detecting broad differences in activity levels, noise from nutritional status, genetic background, population structure, and maternal health swamp the signal of activity patterns at finer grades of resolution.

Lastly, few significant differences exist in levels of humeral asymmetry between Point Hope and other samples. The two temporal periods at Point Hope do differ in the second polar moment of area, with immature individuals from the Tigara period appearing to have lower levels of humeral asymmetry than individuals from the earlier periods. Again, this pattern is suspect due to small sample size. Previous research has indicated that while only very low correlations exist between age and humeral asymmetry, most of the individuals showing high levels of activity-related humeral asymmetry are older children and adolescents (Cowgill, 2010). Closer visual inspection of the data used here shows that there are only four individuals from the early phases of Point Hope and all of them are over the age of eight (Figure 9.8). In addition, the relatively larger number of young individuals from the Tigara subsample may be driving a statistically significant result that does not mirror biological reality.

Conclusions

Immature individuals from the site of Point Hope are characterized by intermediate weight for long bone length, low brachial and crural indices, high levels of postcranial strength, and moderate levels of humeral asymmetry. In many ways, the biological patterns detected in young individuals are similar to those that have been documented in adults. This highlights the fact that both biological and behavioral adaptations occur throughout the lifespan, and that biomechanical and climatic stress are not limited to the adult phases of the life cycle. However, the very early onset of these characteristics, particularly the high levels of postcranial strength detected early in life, hint at more complex mechanisms for the production of long bone strength than have been previously considered under the paradigm of functional adaptation.

References

Alexeeva, T. I., Bogatenkov, D. V. and Lebedinskaya, G. V. (2003). *Anthropology of Medieval Vlakhs in Comparative Study: On Date of Mistikhaly Burial Site.* Moscow: Scientific World (in Russian, summary in English).

Allen, J. A. (1877). The influence of physical conditions on the genesis of species. *Radical Review*, 1, 108–40.

Auerbach, B. M. (2007). *Human Skeletal Variation in the New World During the Holocene: Effects of Climate and Subsistence Across Geography and Time*. Ph.D. Johns Hopkins University School of Medicine.

Auerbach, B. M. (2010). Reaching great heights? Morphological variation and migration on the Great Plains. In *Human Variation in the Americas: The Integration of Archaeology and Biological Anthropology*. Carbondale, IL: Center for Archaeological Investigations, pp. 172–214.

Auerbach, B. M. (2012). Skeletal variation among early Holocene North American humans: Implications for origins and diversity in the Americas. *American Journal of Physical Anthropology*, 149, 525–36.

Bergmann, C. (1847). Ueber die Verhältnisse der wärmeökonomie der Thiere zu ihrer Grösse. *Göttinger Studien*, 3, 595–708.

Bird, D. W. and Bliege Bird, R. (2000). The ethnoarchaeology of juvenile foragers: Shellfishing strategies among Meriam children. *Journal of Anthropological Archaeology*, 19, 461–76.

Bird, D. W. and Bliege Bird, R. (2002). Children on the reef: Slow learning or strategic foraging? *Human Nature,* 13, 269–97.

Bird, D. W and Bliege Bird, R. (2005). Mardu children's hunting strategies in the western desert, Australia. In *Hunter-Gatherer Childhoods: Evolutionary, Developmental and Cultural Perspectives*. New Brunswick, NJ: Aldine, pp. 129–46.

Bird-David, N. (2005). Studying children in hunter-gatherer societies: Reflections from a Nayaka perspective. In *Hunter-Gatherer Childhoods: Evolutionary, Developmental and Cultural Perspectives*. New Brunswick, NJ: Aldine, pp. 92–101.

Blurton Jones, N. G., Hawkes, K. and O'Connell, J. F. (1989). Modeling and measuring costs of children in two foraging societies. In *Comparative Socioecology: The Behavioural Ecology of Humans and Other Mammals*. London: Basil Blackwell, pp. 367–90.

Blurton Jones, N. G., Hawkes, K. and Draper, P. (1994). Differences between Hadza and !Kung children's work: Original affluence or practical reason. In *Key Issues in Hunter-Gatherer Research*. Providence, RI: Berg, pp. 189–215.

Bridges, P. S. (1989). Changes in activities with the shift to agriculture in the southeastern United States. *Current Anthropology*, 30, 385–94.

Cardoso, H. F. V. (2005). *Patterns of Growth and Development of the Human Skeleton and Dentition in Relation to Environmental Quality: A Biocultural Analysis of a Sample of 20th Century Portuguese Subadult Documented Skeletons*. Ph.D. McMaster University.

Carter, D. R and Beaupré, G. S. (2001). *Skeletal Function and Form*. Cambridge: Cambridge University Press.

Chamay, A. and Tschantz, P. (1972). Mechanical influences in bone remodeling: Experimental research on Wolff's Law. *Journal of Biomechanics*, 5, 173–80.

Churchill, S. and Formicola, V. (1997). A case of marked bilateral asymmetry in the upper limbs of an Upper Paleolithic male from Barma Grande (Liguria), Italy. *International Journal of Osteoarchaeology*, 7, 18–38.

Cowgill, L. W. (2010). The ontogeny of Holocene and Late Pleistocene human postcranial strength. *American Journal of Physical Anthropology*, 141, 16–37.

Cowgill, L. W., Warrener, A., Pontzer, H. and Ocobock, C. (2010). Waddling and toddling: The biomechanical effects of an immature gait. *American Journal of Physical Anthropology*, 143, 52–61.

Cowgill, L. W., Eleazer, C. D., Auerbach, B. M., Temple, D. H. and Okazaki, K. (2012). Developmental variation in ecogeographic body proportions. *American Journal of Physical Anthropology*, 148, 557–70.

de Boer, W. F., Blijdenstein, A. F. and Longamane, F. (2000). Prey choice, habitat choice, and timing of people exploiting intertial resources, explained with optimal foraging models. In *Between the Tides: The Impact of Human Exploitation on an Intertidal Ecosystem, Mozambique*. San Clemente, CA: Universal Press, pp. 63–88.

Eschman, P. N. (1992). *SLCOMM Version 1.6*. Albuquerque, NM: Eschman Archaeological Services.

Hawkes, K., O'Connell, J. F. and Blurton Jones, N. G. (1995). Hadza children's foraging: Juvenile dependency, social arrangements, and mobility among hunter-gatherers. *Current Anthropology*, 36, 688–700.

Hiernaux, J. and Froment, A. (1976). The correlations between anthropobiological and climatic variables in sub-Saharan Africa: Revised estimates. *Human Biology*, 48, 757–67.

Holliday, T. W. (1997a). Postcranial evidence of cold adaptation in European Neandertals. *American Journal of Physical Anthropology*, 104, 245–58.

Holliday, T. W. (1997b). Body proportions in Late Pleistocene Europe and modern human origins. *Journal of Human Evolution*, 32, 423–47.

Holliday, T. W. (1999). Brachial and crural indices of European Late Upper Paleolithic and Mesolithic Humans. *Journal of Human Evolution*, 36, 549–66.

Holliday, T. W. (2002). Body size and postcranial robusticity of European Upper Paleolithic hominins. *Journal of Human Evolution*, 43, 513–28.

Holliday, T. W. and Falsetti, A. B. (1995). Lower limb length of European early modern humans in relation to mobility and climate. *Journal of Human Evolution*, 29, 141–53.

Holliday, T. W. and Hilton, C. E. (2010). Body proportions of circumpolar peoples as evidenced from skeletal data: Ipiutak and Tigara (Point Hope) versus Kodiak Island Inuit. *American Journal of Physical Anthropology*, 142, 287–302.

Holliday, T. W. and Ruff, C. B. (2001). Relative variation in the human proximal and distal limb segments. *American Journal of Physical Anthropology*, 116, 26–34.

Holliday, T. W. and Trinkaus, E. (1991). Limb/trunk proportions in Neandertals and early anatomically modern humans. *American Journal of Physical Anthropology*, S12, 93–94.

Holmes, M. A. and Ruff, C. B. (2011). Dietary effects on development of the human mandibular corpus. *American Journal of Physical Anthropology*, 145, 615–28.

Jones, H. N., Priest, J. D., Hayes, W. C., Tichenor, C. C. and Nagel, D. A. (1977). Humeral hypertrophy in response to exercise. *Journal of Bone and Joint Surgery*, 59A, 204–8.

Larsen, C. S. and Kelly, R. L. (1995). *Bioarchaeology of the Stillwater Marsh: Prehistoric Human Adaptation in the Western Great Basin*. New York, NY: American Museum of Natural History.

Larsen, H. and Rainey, F. (1948). *Ipiutak and the Arctic Whale Hunting Culture*. Anthropological Papers of the American Museum of Natural History 42. New York, NY: American Museum of Natural History.

Liversidge, H. and Molleson, T. (2004). Variation in crown and root formation and eruption of human deciduous teeth. *American Journal of Physical Anthropology*, 123, 172–80.

Lovejoy, C. O. and Trinkaus, E. (1980). Strength and robusticity of the Neandertal tibia. *American Journal of Physical Anthropology*, 53, 465–70.

Martin, R. B. (2003). Functional adaptation and fragility of the skeleton. In *Bone Loss and Osteoporosis: An Anthropological Perspective*. New York, NY: Kluwer Academic/Plenum Publishers, pp. 121–38.

Nagurka, M. L. and Hayes, W. C. (1980). An interactive graphics package for calculating cross sectional properties of complex shapes. *Journal of Biomechanics*, 13, 59–64.

Nelson, E. W. (1899). The Eskimo about the Bering Strait. *Annual Report of the Bureau of American Ethnology*, 18, 3–518.

O'Neill, M. C. and Ruff, C. B. (2004). Estimating human long bone cross-sectional geometric properties: A comparison of noninvasive methods. *Journal of Human Evolution*, 47, 221–35.

Pearson, O. M. (2000). Activity, climate, and postcranial robusticity: Implications for modern human origins and scenarios of adaptive change. *Current Anthropology*, 47, 63–99.

Rainey F. 1971. The Ipiutak Culture: Excavations at Point Hope, Alaska. *Addison-Wesley Modular Publications in Anthropology* 8, Reading, MA: Addison-Wesley, pp. 1–32.

Ruff, C. B. (1991). Climate and body shape in hominid evolution. *Journal of Human Evolution*, 21, 81–105.

Ruff, C. B. (1994). Morphological adaptation to climate in modern and fossil hominids. *Yearbook of Physical Anthropology*, 37, 65–107.
Ruff, C. B. (2002). Variation in human body size and shape. *Annual Review of Anthropology*, 31, 211–32.
Ruff, C. B. (2003). Growth in bone strength, body size, and muscle size in a juvenile longitudinal sample. *Bone*, 33, 317–29.
Ruff, C. B. (2007). Body size prediction from juvenile skeletal remains. *American Journal of Physical Anthropology*, 133, 698–716.
Ruff, C. B., Trinkaus, E., Walker, A. and Larsen, C. S. (1993). Postcranial robusticity in *Homo*. I: Temporal trends and mechanical interpretation. *American Journal of Physical Anthropology*, 91, 21–53.
Ruff, C. B., Walker, A. and Trinkaus, E. (1994). Postcranial robusticity in *Homo*. III: Ontogeny. *American Journal of Physical Anthropology*, 93, 35–54.
Sakaue, K. (1998). Bilateral asymmetry of the humerus in Jomon people and modern Japanese. *Anthropological Science*, 105, 231–46.
Saunders, S. R. and DeVito, C. (1991). Subadult skeletons in the Raymond Dart Anatomical Collection: Research potential. *Human Evolution*, 6, 421–34.
Scheuer, L. and Black, S. (2000). *Developmental Juvenile Osteology*. London: Elsevier Academic Press.
Smith, B. H. (1991). Standards of tooth formation and dental age assessment. In *Advances in Dental Anthropology*. New York, NY: Wiley-Liss, pp. 143–68.
Stock, J. and Pfeiffer, S. (2001). Linking structural variability in long bone diaphyses to habitual behaviors: Foragers from the southern African Later Stone Age and the Andaman Islands. *American Journal of Physical Anthropology*, 115, 337–48.
Temple, D. H. and Matsumura, H. (2011). Do body proportions among prehistoric foragers from Hokkaido conform to ecogeographic expectations? Evolutionary implications of size and shape variation among northerly hunter-gatherers. *International Journal of Osteoarchaeology*, 21, 268–82.
Temple, D. H., Auerbach, B. M., Nakatsukasa, M., Sciulli, P. W. and Larsen, C. S. (2008). Variation in limb proportions between Jomon foragers and Yayoi agriculturalists. *American Journal of Physical Anthropology*, 137, 164–74.
Temple, D. H., Okazaki, K. and Cowgill, L. W. (2011). Ontogeny of limb proportions in late through final Jomon period foragers. *American Journal of Physical Anthropology*, 145, 415–25.
Temple, D. H., Cowgill, L. W. and Nakatsukasa, M. (n.d.). Developmental patterning in long bone diaphyseal robusticity among late/final Jomon period foragers. *Anthropological Science*. In press.
Trinkaus, E. (1981). Neanderthal limb proportions and cold adaptation. In *Aspects of Human Evolution*. London: Taylor and Francis, pp. 187–224.
Trinkaus, E. (1997). Appendicular robusticity and the paleobiology of modern human emergence. *Proceedings of the National Academy of Sciences USA*, 94, 13367–73.
Trinkaus, E., Churchill, S. E. and Ruff, C. B. (1994). Postcranial robusticity in *Homo*. II: Humeral bilateral asymmetry and bone plasticity. *American Journal of Physical Anthropology*, 93, 1–34.
Trinkaus, E., Ruff, C. B., Esteves, F. *et al.* (2002a). The lower limb remains. In *Portrait of the Artist as a Child*. Lisbon: Instituto Português de Arqueologia, pp. 435–65.

Trinkaus, E., Ruff, C. B., Esteves, F. *et al.* (2002b). The upper limb remains. In *Portrait of the Artist as a Child.* Lisbon: Instituto Português de Arqueologia, pp. 466–88.
Tucker, B. and Young, A. G. (2005). Growing up Mikea: Children's time allocation and tuber foraging in southwestern Madagascar. In *Hunter-Gatherer Childhoods: Evolutionary, Developmental and Cultural Perspectives.* New Brunswick, NJ: Aldine, pp. 147–71.
Van Gerven, D. P., Beck, R. and Hummert, J. R. (1990). Patterns of enamel hypoplasia in two medieval populations from Nubia's Batn el Hajar. *American Journal of Physical Anthropology*, 82, 413–20.
Van Gerven, D. P., Sheridan, S. G. and Adams, W. Y. (1995). The health and nutrition of a medieval Nubian population: The impact of political and economic change. *American Anthropologist*, 97, 468–80.
Webb, W. S. (1946). *Indian Knoll, Site Oh 2, Ohio County, Kentucky.* The University of Kentucky Reports in Anthropology and Archaeology. Lexington, KT: Department of Anthropology, pp. 111–365.
Weinstein, K. J. (2005). Body proportions in Ancient Andean from high and low altitudes. *American Journal of Physical Anthropology*, 128, 569–85.
Weiss, E. (2003). Effects of rowing on humeral strength. *American Journal of Physical Anthropology*, 121, 293–302.
Winters, H. D. (1974). Introduction to the new edition. In *Indian Knoll.* Knoxville, TN: University of Tennessee Press.
Woo, S. L. Y., Kuei, S. C., Amiel, D. *et al.* (1981). The effect of prolonged physical training on the properties of long bone: A study of Wolff's law. *Journal of Bone and Joint Surgery*, 63A, 780–87.

Part III

Contexts, conclusions, and commentaries

10 *Morphologies from the edge: Perspectives on biological variation among the late Holocene inhabitants of the northwestern North American Arctic*

BENJAMIN M. AUERBACH

Introduction

The past inhabitants of Point Hope – members of the Ipiutak, Norton, and Tigara cultures – present phenotypes associated with other human populations living in high latitudes. This was presumably the result of centuries of occupying the western North American Arctic and common descent from populations that dwelt in the Arctic for millennia. It is unquestionable that the lifeways of these peoples allowed them to adapt (*sensu lato*) to and survive in the extreme climates and subsistence ecologies encountered on the Lisburne Peninsula (Larsen and Rainey, 1948; Van Stone, 1962). Yet, even with the aid of specialized artificial shelter, insulating clothing, carefully controlled fire, and honed subsistence strategies, among other cultural practices, these peoples faced potential environmental pressures that would have shaped their morphology and physiology (see contributions by Cowgill, Hilton *et al.*, and Shakelford, this volume; Edholm and Lewis, 1964; Laughlin, 1979; Moran, 2008; Holliday and Hilton, 2010).

As argued in the contributions to this volume, extensive information may be gleaned about these peoples' acclimatization to life on the Arctic edge through analysis of the skeletal remains recovered from the western edge of the Lisburne Peninsula between 1939 and 1941 by Larsen, Rainey, Giddings, and Shapiro. In addition, the collective research presented in this volume continues to resolve questions concerning the affinities of the two major cultures represented at Point Hope, the Ipiutak and the Tigara. Much of this research is new, as almost

The Foragers of Point Hope: The Biology and Archaeology of Humans on the Edge of the Alaskan Arctic, eds. C. E. Hilton, B. M. Auerbach, L. W. Cowgill. Published by Cambridge University Press.

all analyses of the skeletal remains from Point Hope have only been published over the last thirty years (Tattersall and Thomas, Foreword, this volume). This chapter sets out to synthesize the biological analyses presented in this volume with these previously published studies, while providing some broader context about the phenotypic variation present at Point Hope relative to the rest of North America.

To this end, two sets of questions are asked about the inhabitants of Point Hope in this volume:

(1) The first set is concerned with the origins of the Ipiutak and the Tigara, as well as the genetic relationships between the practitioners of these two cultures. As pointed out by Maley (Chapter 4, this volume), the common recent ancestry of all North American Arctic groups complicates ascertaining affinities between Arctic groups, some of which diverged up to 6,000 years ago (Zlojutro *et al.*, 2006; Crawford, 2007; Crawford *et al.*, 2010), and many of which have much more recent common ancestry (Helgason *et al.*, 2006) or obscured affinities due to gene flow with Siberian populations (Rubicz *et al.*, 2010). Yet ascertaining genetic relationships between the practitioners of the Ipiutak and Tigara cultures is crucial to understanding the general site occupation history at Point Hope. Most publications, such as Charles Utermohle's dissertation (1984) and Holliday and Hilton's analysis (2010), have used morphometric dimensions to examine these relationships. Cowgill's contribution and Maley's contribution to this volume especially add to this area of research.

(2) The second set of questions follows the ascertainment of ancestry, focusing study on the morphological evidence that reflects similarities and differences in the lifeways of the individuals living at Point Hope during the two major cultural phases. This is a much broader undertaking, ranging from examinations of disease and stress (e.g., stature, carious lesions, skeletal lesions, and disease processes) to analyses of trauma (including skeletal signs associated with warfare) to investigations into activity patterns as reflected by skeletal strength. It is notable that most published research to date has focused on these questions, namely, the research by Raymond Costa, Debbie Guatelli-Steinberg, and Gretchen Dabbs, among others (see the Foreword to this volume by Tattersall and Thomas). The preponderance of contributions to this volume, likewise, focus on these topics; contributions by El Zaatari, by Hilton *et al.*, by Krueger, and by Shackelford all are concerned with documenting differences in the lifeways between members of the Ipiutak and Tigara cultures.

The following pages, then, consider these two sets of questions and argue what present biological information has been able to resolve in addressing them. Articulating the results of biological anthropology studies of human remains from Point Hope with archaeological data is difficult, if only because these fields are concerned with evidence that is inherently of different scales (Sassaman, 2010). Thus, this contribution to the volume does not attempt to reconcile archaeological questions with biological data, though original research chapters in this volume do this to varying amounts. The discussion of how the biological data relate to archaeological evidence is addressed by the contributions by Fitzhugh and by Dumond, which follow this chapter. In this chapter, rather, I argue that a substantial amount of knowledge has been gleaned about the past human populations living at the edge of the western Alaskan Arctic by examining the skeletal remains from Point Hope in the context of regional archaeology, as well as in comparison with the skeletons obtained from other sites. While future research is advocated, the resolution obtained to date should productively hone the work of future researchers.

Phenotypic variation and evolutionary forces

It is important to provide an overview of human phenotypic variation in the context of evolutionary forces before reviewing the implications of research into the skeletal biology and morphology of the inhabitants of Point Hope. Many authors (see the following two subsections), including those contributing original research to this volume, invoke evolutionary forces and environmental factors as explanations for phenotypic patterns documented spatiotemporally within and among human groups. Indeed, the differences observed between groups ultimately must be the products of either stochastic or directional evolution, even in the case of population replacement (the superseding population differs due to evolutionary mechanisms encountered ancestrally). Admittedly, any consideration of the influences of evolutionary factors on modern human morphological variation requires more space than this subsection provides. Thus, readers are encouraged to seek out more thorough considerations of the relationship between evolutionary mechanisms, factors, and phenotypic effects.

It is debatable whether the evidence provided, however, warrants some of the arguments made in the broader literature. Studies seldom use testable models of evolutionary forces – namely genetic drift and natural selection (i.e., adaptation) – to explain variation present in the skeleton (Betti *et al.*, 2012; Roseman and Auerbach, n.d.). That is, while morphological differences may be observed between two groups, such as the Ipiutak and Tigara, it is not possible to definitively evoke evolutionary causation without a set of expectations for how drift or selection would shape these phenotypes, followed by examinations of whether the observed phenotypes deviate from expected, modeled patterns.

I raise this caveat here because most current knowledge about the "adaptive" nature of any phenotype has been determined indirectly (Roseman and Auerbach, n.d.). The hypotheses that have been proposed about evolution and human phenotypic diversity are the product of the analysis of modern human morphological variation through metric measurements. These metric differences are related to factors associated with environmental variables and population relationships. Correlations between factors and variation in measurements are assessed as indications of evolutionary relationships, either by direct causation or indirectly, perhaps associated with a latent variable. In none of these cases, however, does the researcher know how strong a correlation must be to signify drift or selection.

However, research that employs explicit genetic[1] models is a more direct test for the effects of genetic drift versus natural selection in shaping phenotypes. Examples of this research from the last decade include Betti *et al.* (2012, 2013), Roseman *et al.* (2010), von Cramon-Taubadel (2009), von Cramon-Taubadel and Weaver (2009), Relethford (2009), Weaver (2009), Roseman and Weaver (2007), and Relethford (2004). (It is important to note that most of these papers have focused on evolution in the cranium; few studies have examined morphologies from the rest of the skeleton.) This approach, unfortunately, is beyond the scope of the research conducted to date within the western North American Arctic, with a notable exception in Maley's contribution to this volume (Chapter 4). Therefore, the definitive evolutionary forces that have shaped biological diversity in the Arctic cannot be invoked at this time, though arguments made by authors about evolutionary forces are a basis for future analyses that take the genetic model approach.

The biology of the inhabitants of Point Hope

Prior arguments about the biology of the past inhabitants of Point Hope

As summarized in the Foreword to this volume, and noted in the preceding pages, there have been few explicit studies into human variation at Point Hope as reflected by skeletal studies. Although research into the origins and lifeways of past groups of humans living at Point Hope has been limited, authors have used the skeletons from the Ipiutak and Tigara cultural period burials to assess more general questions about the morphology of circumpolar peoples. Previous studies that have utilized the skeletons of these two populations as representative of polar human morphology examined body size and shape (e.g., Auerbach and Ruff, 2004; Ruff *et al.*, 2005; Holliday and Hilton, 2010; Auerbach, 2012), though the Point Hope skeletons have not been the only skeletal sample employed for this purpose (e.g., Ruff, 1994; Holliday, 1997; Harvati and Weaver, 2006). Collectively, these studies have demonstrated that both the Ipiutak and Tigara period skeletons exhibit the morphologies typically ascribed to human populations living at high latitudes: wide body breadths, short lower limb lengths relative to torso lengths, and low intralimb proportions (i.e., brachial and crural indices). With the exception of the research presented by Holliday and Hilton (2010), no analysis of differences in body size and proportions directly compared the two major archaeological occupations. Their findings are considered in more detail in the next section ("Point Hope in the context of North American biological variation").

Some previous studies have attempted to ascertain the ancestral relationships of the populations living at Point Hope during the Ipiutak and Tigara periods. Given the overall similarity in their morphologies (i.e., body size and proportions), there is no reason to argue for unique origins for these peoples. Yet, as reviewed by Jensen in this volume (Chapter 2), and as argued by Mason elsewhere (e.g., Mason, 1998, 2000), there is archaeological evidence that the contemporaneous Birnirk cultural peoples (who likely gave rise to the Tigara culture) and Ipiutak peoples were not the same population, and may even have engaged in warfare. This would not have negated gene flow between the groups, but it would have impeded it. The distinction of the two major Point Hope time periods was borne out of the craniometric analyses of Utermohle (1984), though his analysis was unable to argue against biological continuity between the populations associated with the Ipiutak and Tigara cultures. Rather, his study showed a nuanced difference, where the Tigara had a stronger morphological affinity with the Birnirk culture skeletons buried at Point Barrow, Alaska, than the Ipiutak. The Ipiutak, in contrast, had slightly higher affinities with more recent Iñupiat groups from southwestern Alaska, as well as an eastern Siberian population sample, to Utermohle's surprise. His study argued against the significant differences observed in the analyses of Point Hope crania presented in Debetz's 1959 paper. Before the analyses by Maley (see Chapter 4), however, no additional studies sought to determine the biological relationships between the cultural groups who had occupied Point Hope.

The majority of studies on the skeletons from Point Hope, in contrast, have focused on their dentition and mastication. The various studies by Costa (1980a, 1980b, 1982) demonstrated that sex and age differences existed in the incidence of anterior dental attrition, occlusal wear patterns, and caries frequencies between the Ipiutak period skeletons and Tigara period skeletons; in general, the Ipiutak had higher incidences of periodontal disease and more anterior dental attrition and loss. He attributed dietary and possible genetic differences as the causes for these differences. A more recent study by Keenleyside (1998) found the same patterns, though diet was emphasized as the primary factor leading to differences in oral health, with possible contributions from anterior dental crowding. In both cultural periods, based on a broad study by Antón *et al.* (2010), the inhabitants of Point Hope had mandibular corpuses with higher resistances to mechanical bending than other modern human groups, which again argued for influences from diet and/or inherited size and strength. Further evidence provided by El Zaatari's (2008) examination of dental microwear patterns supported the importance of dietary differences between the Ipiutak period inhabitants and the later Tigara period inhabitants, distinctions considered in more detail both in her contribution to this volume and in the following pages. More recently, Holmes and Ruff (2011) measured the cross-sectional geometric properties of the Tigara period juvenile and adult skeletons, comparing these with Coalescent Period Arikara skeletons from the Missouri River drainage basin. Their study results concurred with the strength property patterns reported by Antón *et al.* (2010), but added that the mandibles of the Tigara skeletons were not born with higher loading resistances; rather, individuals living during the Tigara period were argued to show an ontogenetic development of high mandibular strength.[2]

Overall, then, these previous studies of dentition and the mandibles develop the most complete understanding to date of differences between the Ipiutak and Tigara period inhabitants of Point Hope. Both populations experienced higher mechanical loading in association with mastication than other modern preindustrial human groups, though the specific dietary components that contributed to these differed. Most authors did not associate specific dietary distinctions with differences in dental wear patterns or periodontal disease (cf. El Zaatari, 2008), even though archaeological evidence has strongly argued for differences in prey choice and hunting practices between the Ipiutak population and Tigara population at Point Hope (Jensen, Chapter 2, this volume; Mason, Chapter 3, this volume). Parenthetically, it is curious that only Costa (1980b) considered the use of non-dietary mastication as a factor, such as the employment of dentition as a tool for hide working or sinew pulling, though these are difficult to distinguish from deliberate dental ablation (Blakely and Beck, 1984; but see Krueger and Ungar, 2012, and Krueger, Chapter 5, this volume); in fact, before this volume, no author ascribed specific causes to the patterns of dental attrition and wear.

In addition to the groups of studies reviewed here, a handful of analyses examined morbidity and health using skeletons from Point Hope. Few of these explicitly compared the Ipiutak period and Tigara period populations, however. A few studies have reported the presence of skeletal evidence for tuberculosis among the Tigara cultural period inhabitants (e.g., Gossett and Hilton, 2004; Dabbs, 2009, 2011), without reporting evidence for the presence of the infection among the Ipiutak period skeletons (cf. Dabbs, 2009). Lester and Shapiro (1968) reported high incidences of spondylolysis among the Tigara period skeletons but few among the Ipiutak period skeletons. Merbs (1996) echoed this temporal pattern in broader analyses of the Inuit that included more recent Iñupiat Point Hope skeletons (curated at the National Museum of Natural History, Smithsonian Institution). The etiology of this condition, however, and the reasons for its reported higher incidence in more recent populations were not discussed in these studies. Therefore, there is some support for different incidences of disease between the two cultural groups at Point Hope, the relationship of these to specific distinctions in lifeways remains unresolved.

Arguments made in this volume about the inhabitants of Point Hope

The biological contributions to this volume add substantially to the arguments previously made about the foragers of Point Hope. Many of the findings are original. Maley's analysis adds to the previous work by Utermohle (1984), while contextualizing the biological results within archaeological evidence not available thirty years ago when Utermohle published his study. Shackelford is the first to publish examinations of postcranial cross-sectional geometric properties of the limbs and what these reveal about activity distinctions or similarities between the Ipiutak population and Tigara population at Point Hope, as well as between the sexes in each group. Cowgill's novel contribution traces ontogenetic patterns in body proportions and in the cross-sectional geometry of the limbs. Despite the number of publications made previously about the dental anatomy and pathologies at Point Hope, both Krueger and El Zaatari add new data and consider previously unconsidered factors in their analyses. Hilton *et al.* present a comprehensive documentation and analysis of the skeletal pathologies from both the Ipiutak period and the Tigara period skeletons, and importantly place this in the context of the lifeways of other documented foragers. The combined implications of these studies unequivocally add knowledge to current understandings, as well, about past human life in the extreme environment of the Arctic.

As discussed, limited attention has been given previously to the biological affinities and origins of the groups that inhabited Point Hope; Utermohle's 1984 study has remained the primary analysis of biological affinities in the northwestern North American Arctic. Only with Maley's 2011 craniometric and molecular genetic analysis of the ancient DNA from Point Hope skeletons was the biological continuity of these populations addressed further. His study results (Chapter 4) echo Utermohle's conclusions, as well as the archaeological arguments by Jensen and by Mason. Maley supports a case for a Norton origin to the Ipiutak population at Point Hope, which then contributed to the more recent Yup'ik groups living in southwestern and western Alaska. The cranial evidence agrees with previous research that argues the Tigara peoples had descended from Birnirk groups living on the Northern Slope. While the Ipiutak population may not have been the ancestor of the Tigara population, both likely shared an ultimate, common ancestral population that dated to the early or middle Holocene. (How that earlier population then related to the peoples associated with the Norton cultures or the Old Bering Sea cultures cannot be resolved from the available biological data.)

These population affinity patterns have potential importance in understanding other Point Hope population differences reported in this volume. Both Krueger's and El Zaatari's examinations of dental microwear reveal distinctions that likely reflect both dietary and non-dietary influences. Ipiutak individuals and Tigara individuals used their teeth as tools, but Krueger especially shows the Ipiutak population experienced more frequent and intense non-dietary dental activities. That Krueger concludes that the Ipiutak dental patterns are indicative of caribou hide processing, based on ethnographic corollaries, is important and fits with both the archaeological and craniometric arguments for affinities between the Ipiutak and Norton cultures. Sites identified with the Norton culture have been associated with greater dependency on inland mammals than large sea mammals (e.g., Bockstoce, 1979); while population affinity does not necessitate continuity in subsistence practices, similarities in the selection and use of prey choices bolsters cultural connections between the Ipiutak at Point Hope and Norton culture populations elsewhere in the northwestern Alaskan Arctic. (Of course, the similarities in manufacture and style of tools, as well as carvings, have archaeologically linked these cultural groups for some time; see Mason and Jensen's contributions to this volume.) Differences in molar wear, as reported by El Zaatari, complement Krueger's analysis and further support evidence for distinctive prey choices and food (namely meat) preparations between the Ipiutak and the Tigara populations. Indeed, both dental microwear studies corroborate the well-established focus on whaling in the Tigara population, a prey focus that has links to the broader subsistence practices associated with Thule cultures. It is illuminating, though, that Hilton *et al.* further argue for a decline in skeletal health in the later occupants of Point Hope; the Tigara individuals exhibit higher incidence of spondylolysis, infectious disease, and, possibly, trauma. It is tempting to conclude that the Tigara population faced "harder lives" than the Ipiutak population based on these analyses – one might reach this conclusion comparing skeletal lesion frequencies[3] – but the more important revelation is that the lifeways of these two groups were distinctive. These dissimilarities are illuminating, as both groups occupied the same geographic space within a few centuries of each other, but cultural practices and, relatedly, different ancestries (which influenced the unalike cultural practices) shaped distinct niches for the Ipiutak and Tigara peoples.

In light of these findings, then, it is unexpected that the studies of body shape and body proportions do *not* distinguish between the Ipiutak population and Tigara population at Point Hope. Holliday and Hilton's 2010 paper was the first to establish their similarities in body size, brachial index, and bi-iliac breadth. Crural indices are significantly different between the Tigara females (crural index = 82.61) and the Ipiutak females (cural index = 80.60) (Auerbach, 2007; Holliday and Hilton, 2010), with the latter having a lower index and therefore shorter tibial lengths relative to femoral lengths. While crural index has been argued to be more stable than brachial index over time within a population (Auerbach, 2007, 2010), and both indices are established early in development (Cowgill *et al.*, 2012; Cowgill, Chapter 9, this volume), it cannot be established if the slight difference observed between the two populations has bearing on their ancestry or differences in mobility (an argument made by Higgins and Ruff (2011) who examined limb proportions and topography among European groups). Yet, with evidence that these proportions and body size are established early in development (Cowgill, Chapter 9, this volume), the differences in female lower limb indices may be argued to indicate subtle differences in the origins of women in the two time periods; this, in turn, might be important in understanding mating patterns and possible differences in gene flow and/or migration between females and males. Yet Maley (Chapter 4, this volume) does not demonstrate any differences between the sexes in patterns of affinity.

Moreover, the attention given to just one of the morphologies examined by Holliday and Hilton (2010) and by Cowgill (Chapter 9, this volume), when all of the other measures of body size and proportions do *not* differ, obscures the larger picture that these dimensions present about both the Ipiutak population and Tigara population. Both had equally large body masses for their statures, which, in concert with short distal limb element lengths relative to proximal element lengths, is indicative of body shapes that were more efficient at retaining heat than human populations living closer to the equator (see Ruff, 1994); both groups had descended from populations that had lived in the Arctic for some time, an unsurprising but still noteworthy conclusion. Whether this morphology ultimately was a result of genetic drift or adaptation, or both, remains unresolved (see the "Phenotypic variation and evolutionary forces" subsection above). These phenotypic features persisted despite the construction of shelters, advancements in clothing, and well-managed fire, among other cultural adaptations to the arctic environment.

Finally, it is equally striking that studies of the limb bone cross-sectional geometrical properties by Shackelford and by Cowgill in this volume both demonstrated no differences in dimensions associated with bone strength and resistance to mechanical loading. The archaeological argument for the subsistence practices of the Ipiutak population and the Tigara population, as noted herein and by all of the contributors to this volume, makes a case for differences in hunting strategies, mobility, and resource procurement between the two cultures. In short, the Ipiutak were not a seafaring people, and certainly were not whalers like the Tigara peoples, instead depending on hunting that occurred inland (e.g., caribou and smaller mammals) or on the shoreline (e.g., seals and walrus). Thus, both groups might have used their upper limbs in equally mechanically taxing activities (e.g., spearing and bow-and-arrow hunting versus harpooning and rowing). However, the Tigara population would be expected to have indications of less loading in the lower limbs, as they would have not been as terrestrially mobile and would have had comparatively stronger upper limbs, as evidenced in other maritime cultures (Weiss, 2003; Ruff, 2006); the Ipiutak individuals would thus be expected to be more balanced in upper and lower limb strength. There is strong artifact evidence, in addition to the dental analyses by Krueger and by El Zaatari in this volume, that subsistence practices *were* different between both populations. Thus, the Shackelford and Cowgill results reported are cautions against reductionism in the interpretation of cross-sectional geometric patterns compared across archaeological populations. The whaling subsistence practiced by the Tigara population at Point Hope would not have negated other activities that required terrestrial mobility equal to the Ipiutak culture, including but not limited to movement for resource procurement, trade, seasonal migration, and terrestrial subsistence strategies. In conclusion, the Ipiutak population and Tigara population were culturally different, but these differences were not recorded in the cross-sectional geometry of their bones, either because they were not different enough or the analyses performed are unable to register the subtle differences in activity each group practiced.

Overall, these results present an abundance of interesting implications about the Ipiutak population and Tigara population at Point Hope. These groups are arguably of separate recent ancestry, distinguished by craniometric morphologies but, curiously, not by other morphological dimensions except crural indices among females (Holliday and Hilton, 2010; see below) and juveniles (Cowgill, Chapter 9, this volume). Dental microwear, postcranial skeletal lesions, and archaeological evidence collectively support cultural differences in prey choice and subsistence patterns, as well as non-dietary dental use. Comparisons of cross-sectional geometric patterns between the groups are a stark contrast to these lines of evidence, where no limb use differences are apparent by examining dimensions representative of limb strength.

Point Hope in the context of North American biological variation

As discussed at the beginning of this chapter, until this volume most authors incorporating skeletal data from the past inhabitants of Point Hope have used the site as an example of "cold climate" morphologies. Few of these studies recognize the important differences that existed between the two major archaeological occupations of Point Hope, sometimes lumping the Ipiutak period skeletons with the Tigara period skeletons. With the few important exceptions noted in the preceding section (namely, Holliday and Hilton, 2010), a comprehensive study of the variation in body shape, size, and proportions of groups at Point Hope has not been performed.

This section, then, briefly presents a set of analyses to place the morphological variation of the Ipiutak period skeletons and Tigara period skeletons into a broader indigenous North American context. The contributors to this volume place their analyses into a broad regional context (Krueger and Maley), a continental context (El Zaatari), or a global context (Cowgill, Hilton, and Shackelford). These comparative groups were chosen to suit the needs of the authors' respective studies. The following analyses further place morphological variation among the cultural groups that inhabited Point Hope within both a regional, Arctic perspective, as well as a North American context.

Materials and methods

This analysis uses skeletal remains recovered from twenty sites in North America, representing a sampled diversity of morphological variation present in late Holocene North America (see Auerbach, 2007; Auerbach and Ruff, 2010; Auerbach, 2012). These sites are listed in Tables 10.1 and 10.2, and their locations are depicted in Figure 10.1. The Arctic sites were chosen as representative of the western Alaskan groups dating to the time periods of the Ipiutak and the Tigara, and parallel the samples used by Maley in this volume. In addition to these, the Sallirmiut (also called the Sadlermiut), a cultural group that lived on Southampton Island up until their extinction in 1903–4 are included as an Arctic outgroup; there was likely limited gene flow between this group and their ancestors with the inhabitants of Point Hope. In order to contextualize the morphological diversity of these Arctic groups, eleven non-Arctic sites were examined.

Linear dimensions were measured from the skeletal remains. Measurements were taken following Martin (1928) using a portable osteometric board and Mitutoyo Digital calipers, and those used in this analysis were humeral maximum length (HML), radial maximum length (RML), femoral bicondylar length

Figure 10.1. Locations of the sites sampled to examine variation in body size, shape, and proportions. Numbers correspond with site names listed in Table 10.1 and Table 10.2. Note that the Ikogmiut, Kuskowagamiut, and Aleutian Island groups consist of individuals sampled from multiple sites; numbers for these indicate the approximate geographic centroid for those sites.

(FBL), anteroposterior femoral head diameter (FHD), tibial maximum length (TML), and bi-iliac breadth (BIB). All dimensions have minimal (less than 1%) intra-observer measurement error. These dimensions were used to calculate brachial index (RML ÷ HML × 100), crural index (TML ÷ FBL × 100), estimated body mass (following Grine *et al.* [1995] see Auerbach [2010], for the argument behind using this method), and stature (using equations from Auerbach and Ruff [2010]). The means and standard deviations for these are reported for all sampled sites in Tables 10.1 and 10.2.

Statistical analyses were performed using Stata 12.0 for Macintosh. Statistically significant sexual dimorphism occurs in estimated body mass, stature, and brachial indices, and so all analyses were conducted separately by sex. Body mass, stature, and bi-iliac breadth dimensions were compared using ANOVAs; Welch's F' was used in the few instances when the homogeneity of variances among groups was violated. Indices violate the assumptions of parametric statistics, which required alternative methods to be employed to examine

Table 10.1. *Body size, body shape, and body proportions of archaeological North American Arctic groups*

Group [ID]	Date (BP)	Sex[a]	N^b	Estimated body mass (kg)[c]	Estimated stature (cm)[d]	Brachial index × 100	Crural index × 100	Bi-iliac breadth (cm)	Mean femoral head size (mm)	Bi-iliac breadth/ stature × 100
Point Hope Ipiutak [1]	*c.* 1,600–1,050	♀	16	56.9	146.6	71.6	80.2	26.8	41.2	1.82
				(5.1)	(2.8)	(2.2)	(1.5)	(1.2)	(2.3)	(0.06)
		♂	19	67.6	156.3	75.4	81.2	27.5	45.9	1.76
				(3.7)	(6.4)	(2.4)	(2.6)	(1.3)	(1.6)	(0.09)
Point Hope Tigara [2]	*c.* 750–200	♀	22	59.2	148.1	73.2	82.6	26.3	42.2	1.78
				(3.5)	(3.8)	(2.7)	(1.8)	(1.2)	(1.5)	(0.06)
		♂	22	69.5	159.4	74.9	82.9	28.0	47.1	1.76
				(4.4)	(4.9)	(2.3)	(2.0)	(1.1)	(2.7)	(0.08)
Ikogmiut [3]	*c.* 1,000–100	♀	31	57.4	149.2	72.9	80.6	25.9	41.4	1.74
				(4.9)	(3.9)	(2.8)	(1.8)	(1.4)	(2.2)	(0.09)
		♂	29	67.2	157.8	74.7	82.5	26.5	45.7	1.68
				(5.1)	(5.8)	(2.7)	(2.2)	(1.6)	(2.3)	(0.09)
Kuskowagamiut [4]	*c.* 1,000–100	♀	14	57.5	149.2	72.9	80.7	25.5	41.5	1.71
				(5.5)	(3.8)	(2.7)	(1.4)	(1.3)	(2.4)	(0.09)
		♂	14	67.6	157.2	74.7	79.4	26.5	45.9	1.69
				(5.0)	(3.6)	(1.5)	(2.3)	(1.4)	(2.2)	(0.09)
Point Barrow Birnirk [5]	*c.* 1,500–1,000	♀	7	55.6	150.2	73.9	83.0	25.6	40.6	1.76
				(10.1)	(4.3)	(4.6)	(2.3)	(1.6)	(4.4)	(0.06)
		♂	16	69.6	159.3	75.0	82.0	27.5	46.8	1.72
				(4.8)	(4.6)	(2.6)	(2.6)	(1.4)	(2.1)	(0.08)
"Paleo-Aleut" [6]	*c.* 4,000–1,000	♀	15	57.9	147.2	74.5	82.1	27.1	41.6	1.84
				(3.7)	(2.2)	(2.7)	(2.1)	(1.6)	(1.6)	(0.10)
		♂	13	68.0	157.5	77.2	81.6	27.7	46.1	1.76
				(3.6)	(4.8)	(2.8)	(2.1)	(1.0)	(1.6)	(0.06)

(*cont.*)

Table 10.1. (*cont.*)

Group [ID]	Date (BP)	Sex	*N*	Estimated body mass (kg)	Estimated stature (cm)	Brachial index × 100	Crural index × 100	Bi-iliac breadth (cm)	Mean femoral head size (mm)	Bi-iliac breadth/ stature × 100
"Neo-Aleut" [7]	*c*. 1,000–100	♀	20	57.1 (5.5)	146.5 (4.4)	74.6 (2.3)	81.1 (1.7)	25.9 (1.1)	41.3 (2.4)	1.77 (0.07)
		♂	36	66.6 (4.7)	155.9 (5.9)	76.3 (1.7)	81.2 (1.7)	26.3 (1.3)	45.5 (2.1)	1.68 (0.07)
Sallirmiut (Native Point, Southampton Island) [8]	*c*. 950–50	♀	26	62.5 (4.2)	150.7 (5.0)	70.8 (2.0)	80.4 (2.0)	27.1 (1.4)	43.6 (1.9)	1.79 (0.08)
		♂	28	73.2 (5.1)	160.6 (5.1)	72.0 (2.7)	80.5 (1.8)	27.3 (1.3)	48.4 (2.2)	1.71 (0.08)
Koniag[e] [9]	*c*. 700–100	♀	10	57.8 (–)	146.3 (–)	75.9 (–)	80.2 (–)	—	41.6 (2.1)	—
		♂	11	65.6 (–)	154.3 (–)	74.8 (–)	80.8 (–)	—	45.0 (2.1)	—

Reported data are means (standard deviations in parentheses).

See Auerbach (2007) for more information about site groupings, citations for site dates, and details about methods.

See Figure 10.1 for site locations.

[a] Sex was determined using the methods of Phenice (1969) and Bruzek (2002).

[b] Sample sizes reported are the maximum number of individuals measured; some dimensions (especially bi-iliac breadth) were not measureable on the full sample.

[c] Estimated from femoral head diameter using equations by Grine *et al.* (1995); see Auerbach (2007, 2011) for an explanation concerning the use of this equation.

[d] Estimated using the "Arctic" stature estimation formulae in Auerbach and Ruff (2010).

[e] Data calculated from means reported in Holliday and Hilton (2010).

Table 10.2. *Body size, body shape, and body proportions of select archaeological North American non-Arctic groups dating to the late Holocene (last 3,000 years BP)*

Group [ID]	Subsistence	Sex[a]	N[b]	Estimated body mass (kg)[c]	Estimated stature (cm)[d]	Brachial index × 100	Crural index × 100	Bi-iliac breadth (cm)	Femoral head size (mm)	Bi-iliac breadth/ stature × 100
Prince Rupert Tshimshian [10]	Marine forager	♀	18	57.2	146.9	74.9	81.9	26.5	41.3	1.79
				(5.5)	(4.5)	(2.5)	(2.2)	(1.7)	(2.4)	(0.09)
		♂	41	69.5	156.7	77.7	82.8	27.7	46.7	1.76
				(5.3)	(3.4)	(2.4)	(2.1)	(1.0)	(2.3)	(0.07)
Blossom site Windmiller [11]	Forager / horticulture	♀	19	59.5	158.6	76.1	83.7	26.9	42.3	1.71
				(4.5)	(4.1)	(1.7)	(2.2)	(1.3)	(2.0)	(0.08)
		♂	20	70.8	168.0	79.1	84.6	28.0	47.3	1.67
				(6.3)	(6.1)	(2.8)	(2.7)	(1.4)	(2.8)	(0.08)
Channel Island "Canaliño" [12]	Marine forager	♀	17	53.4	150.7	76.8	84.0	25.7	39.6	1.71
				(5.0)	(5.1)	(2.6)	(2.2)	(1.7)	(2.2)	(0.11)
		♂	12	66.3	161.7	79.3	84.8	26.1	45.3	1.61
				(3.8)	(4.0)	(3.2)	(1.7)	(1.2)	(1.7)	(0.06)
Palmer (Weeden Is.) [13]	Marine forager	♀	17	57.9	156.5	76.0	84.7	27.3	41.6	1.73
				(3.6)	(3.7)	(2.0)	(1.3)	(1.4)	(1.6)	(0.08)
		♂	17	65.4	165.2	76.0	84.4	26.7	44.9	1.61
				(3.9)	(3.5)	(2.9)	(2.6)	(1.6)	(1.7)	(0.12)
Irene Mound [14]	Agriculture	♀	19	53.1	155.9	76.2	84.2	25.3	39.5	1.61
				(5.7)	(4.8)	(3.0)	(2.1)	(1.0)	(2.5)	(0.05)
		♂	12	64.3	166.7	78.2	85.6	26.9	44.4	1.59
				(8.4)	(7.1)	(2.5)	(2.5)	(1.7)	(3.7)	(0.09)
Montague site Delaware [15]	Agriculture	♀	11	56.8	157.3	77.8	84.6	25.8	41.1	1.64
				(5.1)	(4.3)	(1.3)	(1.7)	(1.3)	(2.2)	(0.06)
		♂	9	67.4	167.7	78.8	86.0	26.2	45.8	1.56
				(4.6)	(3.3)	(2.3)	(2.5)	(1.2)	(2.0)	(0.08)

(cont.)

Table 10.2. (*cont.*)

Group [ID]	Subsistence	Sex	*N*	Estimated body mass (kg)	Estimated stature (cm)	Brachial index × 100	Crural index × 100	Bi-iliac breadth (cm)	Femoral head size (mm)	Bi-iliac breadth/ stature × 100
Pottery Mound	Agriculture	♀	23	50.9	149.7	77.2	84.5	25.7	38.5	1.73
Pueblo IV [16]				(3.0)	(2.8)	(1.4)	(1.6)	(1.2)	(1.3)	(0.08)
		♂	24	59.6	160.1	78.2	85.1	26.4	42.4	1.66
				(4.7)	(4.1)	(1.3)	(1.8)	(1.0)	(2.1)	(0.08)
Libben (Western	Forager /	♀	17	56.4	158.0	78.9	84.2	26.5	41.0	1.68
Basin tradition) [17]	horticulture			(4.5)	(3.6)	(2.4)	(2.4)	(1.2)	(2.0)	(0.06)
		♂	22	67.6	169.1	79.8	85.5	27.2	45.9	1.61
				(3.2)	(4.1)	(2.5)	(1.8)	(1.8)	(1.4)	(0.10)
Larson site Arikara	Village hunter /	♀	16	57.8	155.9	78.2	85.9	27.0	41.6	1.73
[18]	horticulture			(4.5)	(4.7)	(2.3)	(2.0)	(1.4)	(2.0)	(0.08)
		♂	16	66.5	165.5	79.1	86.7	28.1	45.4	1.69
				(4.6)	(3.8)	(2.9)	(1.8)	(1.3)	(2.0)	(0.06)
Averbuch site	Agriculture	♀	28	56.6	157.2	76.1	83.4	26.6	41.1	1.69
Mississippian [19]				(6.1)	(5.7)	(2.1)	(2.3)	(1.5)	(2.7)	(0.07)
		♂	27	68.1	166.6	78.0	84.0	27.5	46.1	1.65
				(5.5)	(5.2)	(1.5)	(1.5)	(1.6)	(2.4)	(0.07)
Mitchell Ridge	Marine forager	♀	9	55.2	157.2	78.7	84.6	25.9	40.5	1.64
Karankawa [20]				(5.3)	(2.6)	(1.7)	(1.8)	(0.6)	(2.4)	(0.05)
		♂	10	66.0	168.5	78.3	86.0	28.2	45.2	1.66
				(6.8)	(4.5)	(1.9)	(1.8)	(1.4)	(3.0)	(0.06)

Reported data are means (standard deviations in parentheses).
See Auerbach (2007) for more information about site groupings, citations for site dates, and details about methods.
See Figure 10.1 for site locations.
[a] Sex was determined using the methods of Phenice (1969) and Bruzek (2002).
[b] Sample sizes reported are the maximum number of individuals measured; some dimensions were not measureable on the full sample.
[c] Estimated from femoral head diameter using equations by Grine *et al.* (1995).

intralimb indices. Sex-specific linear OLS regressions were calculated for RML against HML, and for TML against FBL, either just among the Arctic samples or among the entire North American sample, depending on the analysis. Unstandardized residuals from these regressions were then compared among the groups using ANOVAs to ascertain significant differences in brachial and crural indices. All statistics used an α of 0.05.

Variation among Arctic groups

Holliday and Hilton (2010) reaffirmed that both the Ipiutak period skeletons and Tigara period skeletons at Point Hope exhibit morphologies associated with human populations living in cold climate environments: wide bi-iliac breadths, relatively short distal limb lengths (i.e., low brachial and crural indices), and high body masses relative to stature. In that paper, Holliday and Hilton did not distinguish between the two Point Hope groups and the Koniag from Kodiak Island in most dimensions (or, incidentally, from their large European sample). Their conclusion was that the groups from Point Hope did not exhibit "hyper-arctic" body size or shape morphologies, despite the location of Point Hope.

Further analyses comparing body breadth, intralimb indices, and body mass among the nine groups summarized in Table 10.1 corroborate the general conclusions drawn by Holliday and Hilton. Results of the ANOVAs comparing dimensions among the groups are presented in Tables 10.3a and b. The only statistical difference between the Ipiutak population and Tigara population is in female crural indices, upholding the findings of Holliday and Hilton (2010). With the exception of significantly wider bi-iliac breadths among Tigara period males in comparison with the "Neo-Aleut" from the Aleutian Islands, neither the Arctic males nor females statistically significantly differ in bi-iliac breadth. The Sallirmiut have higher body masses among both males and females than any of the western Alaskan groups; the Ipiutak males and females are both significantly less massive than the Sallirmiut. Similarly, intralimb indices are significantly lower among the male and the female Sallirmiut in comparison with most of the western Alaskan groups (see Tables 10.3a and b). No statistically significant differences emerge among the Arctic groups in either stature or bi-iliac breadth scaled to stature.

Overall, the Arctic groups do not show statistically significant variation in most dimensions of body shape and size. The principal conclusion drawn from these analyses is that the most evident differences existed among the Southampton Island Sallirmiut, who were generally more massive and had lower intralimb indices than the groups living in the western Arctic. In these dimensions, the Sallirmiut would be considered to exhibit more "hyper-arctic"

Table 10.3a. *Comparisons of body size, shape, and proportions among females from Arctic groups*

Group	Brachial index	Crural index	Bi-iliac breadth	Body mass	Stature	BIB/ Stature
Point Hope Ipiutak [1]		†				
Point Hope Tigara [2]						
Ikogmiut [3]		†				
Kuskowagamiut [4]						
Point Barrow Birnirk [5]	‡					
"Paleo-Aleut" [6]	‡					
"Neo-Aleut" [7]	‡					
Sallirmiut [8]		†		*		
Koniag [9]						

* indicates groups significantly ($p < 0.05$) different from the Ipiutak group; † indicates groups significantly different from the Tigara group; ‡ indicates groups significantly different from the Sallirmiut group.

Table 10.3b. *Comparisons of body size, shape, and proportions among males from Arctic groups*

Group	Brachial index	Crural index	Bi-iliac breadth	Body mass	Stature	BIB/ Stature
Point Hope Ipiutak [1]	‡					
Point Hope Tigara [2]	‡					
Ikogmiut [3]	‡					
Kuskowagamiut [4]	‡	†				
Point Barrow Birnirk [5]	‡					
"Paleo-Aleut" [6]	‡					
"Neo-Aleut" [7]	‡		†			
Sallirmiut [8]		†		*		
Koniag [9]	‡					

* indicates groups significantly ($p < 0.05$) different from the Ipiutak group; † indicates groups significantly different from the Tigara group; ‡ indicates groups significantly different from the Sallirmiut group.

morphologies than the groups that lived in Point Hope. It is noteworthy that the Tigara males did have the widest mean bi-iliac breadths of any of the male Arctic groups analyzed, and both Point Hope groups had among the widest body breadths in the Arctic sample.

This broader study of Arctic variation, then, supports the conclusion that body size, shape, and proportions do not match the population patterning reflected in craniometric variation. There are non-significant trends that indicate more similarity between the Tigara population and Point Barrow Birnirk sample (Table 10.1), as well as more alike dimensions between the Ipiutak population, the Ikogmiut, Kuskowagamiut, and "Paleo-Aleut." While these trends are notable, they do not reach a diagnostic magnitude for population affinities.

Variation in the context of North America

In order to provide a broader, continental context for the Arctic group analyses, additional comparisons were made between the two groups from Point Hope and the eleven non-Arctic groups listed in Table 10.2. Results are presented in Tables 10.4a and b. It should be noted that, while the non-Arctic samples are not an exhaustive representation of the ranges of non-Arctic variation in morphology in North America – large sections of the continent are not represented – they do encompass the range of variation within 1.5 standard deviations of the overall mean in temperate North America of late Holocene sampled in a larger study ($N > 2{,}500$ individuals) of archaeological indigenous North American variation (Auerbach, 2007).

The results of the statistical analyses suggest that the temperate groups differ more in some than in other dimensions of body shape, size, and proportions. Nearly all of the temperate groups have significantly higher brachial indices than the Ipiutak males and females, and a subset of those temperate groups – including the warmer climate sites of Pottery Mound, Irene Mound, and Mitchell Ridge – are also significantly higher in brachial indices than the Tigara males and females. In comparison, it is noteworthy that fewer of the temperate sites exhibit statistically higher crural indices, and that these differ more from both sexes of the Ipiutak population than any from the Tigara period. Generally, however, these patterns match expectations for differences in intralimb proportions between Arctic groups and populations from more temperate locations. That is not the case, however, for either bi-iliac breadth or body mass (see Figures 10.2 and 10.3). Overall, the majority of the temperate, non-Arctic groups have body breadths that are *not* significantly narrower than either Point Hope group, and all of the groups, except for the late Pueblo Pottery Mound population, are equally massive.

As reported elsewhere (Auerbach, 2012), wide bi-iliac breadths are a characteristic of all human populations from the Americas. The evolutionary factors that have contributed to this still await explicit testing (cf. Betti *et al.*, 2012; Roseman and Auerbach, in review), but it is apparent that wide bi-iliac breadths are a common ancestral trait among most indigenous groups from North America. As noted by Ruff (1994), Auerbach (2007, 2010, 2012), and Holliday

Table 10.4a. *Comparisons of body size, shape, and proportions among females from Point Hope with non-Arctic groups*

Group	Brachial index	Crural index	Bi-iliac breadth	Body mass	Stature	BIB/ Stature
Prince Rupert Tshimshian [10]	*					
Blossom site Windmiller [11]	*				* †	
Channel Island "Canaliño" [12]	*	*	*			* †
Palmer [13]	*	*			* †	* †
Irene Mound [14]	*	*	* †		* †	* †
Montague site Delaware [15]	* †	*	*		* †	* †
Pottery Mound Pueblo IV [16]	* †	*	*	* †		* †
Libben [17]	* †	*			* †	* †
Larson site Arikara [18]	* †	* †			* †	* †
Averbuch site Mississippian [19]	*				* †	†
Mitchell Ridge Karankawa [20]	* †	*			* †	

* indicates groups significantly ($p < 0.05$) different from the Ipiutak group; † indicates groups significantly different from the Tigara group.

Table 10.4b. *Comparisons of body size, shape, and proportions among males from Point Hope with non-Arctic groups*

Group	Brachial index	Crural index	Bi-iliac breadth	Body mass	Stature	BIB/ Stature
Prince Rupert Tshimshian [10]						
Blossom site Windmiller [11]	*	*			* †	
Channel Island "Canaliño" [12]	*	*	* †			* †
Palmer [13]		*	†		* †	
Irene Mound [14]	†	*			* †	* †
Montague site Delaware [15]	* †	* †	* †		* †	* †
Pottery Mound Pueblo IV [16]	* †	*	* †	* †		*
Libben [17]	* †	*			* †	*
Larson site Arikara [18]	* †	* †			* †	* †
Averbuch site Mississippian [19]	†				* †	*
Mitchell Ridge Karankawa [20]	* †	* †			* †	* †

* indicates groups significantly ($p < 0.05$) different from the Ipiutak group; † indicates groups significantly different from the Tigara group.

and Hilton (2010), bi-iliac breadth is more stable over time than other morphologies associated with ecogeographic patterns (i.e., intralimb proportions and cormic index), even though the cylindrical model argued by Ruff (1994) demonstrated the importance of body breadth in determining heat dissipation efficiency among humans. The groups that exhibit significantly narrower body

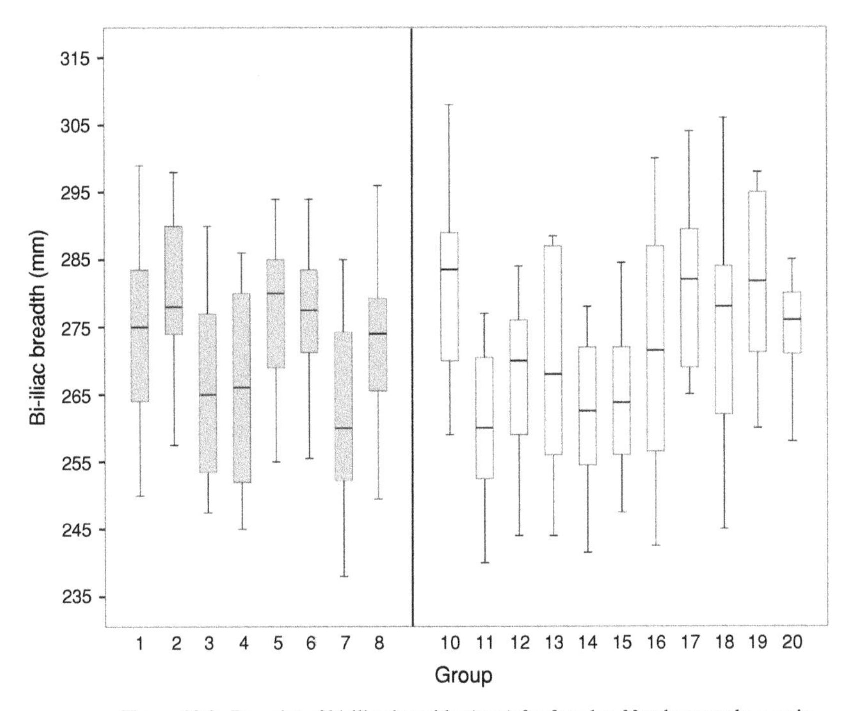

Figure 10.2. Box plot of bi-iliac breadths (mm) for females. Numbers on the x-axis correspond with site names listed in Table 10.1 and Table 10.2. Grey shaded boxes to the left of the vertical line divider designate Arctic groups. Unshaded boxes to the right of the vertical line are non-Arctic sites.

breadths – inhabitants of the Channel Islands off the Californian coast, protohistoric Delaware from modern New Jersey, and Puebloan occupants of Pottery Mound in modern central northern New Mexico – have few climatic factors in common and no recent common ancestry. Thus, it is not possible to disentangle the factors that likely differentiate these groups from the inhabitants of Point Hope.

Yet, it *is* clear that the inhabitants of Point Hope were more morphologically distinguished from groups living in more southern latitudes of North America than any of the groups living in the Arctic. This was evident in intralimb proportions, where most temperate groups have longer distal limb elements relative to proximal elements. It was also evident in stature. Though stature is not associated with efficient thermoregulation (Ruff, 1994), both population history and diet are major contributors to variation in stature (see Auerbach [2011], for a review of factors affecting stature). Except for three groups – the Channel Island population, Tshimshian groups living around Prince Rupert

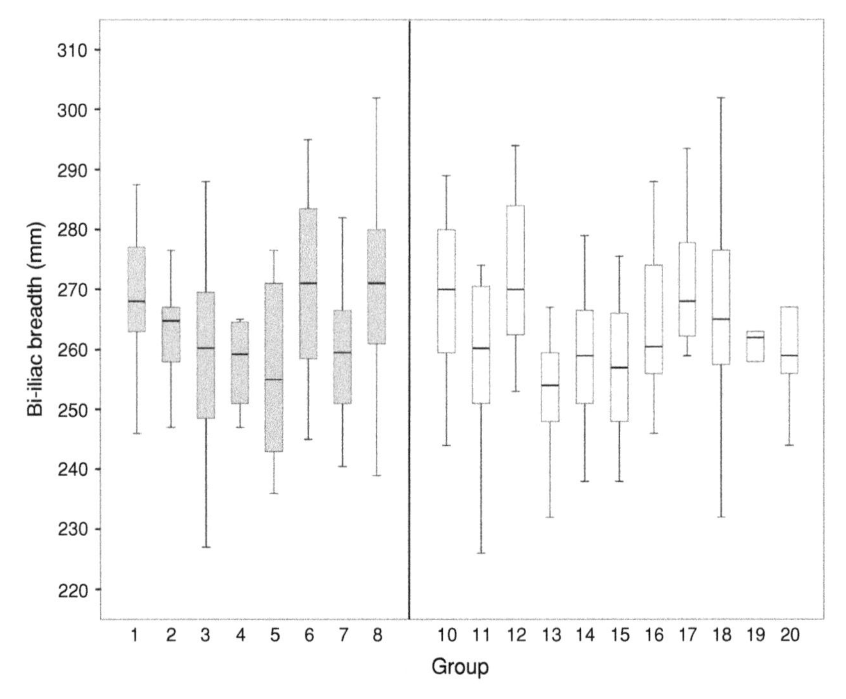

Figure 10.3. Box plot of bi-iliac breadths (mm) for males. Numbers on the x-axis correspond with site names listed in Table 10.1 and Table 10.2. Grey shaded boxes to the left of the vertical line divider designate Arctic groups. Unshaded boxes to the right of the vertical line are non-Arctic sites.

Harbour, and the Pottery Mound Puebloans – all temperate North American indigenous groups were significantly taller than either Point Hope population. Many of these temperate groups were, on average, more than eight centimeters taller than members of either Point Hope culture. This difference in stature is likely the main factor driving differences in overall body shape that are reflected in comparisons of body breadth scaled to stature; most temperate groups were more narrow for their statures than the Arctic groups, matching general ecogeographic expectations (Ruff, 1994). Despite this difference, it is important to remark again on the *lack* of significant differences in estimated body mass among almost all of the groups, both Arctic and non-Arctic.

Implications of North American Arctic morphological variation

One important message that should be taken away from the analyses in this chapter and of this volume concerns the careful selection of human groups as models for studies. Biological anthropologists have arguably moved away from the typological perspectives of the past, wherein North American indigenous skeletal samples from single locations were selected to stand in for all North American indigenous groups (critiqued in Powell, 2005). Yet there is still a tendency to examine the circumpolar groups from North America as representatives for cold climate morphologies, without discriminating among groups based on differences in population history, variation in cultural practices, or dietary disparities. That is, choosing any one archaeological human group from the North American Arctic is *not* representative of *all* North American Arctic human groups. While all of these populations share general, common morphological characteristics (see Tables 10.3a and b), it is crucial to understand that variation exists among the North American Arctic groups in these dimensions, and that some of these phenotypes (e.g., wide bi-iliac breadth and high body mass) are not unique to just indigenous North American groups living in the Arctic. That is, there is an argument to be made that humans living in the high latitudes of North America arrived at *multiple morphological solutions* that reflect a combination of reduced variation retained from ancestors, genetic drift, and adaptation to the demanding environments of the Arctic. Therefore, researchers are strongly cautioned against selecting any high-latitude, circumpolar North American human group as if it were interchangeable with any other North American circumpolar group, without first considering the unique history of the group selected to represent cold climate "adaptations."

Unsurprisingly, information that may be obtained from cranial morphology is subject to different influencing factors than the body morphologies analyzed in this chapter. There is a lack of correspondence between craniometrics and variation in body shape and size in discerning population histories. The dimensions of the skull and postcrania are seldom directly compared in assessing population affinities. Some statistically non-significant patterns in bi-iliac breadth and intralimb indices bear some similarity to the statistically significant craniometric patterns obtained by Maley in this volume, using the same source populations for both sets of analyses. More research should be undertaken in other regions to ascertain the correspondence of these various skeletal morphological sources of evidence for population relationships.

Finally, this study provides further evidence to support the previously reported shared characteristic of wide body breadths and high body masses among North American indigenous populations in general. This conclusion is not meant to imply homogeneity in either dimension among North American groups; variation in these dimensions in North America is both statistically and biologically relevant. Yet the evidence provided in this analysis and elsewhere (Auerbach, 2012; Holliday and Hilton, 2010) argues for reduced variation in these dimensions relative to human groups living in Eurasia and Africa.

Additional studies will be necessary to understand the factors that shape these morphologies, and ultimately what has influenced the reduction in variation observed in the Americas.

Concluding remarks

Returning to the two sets of questions introduced at the beginning of this chapter, what may be resolved about the Ipiutak population and the Tigara population that inhabited Point Hope from the skeletal analyses? The nature of studies such as those presented in this book is that they produce further questions, or may not resolve those put forth. However, the skeletal analyses of this volume add greatly to the understanding of humans foraging on the northwestern edge of the Alaskan Arctic:

(1) The specific population affinities for the inhabitants of Point Hope remain the subject of additional analyses (e.g., isotopic analysis, further ancient DNA analyses, and more morphological analysis). However, the data present a set of working hypotheses:
   - Shared body size, shape, and proportions between the Ipiutak population and Tigara population (with the notable exception of female crural indices), as well as the lack of significant differences among all of the western Arctic groups, supports an argument for an ultimate, common ancestor for all of these groups, or, alternatively, convergence of phenotypes from multiple ancestral groups. Without additional skeletons from deeper Arctic antiquity, differentiating between these models remains unresolved.
   - Craniometric analyses further refine this general perspective. The Ipiutak peoples likely descended, in part or in their totality, from Norton culture practitioners. The fate of the Ipiutak practitioners who lived at Point Hope remains unresolved, though Maley's analysis in this volume suggests continuity (i.e., possible gene flow) with the groups inhabiting the lower Yukon and Kuskokwim River Valleys. If the later Tigara population that occupied Point Hope was related to their Ipiutak predecessors, the craniometric data show no unambiguous evidence for this.
   - Coupling the craniometrics with the difference in crural index among the females of both Point Hope groups, and the non-significant differences in body breadth between the populations, the collective skeletal evidence more likely supports a Point Barrow (or other location) Birnirk origin for the Tigara. This

is an agreement with the preponderance of the archaeological evidence.

(2) Archaeological evidence strongly suggests different subsistence economies and lifeways for the Ipiutak people of Point Hope and the later practitioners of Tigara culture. Only some of the biological data differentiate the nuances of their lifeways:

- Dental microwear studies provide the strongest support for distinctions between the two populations. Both anterior teeth and the molars exhibit differences in scratch and pit features, which together argue for dietary differences, including prey choice and food preparation. Additional examination of the anterior dental wear points toward non-dietary use of teeth in both groups, though this was more pronounced in the Ipiutak population. Such differences may be tied both to prey choice differences and distinctions in the production of items; for example, greater exploitation of caribou by the Ipiutak people would have led to distinct dietary wear compared with whale blubber and meat, in addition to more use of teeth as a tool in preparing hides.
- A comparison of skeletal lesions does distinguish between the two cultural groups. The evidence from Hilton *et al.*'s contribution to this volume (Chapter 7) and previous studies collectively argues for more chronic stress among the Tigara period individuals; higher incidences of lesions associated with tuberculosis (or other infections) may indicate higher population densities at Point Hope during the Tigara period, but, paradoxically (Wood *et al.*, 1992), could also indicate better care and longer term survival for individuals suffering from the infections.[4] The Ipiutak period occupation of Point Hope may have been both less dense, and lacking the cultural buffers that prevent individuals from quickly succumbing to disease and other stresses. This is speculative, however, and calls for further investigation. The higher incidence of spondylolysis in the Tigara population indicates either a higher genetic propensity to vertebral arch fracturing or distinct activities in which the Ipiutak period individuals did not engage. As spondylolysis has been indicated by Merbs (1996) to be common among later period (i.e., Thule culture) North American Arctic groups, which had both separate ancestry and unique activity patterns in relation to populations associated with earlier (e.g., Dorset, Norton, etc.) cultures, either or both causes are possible.
- In contrast with dental and skeletal lesion dissimilarities, the examination of cross-sectional geometric properties of the limbs of the

two groups yields no clear distinctions, in spite of likely different activity patterns between the populations. One conclusion is that individuals in both populations experienced demanding activities, which was more influential on bone strength than population-specific patterns of subsistence and mobility. That is, the highly active lifestyles of the Ipiutak population and Tigara population converged on limb bones with similar strength properties. As shown by Cowgill (Chapter 9), influences from other developmental factors, including diet and ancestry, also would have had a role in shaping this correspondence.

In conclusion, the investigations of this volume have yielded new ideas and refinements to the knowledge of the morphologies of humans living on the edge of the northwestern Alaskan Arctic. Together, they also call for the need for additional study; the past populations of Point Hope await new, additional analyses that follow from these studies. Our understanding of Point Hope would benefit from isotopic analyses, further ancient DNA studies, detailed site demography, and more study into how activities shaped the morphological variation. It is paramount, however, that these studies are conducted in the broader context of human cultural and biological variation in the Arctic. Skeletal biologists, molecular anthropologists, archaeologists, and human biologists must work together to develop a synthetic understanding of life for humans at Point Hope (and elsewhere). The contributions to this volume, however, are a promising step in that direction.

Acknowledgements

The author gives his gratitude to the many curators of collections in North America that permitted access to the skeletons necessary for the data used in this analysis. Special thanks are given to David Hurst Thomas, Ian Tattersall, Giselle Garcia, Ken Mowbray, and Gary Sawyer for access to Point Hope, and their help and support in the many excursions made to the American Museum of Natural History. Thanks as well to my co-editors, Libby Cowgill and Charles Hilton, without whom this volume would not have been possible, and to the contributors to both this volume and the 2006 American Association of Physical Anthropologists Point Hope Symposium in Anchorage, the impetus that set this research into motion. Comments from multiple anonymous peer reviewers greatly improved this contribution. A National Science Foundation Doctoral Dissertation Improvement Grant (BCS #0550673) funded this research.

Notes

1. "Genetic" is used to refer to the mathematical models of population relationships and change over time, and not strictly molecular biology approaches to analyzing these. So long as heritabilities are similar among traits *and* between populations, and patterns of morphological integration are the same, then it is reasonable to use phenotypic data (e.g., osteometric measurements) as a proxy for genetic (e.g., molecular) data. The term "genetic" is not meant to be synonymous with genomics, which focuses more on gene function and molecular structure.
2. This result has a caveat, however. The measurement of juvenile skeletal remains from archaeological contexts automatically necessitates discussion of the Osteological Paradox (Wood *et al.*, 1992), in that the juveniles died before reaching adulthood, and thus their phenotypes may represent maladaptive or less fit variants compared with those individuals who survived to adulthood. Of course, many causes lead to premature death in humans, and not all inherently reflect lower fitness in any particular phenotype. While this problem is present, it does not negate the overall trend reported by Holmes and Ruff (2011). Rather, it is a caution especially to be taken into consideration in studies involving juvenile remains.
3. The Osteological Paradox again is important when considering the presence of skeletal lesions. As Wood *et al.* (1992) explained, individuals presenting lesions survived infections and metabolically stressing disorders. Individuals without lesions had "hidden heterogeneity," meaning that these individuals either died of disease states that did not get expressed in the skeleton, died quickly of disease, or were healthy and died of trauma. This could paradoxically indicate that the Ipiutak people might have experienced higher frequencies of mortality after short-term infection or due to trauma not recorded on the skeleton, while the Tigara people were able to survive longer through infections or other metabolically stressing disorders.
4. It is notable that, in light of Fitzhugh's observations in the next chapter (Chapter 11) of this volume concerning long-distance Asian trade connections to western Alaska, as well as some of the discussion by Mason (Chapter 3) and Jensen (Chapter 2), it is likely that tuberculosis and other infectious diseases might have traveled into locations such as Point Hope. That the Tigara were part of the wide-ranging Thule cultures further argues for a trans-Beringian exchange that led to the transmission of infectious agents.

References

Antón, S. C., Carter-Menn, H. and Deleon, V. B. (2010). Modern human origins: Continuity, replacement, and masticatory robusticity in Australasia. *Journal of Human Evolution*, 60, 70–82.

Auerbach, B. M. (2007). *Human Skeletal Variation in the New World During The Holocene: Effects of Climate and Subsistence across Geography and Time*. Ph.D. Johns Hopkins University School of Medicine.

Auerbach, B. M. (2010). Giants among us? Morphological variation and migration on the Great Plains. In *Human Variation in the Americas: The Integration of Archaeology and Biological Anthropology*. Carbondale, IL: Southern Illinois University Carbondale, pp. 172–214.

Auerbach, B. M. (2011). Reaching great heights: changes in indigenous stature, body size and body shape with agricultural intensification in North America. In *Human Bioarchaeology of the Transition to Agriculture*. London: John Wiley & Sons, pp. 203–33.

Auerbach, B. M. (2012). Skeletal variation among early Holocene North American humans: Implications for origins and diversity in the Americas. *American Journal of Physical Anthropology*, 149, 525–36.

Auerbach, B. M. and Ruff, C. B. (2004). Human body mass estimation: A comparison of "morphometric" and "mechanical" methods. *American Journal of Physical Anthropology*, 125, 331–42.

Auerbach, B. M. and Ruff, C. B. (2010). Stature estimation formulae for indigenous North American populations. *American Journal of Physical Anthropology*, 141, 190–207.

Betti, L., von Cramon-Taubadel, N. and Lycett, S. J. (2012). Human pelvis and long bones reveal differential preservation of ancient population history and migration out of Africa. *Human Biology*, 84, 139–52.

Betti, L., von Cramon-Taubadel, N., Manica, A. and Lycett, S. J. (2013). Global geometric morphometric analyses of the human pelvis reveal substantial neutral population history effects, even across sexes. *PLoS One*, 8, e55909.

Blakely, R. L. and Beck, L. (1984). Tooth-tool use versus dental mutilation: A case study from the prehistoric southeast. *Midcontinental Journal of Archaeology*, 9, 269–84.

Bockstoce, J. (1979). *The Archaeology of Cape Nome, Alaska*. University Museum Monograph 38. Philadelphia, PA: University of Pennsylvania.

Bruzek, J. (2002). A method for visual determination of sex, using the human hip bone. *American Journal of Physical Anthropology*, 117, 157–68.

Costa, R. L. (1980a). Incidence of caries and abscesses in archeological Eskimo skeletal samples from Point Hope and Kodiak Island, Alaska. *American Journal of Physical Anthropology*, 52, 501–14.

Costa, R. L. (1980b). Age, sex, and antemortem loss of teeth in prehistoric Eskimo samples from Point Hope and Kodiak Island, Alaska. *American Journal of Physical Anthropology*, 53, 579–87.

Costa, R. L. (1982). Periodontal disease in the prehistoric Ipiutak and Tigara skeletal remains from Point Hope, Alaska. *American Journal of Physical Anthropology*, 59, 97–110.

Cowgill, L. W., Eleazer, C. D., Auerbach, B. M., Temple, D. H. and Okazaki, K. (2012). Developmental variation in ecogeographic body proportions. *American Journal of Physical Anthropology*, 148, 557–70.

Crawford, M. H. (2007). Genetic structure of circumpolar populations: A synthesis. *American Journal of Human Biology*, 19, 203–17.

Crawford, M. H., Rubicz, R. C. and Zlojutro, M. (2010). Origins of Aleuts and the genetic structure of populations of the archipelago: Molecular and archaeological perspectives. *Human Biology*, 82, 695–717.

Dabbs, G. (2009). *Health and Nutrition at Prehistoric Point Hope, Alaska: Application and Critique of the Western Hemisphere Health Index*. Ph.D. University of Arkansas.
Dabbs, G. (2011). Health status among prehistoric Eskimos from Point Hope, Alaska. *American Journal of Physical Anthropology*, 146, 94–103.
Debetz, G. F. (1959). The skeletal remains of the Ipiutak cemetery. In *Actas del XXXIII Congreso Internacional de Americanistas*, San José, pp. 57–64.
Edholm, O. G. and Lewis, H. S. (1964). Terrestrial animals in cold: Man in polar regions. In *Handbook of Physiology, Section 4: Adaptation to the Environment*. Washington, D.C.: American Physiological Society, pp. 435–46.
El Zaatari, S. (2008). Occlusal molar microwear and the diets of the Ipiutak and Tigara populations (Point Hope) with comparisons to the Aleut and Arikara. *Journal of Archaeological Science*, 35, 2517–22.
Gossett, S. and Hilton, C. E. (2004). Tuberculosis in Iñupiats from Point Hope, Alaska: A possible maritime resource connection. *American Journal of Physical Anthropology*, S30, 133.
Grine, F. E., Jungers, W. L., Tobias, P. V. and Pearson, O. M. (1995). Fossil *Homo* femur from Berg Aukas, northern Namibia. *American Journal of Physical Anthropology*, 97, 151–85.
Harvati, K. and Weaver, T. D. (2006). Human cranial anatomy and the differential preservation of population history and climate signatures. *The Anatomical Record Part A*, 288, 1225–33.
Helgason, A., Palsson, G., Pedersen, H. *et al.* (2006). mtDNA variation in Inuit populations of Greenland and Canada: Migration history and population structure. *American Journal of Physical Anthropology*, 130, 123–34.
Higgins, R. W. and Ruff, C. B. (2011). The effects of distal limb segment shortening on locomotor efficiency in sloped terrain: Implications for Neandertal locomotor behavior. *American Journal of Physical Anthropology*, 146, 336–45.
Holmes, M. A. and Ruff, C. B. (2011). Dietary effects on development of the human mandibular corpus. *American Journal of Physical Anthropology*, 145, 615–28.
Holliday, T. W. (1997). Postcranial evidence of cold adaptation in European Neandertals. *American Journal of Physical Anthropology*, 104, 245–58.
Holliday, T. W. and Hilton, C. E. (2010). Body proportions of circumpolar peoples as evidenced from skeletal data: Ipiutak and Tigara (Point Hope) versus Kodiak Island Inuit. *American Journal of Physical Anthropology*, 142, 287–302.
Keenleyside, A. (1998). Skeletal evidence of health and disease in pre-contact Alaskan Eskimos and Aleuts. *American Journal of Physical Anthropology*, 107, 51–70.
Krueger, K. L. and Ungar, P. S. (2012). Anterior dental microwear texture analysis of the Krapina Neandertals. *Central European Journal of Geoscience*, 4, 651–62.
Larsen, H. and Rainey, F. (1948). *Ipiutak and the Arctic Whale Hunting Culture*. Anthropological Papers of the American Museum of Natural History 42. New York, NY: American Museum of Natural History.
Laughlin, W. S. (1979). Problems in the physical anthropology of North American Indians, Eskimos and Aleuts. *Arctic Anthropology*, 16, 165–77.
Lester, C. W. and Shapiro, H. L. (1968). Vertebral arch defects in the lumbar vertebrae of pre-historic American Eskimos. *American Journal of Physical Anthropology*, 28, 43–8.

Maley, B. (2011). *Population Structure and Demographic History of Human Arctic Populations Using Quantitative Cranial Traits*. Ph.D. Washington University in St. Louis.

Martin, R. (1928). *Lehrbuch der Anthropologie in Systematischer Darstellung mit Besonderer Berücksichtigung der Anthropologischen Methoden für Studierende, Ärtze und Forschungsreisende. Zweiter Band: Kraniologie, Osteologie*. Second Edition. Jena: Gustav Fischer.

Mason, O. K. (1998). The contest between Ipiutak, Old Bering Sea and Birnirk polities and the origin of whaling during the First Millennium A.D. along Bering Strait. *Journal of Anthropological Archaeology*, 17, 240–325.

Mason, O. K. (2000). Archaeological Rorshach in delineating Ipiutak, Punuk and Birnirk in NW Alaska: Masters, slaves or partners in trade? In *Identities and Cultural Contacts in the Arctic*, Publication No. 8. Copenhagen: Danish Polar Center, pp. 229–51.

Merbs, C. F. (1996). Spondylolysis of the sacrum in Alaskan and Canadian Inuit skeletons. *American Journal of Physical Anthropology*, 101, 357–67.

Moran, E. F. (2008). *Human Adaptability: An Introduction to Ecological Anthropology*. Philadelphia, PA: Westview Press.

Phenice, T. W. (1969). A newly developed visual method of sexing the os pubis. *American Journal of Physical Anthropology*, 30, 297–301.

Powell, J. P. (2005). *The First Americans: Race, Evolution, and the Origin of Native Americans*. Cambridge: Cambridge University Press.

Relethford, J. H. (2004). Boas and beyond: Migration and craniometric variation. *American Journal of Human Biology*, 16, 379–86.

Relethford, J. H. (2009). Population-specific deviations of global human craniometric variation from a neutral model. *American Journal of Physical Anthropology*, 142, 105–11.

Roseman, C. C. and Auerbach, B. M. (n.d.) Ecogeography, genetics, and the evolution of human body form. *Journal of Human Evolution*. In press.

Roseman, C. C. and Weaver, T. D. (2007). Molecules versus morphology? Not for the human cranium. *Bioessays*, 29, 1185–8.

Roseman, C. C., Willmore, K. E., Rogers, J. *et al.* (2010). Genetic and environmental contributions to variation in baboon cranial morphology. *American Journal of Physical Anthropology*, 143, 1–12.

Rubicz, R., Melton, P. E., Spitsyn, V. *et al.* (2010). Genetic structure of native circumpolar populations based on autosomal, mitochondrial, and Y chromosome DNA markers. *American Journal of Physical Anthropology*, 143, 62–74.

Ruff, C. B. (1994). Morphological adaptation to climate in modern and fossil hominids. *Yearbook of Physical Anthropology*, 37, 65–107.

Ruff, C. B. (2006). Environmental influences on skeletal morphology. In *Handbook of North American Indians, Volume 3: Environment, Origins and Population*. Washington, D.C.: Smithsonian Institution Press, pp. 685–94.

Ruff, C. B., Niskanen, M., Junno. J.-A. and Jamison P. (2005). Body mass prediction from stature and bi-iliac breadth in two high latitude populations, with application to earlier higher latitude humans. *Journal of Human Evolution*, 48, 381–92.

Sassaman, K. E. (2010). Bridging the empirical divide of human variation research in the New World. In *Human Variation in the Americas: The Integration of Archaeology and Biological Anthropology*. Carbondale, IL: Southern Illinois University Carbondale, pp. 347–64.

Utermohle, C. J. (1984). *From Barrow Eastward: Cranial Variation of the Eastern Eskimo*. Ph.D. Arizona State University.

Van Stone, J. W. (1962). *Point Hope: An Eskimo Village in Transition*. Seattle, WA: University of Washington Press.

von Cramon-Taubadel, N. (2009). Congruence of individual cranial bone morphology and neutral molecular affinity patterns in modern humans. *American Journal of Physical Anthropology*, 140, 205–15.

von Cramon-Taubadel, N. and Weaver, T. D. (2009). Insights from a quantitative genetic approach to human morphological evolution. *Evolutionary Anthropology*, 18, 237–40.

Weaver, T. D. (2009). Out of Africa: Modern human origins special feature: the meaning of neandertal skeletal morphology. *Proceedings of the National Academy of Sciences USA*, 106, 16028–33.

Weiss, E. (2003). Effects of rowing on humeral strength. *American Journal of Physical Anthropology*, 121, 293–302.

Wood, J. W., Milner, G., Harpending, H. C. *et al.* (1992). The Osteological Paradox: Problems of inferring prehistoric health from skeletal samples [and comments and reply]. *Current Anthropology*, 33, 343–70.

Zlojutro, M., Rubicz, R., Devor, E. *et al.* (2006). Genetic structure of the Aleuts and circumpolar populations based on mitochondrial DNA sequences: A synthesis. *American Journal of Physical Anthropology*, 129, 446–64.

11 *The Ipiutak spirit-scape: An archaeological phenomenon*

WILLIAM W. FITZHUGH

Archaeological finds sometimes achieve startling notoriety. Schliemann's discovery of Troy, Carter's entry into the undisturbed tomb of Tutankhamen, and the discovery of Paleolithic art in southwestern Europe rank among the world's greatest finds. Arctic sites have rarely risen to this level of prominence, although a case for Paleolithic cave art could be made since southern Europe lay within the Arctic/Subarctic zone 12,000–27,000 years ago. Post-glacial Arctic archaeology usually generates little attention; but occasionally there have been great rewards when the miracle of frost has preserved treasures that illuminate a spectacular part of human existence, one that never could have been imagined from Arctic societies and cultures as they are known today or in the historical record.

The Ipiutak site of Point Hope, Alaska, clearly falls into a small group of luminous Arctic discoveries that opens a world of understanding about a vanished culture. Ipiutak (the Iñupiaq word for a narrow sand bar separating two lagoons) is not a world we can easily slip into; its fundamental animist beliefs are alien to thousands of years of Western heritage. Even from the more culturally proximate East Asian/Siberian tradition, Ipiutak and its sister cultures of Bering Strait provide uncertain context for scientifically resurrecting its way of life. Many of the most interesting and beautiful Ipiutak creations – such as its swivels, its bone daggers and twisted "pretzels," its grotesque animal forms and rake-like objects, its strange multi-piece burial masks – are as enigmatic to us today as they were to the original excavators and their Iñupiat field assistants. Ipiutak is also unusual in that time and excavation have given us a large body of information from both domestic and mortuary contexts. Even in the absence of wood preservation (except for house floors, log walls, and charcoal) the collection of stone, bone, and ivory objects from Point Hope – more than 17,000 artifacts from 575 houses and 138 graves in the Ipiutak component alone – constitutes an unusually large collection from a single Arctic excavation.

The Foragers of Point Hope: The Biology and Archaeology of Humans on the Edge of the Alaskan Arctic, eds. C. E. Hilton, B. M. Auerbach, L. W. Cowgill. Published by Cambridge University Press.

Equally important, the authors produced one of the most complete archaeological reports ever prepared on any site from a northern region. Helge Larsen and Froelich Rainey's team, consisting of themselves, James L. Giddings (1939 only), Harry Shapiro (1941 only), and just a few others, managed exceptional work over a three-year period under difficult field conditions. Key to their success was the enthusiastic assistance of the Point Hope Iñupiat community who helped instruct them in local customs and subsistence endeavors and assisted in site discovery and excavation. Rainey had experience with Otto Geist on St. Lawrence Island and Larsen had extensive knowledge of circumpolar archaeology and Eskimo ethnography. Both lived in the village of Point Hope for various periods from 1939 to 1942. Rainey hunted with its whaling crews in 1941. Their extended village residence – unusual for archaeologists – greatly enhanced data analysis and interpretation. World affairs also contributed to project legacy. Larsen's alienation from occupied Denmark for much of 1941–5, his residency and professional associations at the University of Alaska, and his employment at the American Museum of Natural History from 1943 to 1945 all contributed to the quality of the final report (Larsen and Rainey 1948). One of the monograph's unusual features – a legacy of John Murdoch's monograph on Point Barrow (1892) and the Danish Eskimology tradition established by Kai Birket-Smith (1929) in his Caribou Eskimo report – is its wide-ranging circumpolar comparisons. Ipiutak house forms, burial types, tool forms, technologies, art, and broad cultural complexes such as shamanism and hunting magic were compared with other archaeological and ethnographic cultures throughout the circumpolar north and beyond, to the Northwest Coast, the American Southwest, and further. Except for the absence of a full osteological report (an interim report on trauma was prepared by Lester and Shapiro [1968] and a study of affinities by Debetz [1959]), *Ipiutak and the Arctic Whale Hunting Culture* (Larsen and Rainey, 1948) was, for its time, one of the most comprehensive archaeological monographs written on any Arctic culture, making a fine companion to the other major monograph of the period, Henry Collins' *Archaeology of St. Lawrence Island* (1937). Our present volume, documenting diverse perspectives on skeletal biology and archaeological interpretation gained from the past half-century, makes Ipiutak an even more important landmark in Arctic archaeology.

This volume presents a timely Ipiutak "science-scape" reassessing the impact of these finds on Arctic anthropology sixty-five years after the original publication. Following Mathiassen's and Frederica deLaguna's collaboration on Inugssuk archaeology in Greenland (when DeLaguna was still a student) and her partnership with Kai Birket-Smith on Chugach ethnology and archaeology in Prince William Sound, Alaska, Larsen's and Rainey's joint work supported by the Danish National Museum and the American Museum of Natural History was the first major international Arctic project to undertake joint fieldwork and publication. They produced the first "complete" reconstruction of a prehistoric Arctic culture including its domestic, technological, mortuary, art, and belief systems. The monograph was the first to demonstrate Asian heritage of an ancient New World culture with specific comparisons in art, shamanism, and burial traditions. Recognition of Ipiutak's Asian heritage was responsible, more than any other excavation or cultural analysis, for establishing the validity of circumpolar anthropology (Fitzhugh, 2010), because it provided the first empirical evidence for long-range connections initially theorized by Gutorm Gjessing (1930) and Waldemar Bogoras (1929; Fitzhugh, 1975, 2010). My contribution to this volume provides a partial reassessment of these contributions in the light of a half-century of new research that was stimulated especially by the opening of the Russian Arctic to collaboration by Russian, Eastern, and Western researchers.

In the early days of anthropology, Arctic cultures were often seen as devolved cultural remnants, a kind of Paleolithic echo of the cultures of Ice Age Europe (Gjessing, 1930). As recently as the 1960s, it was customary to view ancient and recent Eskimo cultures as marginal outliers of the civilized world living in a remote, harsh environment. The views of both ethnographers and archaeologists (e.g., Birket-Smith, 1930; Irving, 1969–70) were tinged with evolutionary and racial baggage that predisposed interpretation of Arctic peoples as exemplars of environmental determinism. By the 1970s–80s those views began to change, both among anthropologists as cultural theory advanced (Schindler, 1985), and among the wider public with the appearance of a series of popular museum exhibitions. The first was the National Gallery of Art's *The Far North* (Collins *et al.*, 1973). This was followed nine years later by the Smithsonian's *Inua: Spirit World of the Bering Sea Eskimo* (Fitzhugh and Kaplan, 1982), and in 1988 by its *Crossroads of Continents: Cultures of Siberia and Alaska* exhibitions (Fitzhugh and Crowell, 1988). These travelling exhibitions brought world-class northern art to widespread public attention for the first time in North America and to Native communities in local "mini-exhibits".

At the same time that the spectacular Ipiutak finds were being discovered and publicized in the West, a parallel discovery was having a similar effect in the Soviet Union. Sergei Rudenko began excavating the frozen Pazyryk burials in South Siberia in the 1920s and returned again in 1947. Although these sites were in the Eurasian steppe and not in the Arctic, the artistic ornamentation – later known as Scytho-Siberian animal style – seen in gold foil-covered wood carvings, embroidered felt, and tattoos on frozen bodies, invited comparison with Old Bering Sea art with which Rudenko was already familiar from his excavations at Bering Strait (Rudenko, 1947). The Pazyryk finds were published in Russian by the Hermitage in 1948 and 1949 (Rudenko, 1970). Larsen and Rainey seem not to have been aware of these finds from the headwaters of the Yenisei River. Instead they linked Ipiutak art to the Bronze and Iron Age cultures represented by Valerii Chernetsov's Ust-Polui finds from sacrificial sites near the mouth of the Ob (Chernetsov, 1935, 1970; Chernetsov and Mozhinska, 1974; Federova, 2003a, b) and in ornamental metal work from Soviet excavations reported by A. M. Tallgren from Perm and the southern Urals (Larsen and Rainey, 1948: 157–61). With limited comparative material available other than Collins' work on St. Lawrence Island, and with nothing similar from the Northwest Coast of America, the authors had to cast a wide net, and they did so in a very detailed manner.

In the 1940s little was known about the complexity and elegance of ancient Eskimo art. The surprising discovery of Old Bering Sea (OBS) art by Diamond Jenness from excavations and collections he made in Bering Strait in 1926, and subsequent explorations by Geist, Rainey, and Collins at Okvik and OBS sites on Punuk and St. Lawrence Island in 1927–9, alerted scholars to the existence of a hitherto unknown set of Arctic cultures and art, but these finds were mostly purchased from Eskimos or came from dwelling site and midden excavations. It was not until Sergei Rudenko, Nikolai Dikov, Maxim Levin, and other Russian archaeologists began systematic excavations on the Bering Strait coast of Chukotka in the 1940s and 50s (Rudenko, 1961) and later at the Ekven and Uelen cemeteries near East Cape in the 1960s–70s (Arutiunov and Sergeev, 2006a [1969], b [1975]) that the full impact of OBS art began to be appreciated. Because most of these publications were in Russian and were available only to a few Western scholars their spectacular results initially brought little attention. The first public viewing of OBS art, *Ancient Eskimo Ivories from the Bering Strait*, that exhibited the beautiful objects that had begun to flood the antiquities market from Native excavations on St. Lawrence Island and other sites, opened at the American Federation of Arts in New York in 1986 (Wardwell, 1986). *Crossroads* (1988) was the first exhibition to present Russian OBS collections in North America. Europeans had seen the Danish share of OBS and Ipiutak art since the 1950s, and OBS materials had been on view at the University of Alaska in Fairbanks, the Smithsonian and the American Museum of Natural History in New York, but the small display cases they were in had little impact. Broader attention by Europeans to the Russian collections came from articles by Mikhail Bronshtein and Patrick Plumet (1994, 1995), and by catalogs and exhibitions in Tübingen, Germany (Leskov and Muller-Beck, 1995) and at the Museum of Oriental Art in Moscow (Bronshtein *et al.*, 2007). A comprehensive U.S. exhibition initiated by Julie Hollowell and organized by the Princeton University Art Museum in 2009 highlighted Okvik, OBS, and Punuk art borrowed from museum and private collections around the world, by far the largest number coming from the Museum of Oriental Art in Moscow

(Fitzhugh *et al.*, 2009; Bronshtein, 2009). Concurrently, an exhibition organized and curated by Edmund Carpenter, *Upside Down: Les Arctiques* (Carpenter *et al.*, 2008) opened at the Branly Museum in Paris in 2008 and later as *Upside Down: Arctic Realities* (Carpenter *et al.*, 2011) at the Menil Collection in Houston in 2011. The Carpenter exhibitions included many fine objects of Okvik, OBS, and Punuk art from the Carpenter/Menil and Moscow Museum of Oriental Art collections; it also included Dorset art from the Canadian Museum of Civilization, and Ipiutak art from the Danish National Museum. Finally, to assist in assessing Larsen and Rainey's interpretations we now have available the translated reports of Chernetsov and Mozhinska's work on Yamal (Chernetsov and Mozhinskaya, 1974) and catalogs of materials excavated by Fedorova from exhibits in St. Petersburg and Salekhard (Fedorova, 2003a, b) that present Bronze and Iron Age art of Western Siberia. Because papers by Mason (Chapter 3) and Dumond (Chapter 12) in this volume primarily concern culture history, especially on the Alaskan side, the following discussion focuses on Ipiutak art and ritual, the elements of Ipiutak culture that Larsen and Rainey used to support their theories of Asian relationships.

Ipiutak had a pivotal position in the field Danish scholars coined as "Eskimology" early in the twentieth century. Once it was understood that Eskimo origins and its most central developments took place in the Bering Strait region, as demonstrated conclusively by Collins (1937), the discoveries of fine art and hundreds of Ipiutak houses displayed in ranks that gave the appearance of "town-planning" at Point Hope startled scholars. Further, the site's large well-preserved cemeteries linked to its different settlement periods, presented a dramatic contrast with earlier multi-component village midden excavations conducted previously by Jenness, Geist, Rainey, and Collins. How could such sites have been created by Eskimos who until that time had been represented primarily by climate-stressed Inuit peoples from the Eastern Arctic and Greenland? Equally influential was the surprising artistry of the Ipiutak artifacts. Finds from the nearby sites of Jabbertown and Old Tigara – the initial target of Larsen and Rainey's attention owing to Knud Rasmussen's earlier reports – resembled those of the Eastern Arctic Thule and historic Inuit cultures. The materials excavated from Ipiutak houses and graves produced a wealth of data that provided the first large-scale cultural interpretation of a pre-Thule Western Arctic Eskimo culture. Ipiutak was revealed as a culture profoundly different from contemporary eastern Inuit or from previous excavations in Bering Strait or western Alaska; in particular, their artistic themes echoed Rudenko's Scythian-related Pazyryk finds. Ipiutak people lived in wood houses, used open fires without soapstone lamps or pottery; had few dogs and no dog-drawn sleds, used chipped stone rather than ground stone tools, hunted with harpoons lacking drag-floats, and were peculiarly attached to death rituals and ornaments unlike any found at other Eskimo sites in North America. Even compared with sites from Bering Strait, Ipiutak and its closely related Near Ipiutak village (now known to be an Ipiutak predecessor rather than, as Larsen and Rainey supposed, a descendant) appeared as an enigmatic early Eskimo culture, if it was Eskimo at all.

Art and ritual

The most evocative and enigmatic finds are Ipiutak's art and ritual materials, which occurred in both domestic and mortuary contexts. The volume of this material is immense. Artistic treatment was applied to items as prosaic as arrowheads and as iconic as its famous masks, bands, tubes, and animal carvings. Sculptural masterpieces include animal carvings, strange bear-imaged combs, miniature human and animal heads, and hundreds of intricately fashioned open work ivory and bone carvings made in a multitude of shapes: "swivels," twisted pretzel-shaped pieces, chains, and links ornamented with animal heads or faces, often of loons. Even though soil acids have compromised the beauty of many objects, the high quality of workmanship is striking.

At the beginning of their discussion of art Larsen and Rainey state:

> ... the Ipiutak [people] were not only extremely skillful carvers but possessed an original imagination and ingenuity with few counterparts in the world. These are clearly demonstrated in the openwork carvings, almost every example of which is unique. Whether we designate the openwork carvings as art objects or as the unique outcome of eccentricity is a matter of opinion. However, even if we omit these carvings from consideration, we still have numerous evidences of their artistic ability. In their sculpture they display an appreciation of beauty of form; in their decorative art they demonstrate an unusual ability to adapt a composition to the shape of the surface. Above all, in all their handicraft they demonstrate a delight in nicely shaped artifacts, which, if possible, were given a final decorative touch: a simple line design, an elaborate pattern, or perhaps an animal head. In no phase of Eskimo culture do we find as great a quantity and as diverse a range of decorated artifacts as in the Ipiutak culture ... the closest parallel may be found in the Okvik phase of the Arctic Whale Hunting culture. In the Old Bering Sea phase, Eskimo decorative art attains its climax, especially in the extreme elaboration of surface ornamentation, but the decoration is limited to a few types of artifacts, such as harpoon heads and socket pieces, needle cases, scrapers, and winged figures. In later phases of the Arctic Whale Hunting culture, ornamentation is simpler and occurs less frequently. Of the historic Eskimo only the Bering Sea Eskimo and the Angmagssalik Eskimo, far to the east, are comparable in craftsmanship and artistic achievement to the Ipiutak people. (Larsen and Rainey, 1948: 135)

These comments are as prescient today as they were when written in the 1940s. Even though much has been learned about Ipiutak culture and its southern distribution and links with Norton culture, as noted by Mason (Chapter 3) and Dumond (Chapter 12) herein, it is useful to consider the Norton and Bering Sea (i.e., Yup'ik) Eskimo connection as it relates to art, religion, and ritual, as this provides a clearer understanding of Ipiutak and corrects the mis-impression that the authors emphasized only the Asian component in explaining Ipiutak's peculiarities within the Eskimo mainstream. As the original authors pointed out, and as Mason and Dumond affirm in this volume, there is a pronounced discontinuity, at least in northwest Alaska, following Ipiutak, before the modern Arctic whaling culture sequence begins with Punuk, a century or so before 1000 CE. So many changes occur that Mason has argued strongly, here and elsewhere, for warfare and population replacement by an Asia-derived Punuk force, at least in the whaling territories north of Bering Strait. Nevertheless, continuities to southern parts of Alaska are also present.

What is most striking about this Punuk shift is the replacement of an earlier underlying belief system that was rooted in the idea that hunting was a spiritual contract between individual hunters and the animals they pursued. A semblance of this philosophy was described by Edward Nelson (1899) in his ethnography of the Yup'ik peoples of the lowland Yukon-Kuskokwim delta. Nelson's collections reveal the pervasiveness of this concept as expressed in material culture, ceremony, stories, ritual, and social life (Fitzhugh and Kaplan, 1982; Fitzhugh and Crowell, 1988) and became even more evident in ethnography conducted since the 1970s by Anne Fienup-Riordan (1988, 2005) and her Yup'ik associates. The belief that all objects, animate and inanimate, had sentient souls and that humans needed to maintain personal relationships with these beings in order to sustain life as hunters and prey, or users of worldly materials, was fundamental to Yup'ik survival. When things went wrong, various rituals might set things right; for instance, shamans might be dispatched to the controlling master spirits to bring food, change weather, or stem disease or conflict. Among early cultures of the Beringian region, including the Ainu, these views led to a proliferation of individualized artistic expression seen in the meticulous preparation of clothing, and manufacture of tools, weapons, and items of everyday use. Logically, hunting equipment became a prominent place for expressing one's respect for the spirits of the game and their controllers by making one's implements beautiful and pleasing to animals as well as to gain the respect of other people. The motivation for artistic elaboration seen in early Eskimo and ethnographic art of Northwestern America was probably driven by similar belief and practice. The artistry of material culture was created from cultural templates that were personalized by individual craftsmen and women. Replication and commoditization were eschewed, and pride in artistic creativity was a general characteristic. Ipiutak is a striking archaeological example of such a belief system as it applied to material culture. We might even go so far as to describe this approach to art as a kind of cosmological "spirit-scape" in which artifacts – like human and animal spirits – serve multiple purposes and have multiple shapes and forms; the artifact as object and the artifact as a vehicle for communicating with the spirit world.

Larsen and Rainey several times (e.g., p. 136 in Larsen and Rainey, 1948) express surprise and frustration at not being able to define a specific "Ipiutak" art style, as Collins had successfully done for OBS I (Okvik), OBS II, OBS III, and Punuk. While a considerable discussion has recently emerged over the original interpretation of these styles as culturally and chronologically specific designations (Gerlach and Mason, 1992) and the extent to which there may be little-understood regional, ethnic, or social components ("polities," in Gerlach and Mason parlance) within these categories, these styles have withstood scrutiny of more than half a century. Most of this new evidence, however, comes from St. Lawrence Island and Chukotka, where it substantiates and to some extent clarifies the dating of these styles (but see Gerlach and Mason, 1992; Mason, 1998, 2000, 2006, 2009a, b; Chapter 3, this volume; and Dumond, Chapter 12, this volume; Fitzhugh, 2009a). What has been less clear, and what was less clear to the original authors, and still remains so today, is how Ipiutak art relates to the history of Eskimo cultures in western and southwestern Alaska, and, in particular to Okvik and Norton and its presumed Yup'ik Bering Sea Eskimo descendants.

The problem with defining an Ipiutak style is peculiar in that most who have studied this material agree that "you know it when you see it." If so, why has it been difficult to describe? The inherent problem is that there is no unique design style applied throughout the Ipiutak assemblage of diagnostic artifact types such as multi-spurred harpoon points, openwork carvings, masks and maskoids, and other items. When its decorative art is rolled out, it can be seen as obviously related to Okvik (e.g., one of the finest harpoon heads from the site (Figure 11.1)), is "so Okvik" that Larsen and Rainey thought it might be intrusive, the result of inter-polity trade or spoils of war – and that may indeed be the case). Ipiutak decorative art consists of deeply cut lines, often accompanied by equally deep convergent forms and more lightly engraved secondary or tertiary lines conforming to the shape of the parent object. These lines are often, as in Okvik, decorated with lighter secondary lines parallel to the first and ornamented with delicate tooth-like spurs or more deep triangular spurs. Intersecting lines are often marked with nucleated circles, circle-dots, ovoids, or ellipses, often with internal spoke-like elements or spurs. Many of these elements may be seen as animal "joint-marks." Most of the decoration is linear, but in rare instances, as in a pair of snow goggles (Larsen and Rainey, 1948:

Figure 11.1. Okvik style harpoon head from House 69 at the Ipiutak site (cat. No. AMNH P4366). The image is reproduced from Larsen and Rainey (1948: fig. 14) with permission from the American Museum of Natural History.

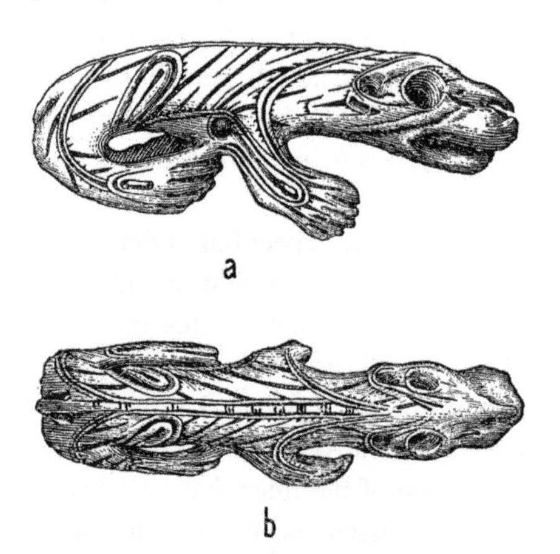

Figure 11.2. Ivory carving of a young walrus. The image is reproduced from Larsen and Rainey (1948: fig. 31) with permission from the American Museum of Natural History.

Figure 27) and two elegantly decorated flat ivory bands (Figure 11.5) a strong curvilinear design prevails, giving such pieces an OBS II/III aspect. The baby walrus (Figure 11.2) combines dominant lines whose curves follow the shape of the animal's form (Figure 11.5) with large spike-like incisions representing ribs attached to a segmented vertebral spine. A number of other Ipiutak specimens also feature skeletal art (e.g., a bear, Larsen and Rainey, 1948: Figure 30). Raised bosses and circular or ellipsoid joint marks are also seen on a walrus and a few other carvings but are rare compared to their prominence in OBS III art. The dominant lines that determine the basic structure of Okvik and OBS decorative patterns can be described as "form-lines," a term Bill Holm used to analyze Northwest Coast art (Holm, 1965), but seems equally valid for OBS.

Larsen and Rainey's concept of uniqueness of form and decoration is one of the features that makes Ipiutak art difficult to classify. Although the surface decoration described above can be seen as a backdrop, larger decorated objects such as bands, plaques, death masks, and antler tubes (Larsen and Rainey, 1948: Figures 28 and 45) contain schematic human and cryptic animal elements. The gaping mouth of an antler tube (Figure 11.3a) is carved into an elegant fanged wolf-like creature whose two form lines trail behind, ending in caribou hoof motifs (Figure 11.3b). Multi-iconic sculptures and surface designs feature strongly in the Ipiutak spirit-scape. A close parallel collected from the Tlingit consists of an ivory tube with open-mouth beasts carved at each end (Fitzhugh and Crowell, 1988: Figure 451). While carved in distinctive Tlingit style with ovoids and form-lines, its use as a shaman's sucking tube for curing rituals suggests a similar function for the Ipiutak tube. This fine artifact came from Burial 21, whose legs had been purposefully disarticulated; it was probably the burial of a shaman whom the survivors did not want to have "walking around" causing mischief or misfortune. In addition to this tube the burial contained four other incised bone tubes, and an ivory rod carved with a human hand at its proximal end (Larsen and Rainey, 1948: Plate 73.1) was threaded through several of the individual's lumbar vertebrae. This adult had a child buried between its legs.

Throughout the Ipiutak assemblage one is struck by the individualized treatment given to many artifact types. Sometimes this is seen in the diversity of sculptural forms in a class of objects like animal heads; in others it is the surface decoration of arrowheads or harpoon heads. Small carvings of human heads were also common finds in houses and burials. Larsen and Rainey likened these to the small death-heads common in Lamaist art, but they are also found in Kachemak and Koniag sites from Kodiak Island (Crowell *et al.*, 2001). Many other elements of the Ipiutak assemblage are also found in Bering Sea Yup'ik Eskimo and Alutiiq ethnographic collections and their archaeological predecessors, including carved ivory ornaments, charms, and small ritual items that have near-identical parallels in Ipiutak (e.g., antler-hoof charms, animal-themed line-attachers, loon skulls with inlaid jet eyes, and a wide variety of decorative forms that probably were mounted on bent-wood hunting hats). Many of these types of objects have not yet been found at Norton sites in the Yukon-Kuskokwim delta, the least well-known archaeological region of coastal Alaska. It is here that early versions of Okvik art and materials similar to Ipiutak could be expected. An undated beast-headed bone quiver stiffener collected by E. W. Nelson from

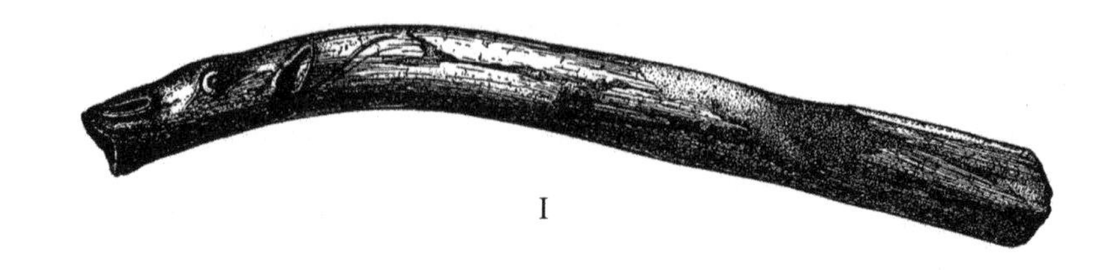

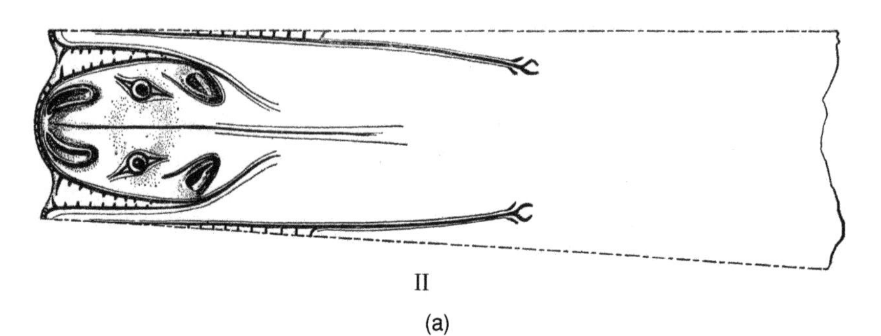

(a)

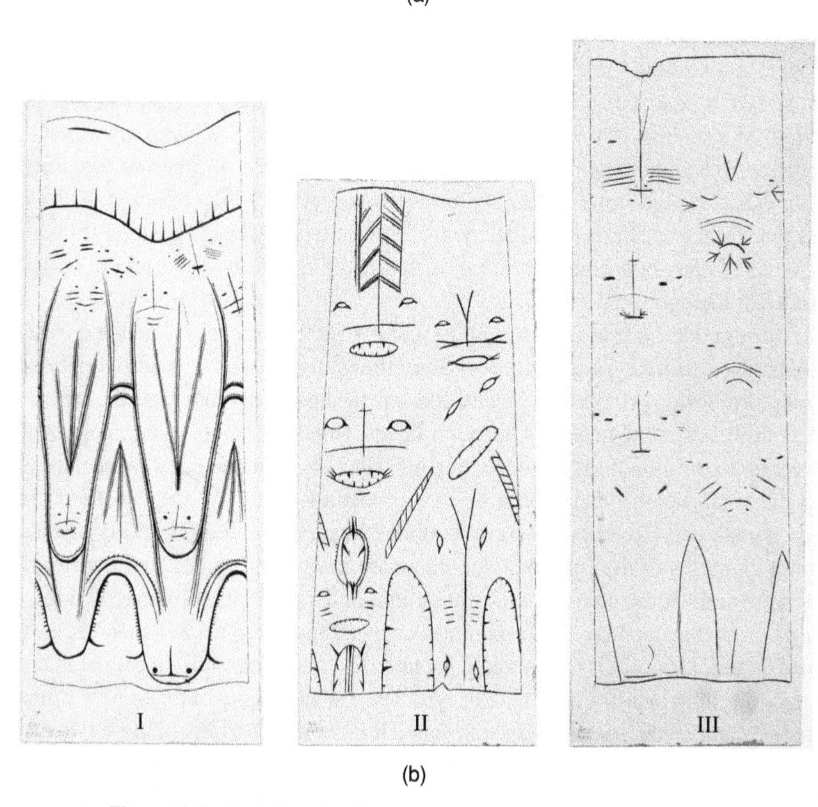

(b)

Figure 11.3. (a) Antler tube with engraved image of a wolf-like beast (AMNH cat. no. 60.1-7453, Burial 21) reproduced from Larsen and Rainey (1948); (b) Decorated antler tubes (AMNH cat. nos. 60.1-7452, Burial 21; 60.1-7455, Burial 21; 60.1-7451, Burial 21). All reproduced with permission from the American Museum of Natural History.

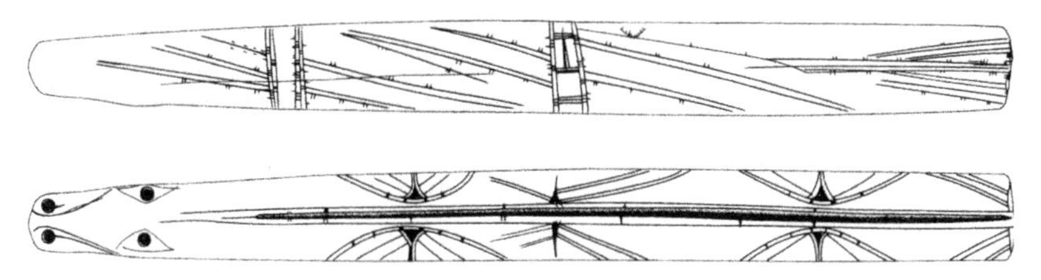

Figure 11.4. Bone quiver stiffener from Chalitmut, Yukon-Kuskokwim Delta (Smithsonian Institution Ethnology cat. no. 36396).

Chalitmut, noted by Collins (Figure 11.4; Collins, 1959; Fitzhugh and Kaplan, 1982), could well be an indicator of Okvik-like discoveries yet to come from the Yukon-Kuskokwim delta.

What these Ipiutak materials display is precisely what is missing from the later prehistory of Arctic regions from Chukotka to Greenland. Larsen and Rainey recognized these later whale hunting cultures had little in common with Ipiutak, whose spiritual life and art shared similarities with southern Eskimo and even Northwest Coast cultures. Today, with more appreciation for the diverse art of the Alaskan Yup'ik, Alutiiq (Koniag), and Unangax (Aleut) peoples, and better understanding of ethnography and belief as documented by Anne Fienup-Riordan and others, these similarities appear more central to the underlying culture "stratum" that produced Ipiutak. Even so, a number of items are strangely absent. Dolls, so far missing at Ipiutak sites, are common in the Yup'ik tradition and in archaeological and ethnographic collections from Kodiak, the Aleutians, and the Bering Sea coast. Small figures or images of humans served as house guardians among the Unangax and almost certainly had a spiritual dimension among the Yup'ik, but are rarely found in OBS sites on St. Lawrence Island and in Chukotka, although they are common in Okvik and Punuk contexts.

Another interesting absence is the mysterious winged object; but before discussing this, a technical digression into the Bering Sea harpoon system is needed. Early excavations in Bering Strait turned up many examples of finely decorated harpoon heads and socket-pieces. The mechanics of the harpoon head, fore-shaft, and socket-piece is well established (Arutiunov, 2009a). The size of the head was proportional to the game; the foreshaft, which detached from the harpoon head and socket upon impact, insured the harpoon head penetrated deeply beneath hide and blubber; and the socket-piece had the dual function of cushioning the foreshaft as well as providing an inertial punch for deep penetration. Socket-pieces for large game, such as walrus and whales, had to be heavy. The combination of these ivory parts produced a missile with a heavy front end, a shaft at least two meters long, and a hand-hold near its front end for proper balance. Whaling harpoons used by Thule culture and later Eskimos followed this construction and were launched from open skin boats by hunters who threw standing up. Harpoon-darts were lighter weapons that could be launched from kayaks, but were used only for hunting small seals.

Along with large harpoons with heavy socket-pieces suitable for hunting walrus, excavations at Okvik and OBS middens in Bering Strait recovered wing-shaped objects of various shapes and decorative styles. Collins showed these elegant objects could be arranged into a series that began with broad butterfly-like wings followed by objects whose wing size was gradually reduced until it formed a trident, and then a crown-like crest. Later research turned up pieces that showed the series ended with the crest being reduced to a single post. The sequence was ordered chronologically by style changes that began with Okvik (Collins' OBS I equivalent) and proceeded through OBS II, and OBS III, and ended with tridents, crowns, and posts in Punuk. Throughout their evolution two features remained constant: a small dimple was always present at the distal end and a countersunk hole for insertion of a wood shaft at the proximal end. For decades the objects were thought to have served as ornaments on ceremonial staffs like those used by Tlingit orators. The matter was resolved when burials from Ekven and Uelen showed winged objects fastened to the butt ends of wooden harpoon shafts whose front ends had heavy socket-pieces and walrus harpoon heads. The shafts were too short to have been used as spears, and the dimples indicated that these ivory-tipped missiles were thrown by hunters in kayaks who used throwing boards with hooks that engaged the indentations in the rear of the winged objects. Such a front-loaded dart could not be launched without a counterweight to balance the lighter rear end of the shaft. Although we now understand the technology, the origin of this mechanical system – particularly of the counterweight – still remains unknown, although Russian scholars have proposed it may have begun by employing a walrus atlas vertebra , since some of the early counterweights have a similar shape (Arutiunov, 2009b: Figure 8).

Equally mysterious was the significance of the carving and decoration on these harpoon parts. More recently, several explanations for its complex iconography have been proposed. Arutiunov (2009b) believes socket-pieces representing the face of a toothy predator symbolized the polar bear spirit, the master maritime predator whose power could enhance the hunter's quest, while the counterweight (as axis vertebra) signified the location of the quarry's spirit or life force. Fitzhugh (2009b) proposes the harpoon may harbor a host of spirit helpers, or *tunrat,* according to modern Yup'ik belief: the often feathered appearance of the harpoon head representing a master bird of prey; the socket-piece, a master predator of land or sea; and the stabilizer, the master spirit of the universe or *tunghak*, who nineteenth century Yup'ik believed dwelled in the moon. Ipiutak socket-pieces carry equally ornate sculptures of predators which grip the harpoon foreshaft in their jaws (Larsen and Rainey, 1948: Figures 15 and 47; Plate 38).

Poor preservation of the surfaces of most Ipiutak specimens makes comparison with OBS engraving difficult, but the socket-pieces clearly represent animal predator spirits. Ipiutak harpoons less often display bird-like imagery, and counterweights are completely absent. This absence must be important since it is so dominant in the Okvik–OBS–Punuk tradition. It is particularly notable since the dominant iconography on counterweights in the Bering Strait sequence features a semi-human beast whose prominence would appear to rank this being as a supreme master spirit. Thus, while similar iconography exists, Ipiutak technology and engraved art related to the hunting of large sea mammals differed in some significant details. In this regard, as in other categories noted above, Ipiutak seems more closely related to the southern Alaskan tradition. These comparisons between Ipiutak and southwest Alaskan cultures may prove to be only the tip of a cultural "iceberg" once more information becomes available from Alaska's geologically subsiding frozen coast south of Bering Strait.

The image of the OBS counterweight spirit serves to introduce a final category: masking. At Ipiutak, masking is known from its famous composite death masks, from a series of small human-heads and death-head carvings, so-called "snow goggles", and images of stylized human faces found on ivory plaques and tubes. The stylized faces are interesting because of their total abstraction, showing eyes as single dots, and nose and mouth by simple fine, straight lines; sometimes a toothy open mouth (singing, sucking, blowing?) is also shown. Despite their small size and simplicity, these images sometimes provide insight into facial decoration: short lines at the corners of the mouths, across the nose and cheekbones, slanting parallel lines on the cheeks, single and Y-shaped lines between the eyes and on the forehead, and marks suggesting labrets are shown in positions indicating male gender (by ethnographic example), almost certainly representing tattoos. On antler tubes they appear to float as a congregation of spirits that may assist the shaman during curing rituals – the Ipiutak spirit-scape in action. Similar stylized faces appear in petroglyphic art at Cape Alitak on Kodiak Island and on the incised pebbles often found at *c*. 1300 CE Koniag culture sites on Kodiak Island. Similar faces are also, in rare instances, incised on Dorset harpoons in the Canadian Arctic.

Similar but more elaborate faces form the central panel of the famous incised bands from Burial 61 (Figure 11.5). In this case, one band (Larsen and Rainey, 1948: Figure 38a, published upside-down) shows the classic stylized face with cheek and nose-bridge tattoos and a vertical line at the center of the chin, accompanied by two circles (labrets?). This face is set within a larger face

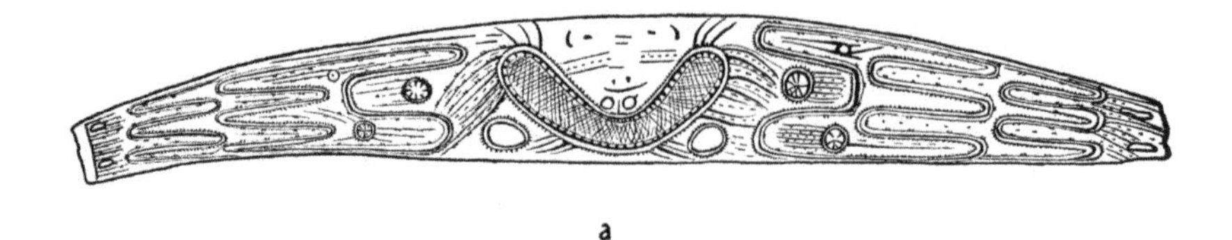

a

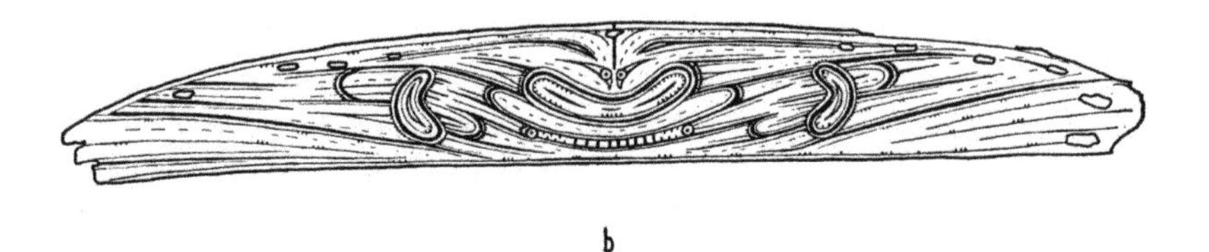

b

Figure 11.5. Decorated bands from Burial 61 (AMNG cat. nos. 60.1-7703, B61; 60.1-7702, B61), reproduced from Larsen and Rainey (1948: fig. 38) with permission from the American Museum of Natural History.

dominated by a whiskered or tattooed mouth with cross-hatched interior and two labrets below the corners of the mouth. This panel is flanked by wavy bands and spurred circles that might represent the heavens. The second band (Larsen and Rainey, 1948: Figure 38b, also published upside-down) is rendered in an evolved Okvik style reminiscent of OBS II, with multiple mouths, spurred nostrils and eyes, and a vertical line in the middle of the forehead. Instead of wavy lines and celestial bodies, this image is flanked by kidney-shaped ovoids (labrets?) set in a burst of straight form lines and lighter spurred lines and dashes. It is hard not to imagine these exquisite bands as representing dual master spirits set amidst the cosmos (perhaps the sun and the moon?) akin to the images of OBS stabilizers, but related more closely, pictorially, to Yup'ik and Alutiiq face art traditions (e.g., Crowell *et al.*, 2001: Figure 187). The same may be said of human-head renditions, some of which resemble the grotesque exaggerated human-faced burial masks known from Unga and other caves in the Aleutian Islands.

Most famous are the two ivory composite masks found with Burials 64 and 77 (Larsen and Rainey, 1948: Plates 55, 54). Larsen and Rainey believed these masks were used, as in archaeological and ethnological examples in Siberia, to cover the faces of the deceased between the time of death and burial, when they were removed and, as in the case of the larger Ipiutak one, disassembled. The smaller mask was found over the body of a child lying between the legs of an adult male in Burial 77; its pieces were intact, fastened to a piece of wood. Jet inlays for labrets below the corners of the mouth, jet eyes, and ivory nostril plugs were present. The larger and better preserved mask (Figure 11.6; Larsen

Figure 11.6. Composite ivory mask from Burial 64 (AMNH cat. no. 60.1-7713a-k) reproduced from Larsen and Rainey (1948: fig. 39) with permission from the American Museum of Natural History.

and Rainey, 1948: Figure 39, Plate 55) is more elaborate and is decorated with wavy incised lines punctuated by nucleated circles with radial, spoke-like lines, an animal (seal?) face at the top, two humanoid tattooed faces with labrets at the bottom corners, jet labrets, double-ended larvae-like creatures (botfly larvae in Owen Mason's interpretation) at the corners of the mouth, holding the mouth and nose pieces in place. Finds from other graves included decorated mouth-covers that appear to have been sewn onto the deceased or a lost wooden face cover, and jet inlays or pegs for human and loon head eyes. Larsen and Rainey compared the composite masks to Evenk, Yakut, and Finno-Ugric burial customs that required covering the faces of the dead and plugging body orifices during pre-burial rituals. Jenness (1952) proposed similarities to Han burial customs, and Collins (1971) identified Chinese Shang and Eastern Zhou Dynasty multi-piece jade composite burial masks as the most likely inspiration.

Also within the mask category are several sets of snow goggles, items also rare in the OBS tradition but common in Yup'ik ethnographic culture, where they often take on the aspect of a masked creature. Ipiutak goggles are unusual in having round eye holes and curvilinear decoration, perhaps having served as ceremonial masks rather than as functional snow goggles. Another rare mask-like implement consisted of two toothed ivory rakes or combs carved with ornate bear and seal imagery with jet eye inlays, found in burials along with hunting equipment. These pieces became the authors' prime suspects in attributing Asian animal style art influences at Ipiutak. The most elaborate example of this type comes from Point Spencer on Seward Peninsula (Collins *et al.*, 1973: Figure 21).

Ritual and society

The social context for ceremonial life and death rituals at Ipiutak is largely a matter of speculation. Despite the presence of a few dwellings that were larger than the typical domestic house – perhaps functioning as *qargis* (a men's house/workshop) – these dwellings are not large enough for major community- or clan-based gatherings; nor did their contents signal anything other than the production of men's hunting gear and equipment. Amplification of this view from Ipiutak comes from Deering, on the north side of Seward Peninsula, where Helge Larsen in 1950 excavated an 8 × 12 meter structure with an 8 × 4 m anteroom next to an 8 × 8 meter area with a central hearth and side benches. The large size of this structure and its contents suggested it functioned as a men's house or *qargi* (Larsen, 2001). Radiocarbon dates on wood range from *c.* 1,200 to 1,500 years BP, in line with new dates run on wood from several Ipiutak houses (Larsen, 2001: 80). Considering the good quality of wood preservation, it is strange that little ceremonial or ritual gear, and not a single fragment that could be interpreted as a dance mask or mask part, was found at Deering. Ornamented artifacts replicate types known from Ipiutak, including two ivory flint flaker handles with different ornamentation styles, one classically linear Okvik with nucleated junctions and the other with sinuous wavy designs. The designs on these two artifacts parallel the same two styles found on the Ipiutak engraved ivory bands. Other interesting finds included engraving tools (Figure 11.7), a remarkable two-pronged salmon spear (kakivak), a harpoon socket-piece with a twisted ("man-in-the-moon") human face, fishing rod ferrules, tattooed human face engravings, and many animal-headed utensils and amulets. Snowshoe and sled parts and arrowheads speak well for winter caribou hunting, and faunal remains suggest seals, walrus, caribou and fish were economic mainstays. A remarkable knife handle was recovered with a complicated design of interlocking ovals and circles that appears almost notational or narrative-like, defying the normal stylistic conventions but reminiscent of the more "storied" graphics of the Ipiutak bands and tubes. Sled parts and dog remains were present, but no harness parts.

In the absence of dance masks one is tempted to see Ipiutak social and ceremonial life as transitional, evolving toward the larger dwellings known from Punuk and later times in Bering Strait, and in southern Alaska and the Aleutians. Lacking drum parts at both Ipiutak and Deering, communal dancing

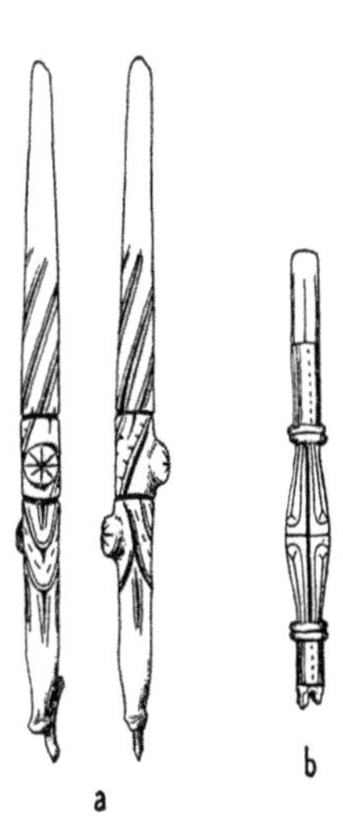

Figure 11.7. Ivory engraving styluses from House 43 and Burial 54 (AMNH cat. nos. P3724; 60.1-7689) reproduced from Larsen and Rainey (1948: fig. 18) with permission from the American Museum of Natural History.

and pot-latching seems not to have been practiced. This is difficult to square with Larsen and Rainey's belief that a cult of Asian shamanism appeared for the first time at Ipiutak. One thing is clear: the quality and individuality of carving and the artisanal care in the making of stone tools indicate that Ipiutak shared the belief that one's hunting success depended on more than technical prowess; the concept that animals would be attracted to hunters whose weapons and equipment were finely made set a high standard for art and individual creativity that was a widespread belief among non-whaling pre-Punuk cultures, and which persisted, to a large degree among the Eskimo cultures south of Bering Strait, even where whaling was sometimes practiced. Nevertheless, compared with Norton predecessors, social complexity and hierarchies were beginning to emerge even though reinforced bows and slat armor were absent.

Circumpolar and Asian connections

Much scholarly attention has been given to Larsen and Rainey's interpretations about the source of innovations seen at Ipiutak. In their day little was known about pre-Thule cultures. The Denbigh Flint Compex (a.k.a. Arctic Small Tool tradition) was not yet discovered, and Choris, Norton, and a host of other cultures were still to be found. For these reasons Larsen and Rainey believed that they had discovered the earliest Alaskan Eskimo culture then known. They considered Near Ipiutak to be a late Ipiutak variant rather than its precursor, as it is now understood. The presence of an extensive flint-working tradition and the belief that caribou hunting was the sustaining economic pattern was in keeping with concepts of the day for a relatively recent inland origin of Eskimo culture. Absence of pottery and soapstone vessels and a domestic life based on open fires helped support their interpretation. Today, archaeological opinion is quite different: Ipiutak is seen not as a summer camp on the Chukchi Sea but as a winter–spring occupation by people who subsisted largely on seals and walrus. Notwithstanding evident ties to south Alaska and echoes of the Northwest Coast, there is still a need to explain more fully features that to the authors argued for Asian inspiration – elements such as the raised bosses, joint marks, skeletal markings, as well as animal transformations and griffin-like motifs that harked to Scytho-Siberian Animal Style art in bronze and felt of Central Asia. Other elements seemed to call for prototypes from the Ob River basin and the animal art of the Iron Age Urals, where polyiconic art was also present in bone, ivory, and bronze. Given the lack of archaeological data from coastal Northeast Asia and Arctic coasts of Russia these views seemed plausible, considering Eskimo-like adaptations brought to light by Chernetsov's work in the Yamal region (Chernetsov, 1935, 1970; Chernetsov and Mozhinskaya, 1974).

Subsequent work in West Siberia and along the Russian Arctic coast (Fitzhugh 1998, 2002, 2010; Fedorova, 2003a, b; Pitul'ko, 1999; Khlobystin, 2005) removed this area from consideration, but Scytho-Siberian links seen in Rudenko's Pazryk art have continued to intrigue archaeologists. Larsen and Rainey saw links between the Ipiutak death masks and the Chinese tradition of masking the dead and plugging their eyes and orifaces to bar evil spirits. Later, Collins (1971) provided more specific links between these masks and Chou Dynasty jade death masks. Even stronger Asian connections were suggested by the hundreds of twisted ivory chains, swivels, animal-headed links, and other forms that seemed to replicate metal ornaments and bone and leather spirit-figures found on Siberian shaman costumes, and which were also present on everyday dress among the Evenk (Tungus), Even, Amur tribes, Chukchi, Yakut (Sakha), and other historically known Siberian peoples. Even the surface decoration on OBS, Ipiutak, and Punuk harpoon tools could easily be a version of the Bronze and Iron Age tattooing found on the frozen bodies of Pazyryk warriors (Rudenko, 1970), on Late Jomon ceramic figurines in Japan, and on the clothing of Ainu and other Northeast Asian peoples in the historical period where these designs served to bring good luck and protect wearers from malevolent spirits. Despite preliminary research to clarify Ipiutak's Asian connections, initially by Schuster (1951; Schuster and Carpenter, 1996), who pointed out general similarities in joint marks, skeletal art, transformation figures, and other motifs, by Fitzhugh (2002, 2004), who researched artistic parallels seen in 3,000-year-old Mongolian deer stones, and continuing research by Mason (2009a, and Chapter 3, this volume), firm evidence has been hard to find, largely because of the paucity of archaeological data from coastal and interior regions of Northeast Asia during the crucial period, *c.* 1,000 BC–1,000 CE.

The question of how Ipiutak might have received these ideas from Asia – whether by cultural diffusion, migration, or some other means – may eventually be illuminated by skeletal and DNA studies. Debetz, who made a preliminary study of Ipiutak crania (1959, 1986), saw significant differences between Ipiutak and other Eskimo groups and a stronger biological relationship with Siberia's Northern Mongoloids as represented by Reindeer Tungus (Evenk and Even) and Yukagir. But he presented these results as tentative and needing more extensive analysis. These connections are strengthened by linguistic observations by Michael Krauss (2005) who has identified traces of North Alaskan Inupiaq language in place names along the northern coast of Chukotka as far west as the Kolyma River. Such a connection, reaching from Point Hope to the Kolyma via Wrangel Island, might help explain Ipiutak's Asian heritage as well as features separating them from Bering Strait and Yup'ik peoples.

Today, Larsen and Rainey's measured belief that Ipiutak culture formed from a south Alaskan tradition stimulated by art and ritual concepts introduced from East and Northeast Asia is still a viable hypothesis. It seems more and more likely that Ipiutak emerged from a spiritual-dominated hunting and fishing culture in northwestern North America that was influenced by the spread of Asian peoples and ideas resulting from the expansion into Northeast Asia of Turkic-speaking reindeer-herding and horse-breeding peoples following the sixth century CE. Western Siberia could not have been the source of new inspiration, which must have come from geographically more proximate Northeast Asia, especially from northern Japan, the Amur Basin, the coasts of the Okhotsk Sea, and perhaps from the Arctic coast whose river headwaters lay close to these northwest Pacific rim regions. The absence of drum parts in early Alaskan sites suggests that the spread of a new religious culture, accompanied by a more politically active form of shamanism may have been important in transmitting some of these ideas. But Larsen and Rainey may have over-stressed this as the specific transformative mechanism. More likely the so-called Ipiutak "death cult" appears unique because of a new form of burial that was introduced, most likely from Asia, which included the burial of individuals accompanied by a full complement of earthly goods and charms to protect their lives in the next world. Fortunately for archaeologists and general knowledge, their graves and homes were preserved, giving us a rare glimpse of a people whose spirit-filled world was about to be changed forever after the arrival of Punuk culture and people: Siberians who had perfected a new economy, a more hierarchical social order, belief systems more tuned to corporate social life, more wide-ranging contacts and trade by dogsled and umiak, powerful new sinew-backed bows and slat armor, and a major escalation of warfare and political confrontation.

Conclusion

Larsen and Rainey joined forces in the field to explore the question of a "Paleoeskimo substratum" that emerged from Mathiassen's and Jenness' work on Dorset in northern Hudson's Bay and Solberg's Greenland "stone age." Their work at Ipiutak provided the first glimpse of a Western Arctic culture, comparable in many ways – especially as seen in the stone technology and in wood working at Deering – with Dorset, but very different from Punuk and the later Arctic whaling cultures. Stylistically affiliated, but distinct from the Okvik and Old Bering Sea cultures, Ipiutak suggested the presence of a contemporary third tradition, now confirmed by ^{14}C dating. Gerlach and Mason argue for an ethnic and political basis for these different traditions, based on occasional sharing of artifact types and styles appearing in sites of different traditions, confirming the movement of artifacts and individuals between these polities, which were trading, fighting, and socializing to varying degrees. Advances of archaeology in Alaska and Chukotka enable us to see these traditions in broader contexts or polities rather than as isolated cultures. Underlying traditions and structures argue for Ipiutak's emergence from a southern/western Alaskan culture base to which many new ideas were grafted from northern and east-central Asia. At a time when most archaeologists were defining their cultures in rigid chrono-spatial terms, Larsen and Rainey had set their sights on a wider horizon, drawing inspiration and comparisons from the entire circumpolar region as well as the circum-North Pacific, Siberia, and East Asia. They tempered their work with thoughtful ethnography and first-hand Arctic experience; they showed great knowledge of their colleagues' work, and for the Point Hope community that made their research possible and enjoyable.

The chapters presented in this volume constitute the "final chapter" needed to round out the obligation required of all archaeologists: that the fullest potential of research and publication should be accomplished to justify the intrusion of excavation and requirement for professional curation. In the case of world-class collections like those from Point Hope, especially in an era of repatriation, this obligation is even more stringent. These collections have been curated at the American Museum of Natural History and the Danish National Museum for seventy-five years. It is to the credit of these institutions and their curators and administrators that the collections have been cared for and that many of the finest objects have been made available for publication and display. Knowledge gained from the studies published here helps ensure that these collections are respected and will be preserved far into the future. Reconstructing history from archaeology is a risky enough business without losing large pieces of a tapestry that has already been ravaged by loss, decay, and the idiosyncrasy of discovery. Point Hope is a geologically fragile environment whose existence is threatened today more than ever by global warming and rising seas. Hopefully these studies will enhance the collaboration that made the Point Hope collections possible in the first place and can lead to further studies, such as ancient DNA work, that will inform the mysteries that still prevail over the history of the Ipiutak people and their culture.

References

Arutiunov, S. A. (2009a). The Eskimo harpoon. In *Gifts from the Ancestors: Ancient Ivories from Bering Strait*. Exhibition catalog. Princeton, NJ: Princeton University Art Museum, pp. 52–7.

Arutiunov, S. A. (2009b). The enigma of ancient Bering Strait art. In *Gifts from the Ancestors: Ancient Ivories from Bering Strait*. Exhibition catalog. Princeton, NJ: Princeton University Art Museum, pp. 126–37.

Arutiunov, S. A. and Sergeev, D. A. (2006a [1969]). *Ancient Cultures of the Asiatic Eskimos: The Uelen Cemetery*. Anchorage, AK: Shared Beringian Program, National Park Service. [Translation by Richard L. Bland of *Drevnie kul'tury aziatskikh eskimosov (Uelenskii mogil'nik)*. Akademiia Nauk, SSSR, Institut Etografi Imeni, N.N. Moscow: Milkukho-Maklaia.]

Arutiunov, S. A. and Sergeev, D. A. (2006b [1975]). *Problems in the Ethnic History of the Bering Sea: Ekven Cemetery*. Anchorage, AK: Shared Beringian Program, National Park Service. [Translation by R. L. Bland of *Problemy etnishiskoi istorii Beringomoria: Ekven mogil'nik*. Akademiia Nauk, SSSR, Institut Etografi Imeni, N.N. Moscow: Milkukho-Maklaia.]

Birket-Smith, K. (1930). The question of the origin of Eskimo culture: A rejoinder. *American Anthropologist*, 32 (4), 608–24.

Bronshtein, M. M. (2009). Early Eskimo art from from Ekven and Bering Strait. In *Gifts from the Ancestors: Ancient Ivories from Bering Strait*. Exhibition catalog. Princeton, NJ: Princeton University Art Museum, pp. 138–61.

Bronshtein, M. M. and Plumet, P. (1994). La préhistoire des Eskimaux. *La Recherche*, 25 (269), 1004–10.

Bronshtein, M. M. and Plumet, P. (1995) Ekven: l'art préhistoire béringien et l'approach russe de l'origin de la tradition culturelle esquimaude. *Etudes/Inuit/ Studies*, 19 (2), 5–59.

Bronshtein, M. M., Dneprovsky, K. and Sukhorukova, E. A. (Editors) (2007). *The World of Arctic Maritime Hunters: Steps into the Unknown*. Exhibition catalog. Moscow and Anadry: Russian State Museum of Oriental Art.

Carpenter, E. S., Sutherland, P., McGhee, R., Bronstein, M. and Dneprovsky, K. (2011). *Upside Down: Arctic Realities*. Exhibition catalog. Houston, TX: Rock Foundation and Menil Collection. (Published in French in 2008 as *Upside Down: Les Arctiques*, Quai Branly Museum and Rock Foundation.)

Chernetsov, V. N. (1935). An early maritime culture of the Yamal Peninsula. *Sovietskaya Etnografiia*, 4–5, 109–33.
Chernetsov, V. N. (1970). On the problem of the ancient substratum in the cultures of the circumpolar region. In *Proceedings of the Seventh International Congress of Anthropological and Ethnological Sciences*, Volume 10, Moscow, pp. 260–7.
Chernetsov, V. N. and Mozhinskaya, W. (1974). *Prehistory of Western Siberia.* Anthropology of the North: Translations from Russian Sources, Volume 9. Montreal and London: Arctic Institute of North America.
Collins, H. B. (1937). *Archaeology of St. Lawrence Island, Alaska.* Smithsonian Miscellaneous Collections 96 (1). Washington, D.C.: Smithsonian Institution Press.
Collins, H. B. (1959). An Okvik artifact from Southwest Alaska and stylistic resemblances between Early Eskimo and Paleolithic art. *Occasional Papers of the Stefansson Collection.* Hanover, NH: Dartmouth College, pp. 13–27.
Collins, H. B. (1971). Composite masks: Chinese and Eskimo. *Anthropologica*, XIII, 271–8.
Collins, H. B., de Laguna, F., Carpenter, E. S. and Stone, P. (1973). *The Far North: 2000 Years of American Eskimo and Indian Art.* Exhibition catalog. Washington, D.C.: National Gallery of Art.
Crowell, A. L., Steffian, A. F. and Pullar, G. L. (Editors) (2001). *Looking Both Ways: Heritage and Identity of the Alutiiq People.* Fairbanks, AK: University of Alaska Press.
Debetz, G. F. (1959). The skeletal remains of the Ipiutak cemetery. *Proceedings of the International Congress of Americanists*, 31, 57–64.
Fedorova, N. (2003a). *Sokrovitsa Pruob'ia. Zapadnaya Sibiri na torgovikh putyakh srednevekoviya* [*Treasures of the Ob: Western Siberia on the Medieval trade routes*]. Exhibition catalog. Salekhard and St. Petersburg: Yamal Nenets Autonomous Okrug Administration, State Hermitage Museum, and Institute of History and Archaeology of the Ural Division of the Russian Academy of Sciences.
Fedorova, N. (2003b). *Ust'-Polui: 1 bek do N.E.* [*Ust-Polui: 1st century B.C.*]. Exhibition catalog. Salekhard and St. Petersburg: Yamal Nenets Autonomous Okrug, Museum of Anthropology and Ethnography, Shemanovsky Museum, and Institute of History and Archaeology of the Ural Division of the Russian Academy of Sciences.
Fienup-Riordan, A. (1988). Eye of the dance: Spiritual life of the Bering Sea Eskimo. In *Crossroads of Continents: Cultures of Alaska and Siberia.* Washington, D.C.: Smithsonian Institution Press, pp. 256–70.
Fienup-Riordan, A. (2005). *Yup'ik Elders at the Ethnologisches Museum Berlin: Fieldwork Turned on its Head.* Seattle, WA: University of Washington Press.
Fitzhugh, W. W. (1998). Searching for the grail: Virtual archeology in Yamal and circumpolar theory. *Publications of the National Museum, Ethnographic Series*, 18, 99–118.
Fitzhugh, W. W. (2002). Yamal to Greenland: Global connections in circumpolar archaeology. In *Archaeology: The Widening Debate.* Oxford: Oxford University Press, pp. 91–144.

Fitzhugh, W. W. (2004). The deer stone project: Exploring northern Mongolia and its Arctic connections. In *The Deer Stone Project: Anthropological Studies in Mongolia 2002–2004*. Washington, D.C., and Ulaanbaatar: Arctic Studies Center, Smithsonian Institution, and the National Museum of Mongolian History.

Fitzhugh, W. W. (2009a). Notes on art styles, cultures and chronology. In *Gifts from the Ancestors: Ancient Ivories from Bering Strait*. Exhibition catalog. Princeton, NJ: Princeton University Art Museum, pp. 88–93.

Fitzhugh, W. W. (2009b). Eagles, beasts, and gods: Art of the Old Bering Sea hunting complex. In *Gifts from the Ancestors: Ancient Ivories from Bering Strait*. Exhibition catalog. Princeton, NJ: Princeton University Art Museum, pp. 162–89.

Fitzhugh, W. W. (2010). Arctic cultures and global theory: Historical tracks along the circumpolar road. In *A Circumpolar Reappraisal: The Legacy of Gutorm Gjessing (1906–1979)*. Proceedings of an International Conference held in Trondheim, Norway. Edited by Christer Westerdahl. BAR International Series S2154. Oxford: Archaeopress, pp. 87–109.

Fitzhugh, W. W. and Crowell, A. (Editors) (1988). *Crossroads of Continents: Cultures of Siberia and Alaska*. Exhibition catalog. Washington, D.C.: Smithsonian Institution Press.

Fitzhugh, W. W. and Kaplan, S. A. (1982). *Inua: Spirit World of the Bering Sea Eskimo*. Washington, D.C.: Smithsonian Institution Press.

Fitzhugh, W. W., Hollowell, J. and Crowell, A. (Editors) (2009). *Gifts From the Ancestors: Ancient Ivories from Bering Strait*. Exhibition catalog. Princeton, NJ: Princeton University Art Museum.

Gerlach, C. S. and Mason, O. K. (1992). Calibrated radiocarbon dates and cultural interaction in the Western Arctic. *Arctic Anthropology*, 29 (1), 54–81.

Irving, W. N. (1969–1970). The Arctic small tool tradition. In *Proceedings of the 8th International Congress of Anthropological and Ethnological Sciences*. Tokyo, pp. 340–2.

Jenness, D. (1952). Discussion of H. Larsen's "The Ipiutak Culture: its origin and relationships." *Proceedings of the 29th International Congress of Americanists 3*, Chicago, IL: University of Chicago Press, pp. 30–4.

Khlobystin, L. P. (2005). *Taymyr: The Archaeology of Northernmost Eurasia*. Contributions to Circumpolar Anthropology 5. Translation from the Russian by L. Vishniatsky and B. Grudinko. Washington, D.C.: Arctic Studies Center, Smithsonian Institution.

Larsen, H. (2001). *Deering: A Men's House from Seward Peninsula, Alaska*. Publications of the [Danish] National Museum, Ethnographical Series, Volume 19. Department of Ethnography, SILA – Greenland Research Centre, Copenhagen: National Museum of Denmark.

Larsen, H. and Rainey, F. (1948). *Ipiutak and the Arctic Whale Hunting Culture*. Anthropological Papers of the American Museum of Natural History 42. New York, NY: American Museum of Natural History.

Leskov, A. M. and Muller-Beck, H. (Editors).(1995). *Arktische Waljäger vor 3000 Jahren: Unbekannte Sibirische Kunst* (*Arctic Whale Hunters 3,000 Years Ago: Unknown Siberian Art*). Second Edition. Exhibition catalog. English translation by R. L. Bland. Mainz: V. Hase and Koehler.

Lester, C. W. and Schapiro, H. L. (1968). Vertebral defects in the lumbar vertebrae of pre-historic American Eskimos. *American Journal of Physical Anthropology*, 28, 43–8.
Mason, O. K. (1998). The contest between Ipiutak, Old Bering Sea and Birnirk polities and the origin of whaling during the first millennium A.D. along Bering Strait. *Journal of Anthropological Archaeology*, 17 (3), 240–325.
Mason, O. K. (2000). Archaeological Rorshach in delineating Ipiutak, Punuk and Birnirk in NW Alaska: Masters, slaves or partners in trade? In *Identities and Cultural Contacts in the Arctic*, Publication No. 8. Copenhagen: Danish Polar Center, pp. 229–51.
Mason, O. K. (2006). Ipiutak remains mysterious: A focal place still out of focus. In *Dynamics of Northern Societies*. Proceedings of a Symposium. Copenhagen: Danish National Museum and Danish Polar Center, pp. 106–20.
Mason, O. K. (2009a). Art, power, and cosmos in Bering Strait prehistory. In *Gifts from the Ancestors: Ancient Ivories from Bering Strait*. Exhibition catalog. Princeton, NJ: Princeton University Museum of Art, pp. 112–25.
Mason, O. K. (2009b). Flight from Bering Strait: Did Siberian Punuk/Thule military cadres conquer Northwest Alaska? In *The Northern World AD 900–1400*. Salt Lake City, UT: University of Utah Press, pp. 76–130.
Nelson, E. W. (1899). The Eskimos about Bering Strait. *Bureau of American Ethnology Annual Report*, 18, 1–518. Washington, D.C. [reprinted 1983 with introduction by W. W. Fitzhugh, Classics of Smithsonian Anthropology Series.]
Pitul'ko, V. V. (1999). Archaeological survey in central Taymyr. In *Land–Ocean Systems in the Siberian Arctic: Dynamics and History*. Berlin; Springer, pp. 457–67.
Rudenko, S. I. (1961). *The Ancient Cultures of the Bering Sea and the Eskimo Problem*. [Translation by P. Tolstoy.] Anthropology of the North. Translations from Russian Sources, 1. Toronto: Arctic Institute of North America.
Rudenko, S. I. (1970). *Frozen Tombs of Siberia: The Pazyryk Burials of Iron Age Horsemen*. Berkeley, CA: University of California Press.
Schindler, D. (1985). Anthropology in the Arctic: A critique of racial typology and normative theory. *Current Anthropology*, 26 (4), 475–500.
Schuster, C. (1951). A survival of the Eurasiatic animal style in modern Alaskan Eskimo art. In *Selected Papers of the 29th Congress of Americanists*, New York, NY, and Chicago, IL: University of Chicago Press, pp. 35–45.
Schuster, C. and Carpenter, E. (1996). *Patterns that Connect: Social Symbolism in Ancient and Tribal Art*. New York, NY: Harry N. Abrams.
Wardwell, A. (1986). *Ancient Eskimo Ivories of the Bering Strait*. New York, NY: American Federation of Arts.

12 *Point Hope in certain contexts: A comment*

DON E. DUMOND

The corpus of information on which so much of this volume is based resulted from excavations that Helge Larsen and Froelich Rainey conducted at Point Hope from 1939 to 1941 (Larsen and Rainey, 1948). The excavators especially emphasized the discovery of a previously unknown culture, designated by them "Ipiutak," while what was much less clear to them at the time was that they had produced evidence of a progression of archaeological cultures that is now known to be basic to northwest Alaska. Presently recognized to begin with what Larsen and Rainey called "Near Ipiutak," this progression runs through Ipiutak and later Birnirk to so-called Western Thule, on to the Tigara cultural deposits at Point Hope, and to recognized late prehistoric Iñupiaq Eskimo. The significance of only portions of this was evident to Larsen and Rainey at the time of their work, however.

Their approach to Point Hope – where recent Iñupiat had first been met and reported by the crews of European ships in the second decade of the nineteenth century – was based on a description by Knud Rasmussen, who as organizer of the Danish Fifth Thule Expedition (1921–4) had traversed Arctic America by dog sled from Baffin Island in the east to north Alaska in the west, and who suggested that the "old village of Tigara [i.e., Tikeraq] at Point Hope was one of the largest and most interesting sites along the Arctic Coast" (Larsen and Rainey, 1948: 5; see Rasmussen, 1999: 329–31). At the time Larsen and Rainey made their choice, there was already available to them a certain amount of local historical information regarding far northwest Alaska; this is summarized by Jensen (Chapter 2, this volume) in her outline of north Alaskan archaeology.

Specifically, early in the twentieth century the ethnologist-traveler Valhjalmur Stefánsson had purchased archaeological collections in northern Alaska. And it was after work by Rasmussen's Fifth Thule Expedition, which was focused especially on areas in northern Canada, that expedition archaeologist

The Foragers of Point Hope: The Biology and Archaeology of Humans on the Edge of the Alaskan Arctic, eds. C. E. Hilton, B. M. Auerbach, L. W. Cowgill. Published by Cambridge University Press.

Therkel Mathiassen concluded that the archaeological culture he had designated "Thule" in north Canada, harking to finds from the Thule district in northwest Greenland, had actually originated in northernmost Alaska (e.g., Mathiassen, 1930). Fueled by some of this knowledge, in the 1920s the Canadian anthropologist Diamond Jenness conducted excavations in the Bering Strait region (at Wales and on the Diomede Islands), the results of which led him to conclude that Bering Strait had been home to an especially ancient Eskimo culture (Jenness, 1928). This in turn was among the attractions that led the Smithsonian Institution's Henry B. Collins, Jr., to excavate on St. Lawrence Island from 1928 to 1931 (Collins, 1937). Meanwhile, in a partly coordinated thrust in the early 1930s, archaeologist James Ford moved on northwest Alaska, excavating at the Birnirk site near Point Barrow in 1936. Although Ford's results were not to see print for some years (Ford, 1959), they were evidently well known to Collins when he announced his own conclusions to the effect that the origin of the Thule culture of north Alaska, Canada, and Greenland had in fact been in the Bering Strait region. This development he saw to have proceeded through cultural manifestations designated by himself and others as Okvik, Old Bering Sea, Birnirk, and Punuk, thus leading to the Thule culture as known from Canada and points east (Collins, 1937).

It is no doubt significant that Collins' own published results appeared before Larsen and Rainey made their approach to Point Hope. Further, it is noteworthy that the Larsen and Rainey results at Point Hope were to be published (1948) the year before the first appearance of the basic explication of dating by means of radioactive carbon (Libby *et al.*, 1949). Without the support of a dating method essentially independent of the cultural sequence itself, they perforce proceeded on the basis of a formally defined sequence of artifacts – especially harpoon heads – and comparisons with other such sequences known from the larger region, which included Bering Strait.

In essence, Larsen and Rainey adopted the Collins scheme of cultural development, with two exceptions to accommodate their own finds. First, the spectacular and evidently shaman-induced art work of the Ipiutak burial culture led them to propose that Ipiutak might well represent the original Eskimo culture, as the product of a shamanistic people whose forebears had migrated from northern Asia not long before. Whereas the presence of certain artifacts in the Ipiutak collection seemed clearly to be of (Bering Strait) Okvik-culture origin, suggesting some degree of contemporaneity between Ipiutak and Okvik, from their Asia-origin hypothesis Larsen and Rainey had to presume that Ipiutak was the earlier of the two. Second, their designation of a "Near Ipiutak" culture was based on grave lots and midden remains that were *almost* like those of Ipiutak, but not quite. A major discrepancy was the presence in the Near Ipiutak excavation units of pottery and small stone vessels showing use as oil-burning lamps, in marked contrast to the many more units ascribed to Ipiutak, all of which were non-ceramic and lamp-less. Knowing of the consistent ceramic and lamp use by virtually all later Alaskan Eskimo, they concluded that "Near Ipiutak" must follow Ipiutak. Both of these chronological conclusions have been soundly rejected in later years with support from radiocarbon dating. A significant part of this rejection has also related to demonstrations that "Near Ipiutak" was a form of the Norton culture first identified by J. L. Giddings in Norton Bay (at the southeast corner of the Seward Peninsula) only in the late 1940s (Giddings, 1949, 1964), and shortly shown with radiocarbon evidence to be centuries older than Ipiutak (see Larsen, 1982), and almost certainly older than Okvik (e.g., Giddings, 1960; see also Dumond, 2008).

Further, the attention of Larsen and Rainey was drawn to the obvious whaling-focused culture of modern residents of Tikiġaq and to the plentiful evidence of comparable whaling by people of the late prehistoric Tigara culture, evidence strikingly absent from the Ipiutak-attributed cultural remains. In contrast, with plentiful evidence in Ipiutak deposits of caribou remains, especially in the form of extensive projectile tips and other items made of caribou antler, they concluded that these offspring of recent Asian immigrants were in fact a predominantly inland people who lived only seasonally at Point Hope – in spring and summer – where they subsisted for that period largely on sea mammals such as seals and walrus. This conclusion was reached, however, in the absence at the time of any evidence of Ipiutak-like sites in the Alaska interior.

With this they also concluded that several centuries separated the occupations of Ipiutak and Tigara at Point Hope. This was in view of at least slight indications of the presence there of people of Birnirk and Western Thule cultures, which by reference to the Bering Strait cultural sequence must be considered to have intervened in time between the two.

These results from the Larsen and Rainey work included four primary sets of suppositions to be examined by later researchers. First, there was simply the question of the dating of the Point Hope series relative to other north Alaskan prehistoric sequences – especially that at Bering Strait – and the relationships implied between peoples represented in these two sub-regions: Bering Strait and northwest mainland Alaska. Second, and looking in the other direction, there was the relationship of Ipiutak of Point Hope to any contemporary sites of the north Alaskan interior to the east. Third, there were their conclusions regarding the comparative subsistence regimes of Ipiutak and Tigara; and fourth, the question of the degree to which the latter, Tigara, may have descended from the former.

These questions are dealt with in various ways within the chapters of the present volume.

Dating

By the 1950s, radiocarbon dating was coming into prominence as a "culturally independent" means of assessing archaeological chronology. As the years moved ahead and more and more radiocarbon measurements were reported, however, it became evident that the dating method itself was in need of correction. Whereas the radiocarbon pioneers had assumed that the amount of radioactive carbon in the atmosphere was constant from year to year and decade to decade – and so would be ingested at equal rates by all living things now and in the past – it became clear that various environing conditions caused proportions of this atmospheric carbon to vary from time to time, although with relative uniformity worldwide. As a result, for instance, radiocarbon dating results became recognized to vary when compared to ages based on enumerations of regularly progressing growths such as in tree rings. For the sake of accuracy, then, corrections – many based on ages assessed by means of long-lived trees – were devised that would allow the newer radiocarbon calendar to be corrected or "calibrated" in relation to the solar calendar. This, then, has been the basis of recent more comprehensive attempts to date the Point Hope sequence in comparison to the rest of the northwest Arctic.

A major attempt in this regard was that of Gerlach and Mason (1992), who presented a battery of determinations, as measured and as calibrated, for north Alaska and the Bering Strait region. By this time, also, there had become known other coastal Ipiutak locations with constructed habitations, especially that at Cape Krusenstern, some distance south of Point Hope. Houses there were not in such numbers as at Point Hope, but nevertheless comprised several settlements of relatively substantial size. There had also been enough research in segments of northern Alaska to show that there did seem to be a number of inland sites that were apparently Ipiutak related, although without exception these appeared to be very restricted in size as compared to Point Hope and Cape Krusenstern. These advances made it possible for Gerlach and Mason to assess ages of Ipiutak occupations at both coastal and interior sites, and to compare them with ages from the Bering Strait sequence, which by the time of their work had come also to include results from sites on the coast of northeast Asia. These latter included relatively extensive cemeteries near East Cape, especially those sites designated Uelen (Arutiunov and Sergeev, 2006a) and Ekven (Arutiunov and Sergeev, 2006b), in which graves of people of the Okvik, Old Bering Sea, Punuk, Birnirk, and western aspects of the Thule culture were represented in some profusion and – in some cases – with interred displays of apparent wealth objects. Significant results included the indication that, beginning by the fourth century CE and enduring until 1,150 or 1,050 years BP, American Ipiutak was in very large part contemporary with Bering Strait Okvik and Old Bering Sea, with evidence of Birnirk appearing in both regions in still later years (Gerlach and Mason, 1992).

In brief and with hindsight, the early-to-late progression of material collections from Point Hope (with recently presented approximate dates) now appear to be as follows:

"Near Ipiutak" – This was defined at Point Hope on results from sixteen burial lots, one possible house assemblage, and several midden finds, and is now recognized as Norton culture, which in north Alaska is dated to the latest centuries BCE and early centuries CE – possibly as early as 2,450 years BP, certainly as early as carbon ages that should calibrate to 2,250 years BP (e.g., Giddings and Anderson, 1986: 30). Far to the south, on the Alaska Peninsula, dates on Norton deposits have been calibrated at least as early as 2,250 years BP (e.g., Dumond, 2011: Table 5.2).

Ipiutak – This was defined at Point Hope on the basis of more than 105 burial lots and some seventy house collections, and has been dated by Gerlach and Mason (1992) by means of twenty-three carbon ages from four coastal sites, and forty-three determinations from eight inland sites. The central tendencies for the coastal sites are between 1,500 years BP and 1,100 years BP, with the earliest reasonably possible at about 1,750 years BP. The central tendencies for the inland sites are between about 1,350 and 1,000 years BP, with the earliest reasonably possible also at about 1,750 years BP. Here, however, it must be noted that when it comes to small sites with small artifact samples (as are most in the interior) the discrimination of "Ipiutak" from "Norton" may be virtually impossible.

Birnirk – This was defined at Point Hope on the basis of three burials, and dated in sites near Point Barrow at around 1,150 years BP, possibly 1,050 years BP (Gerlach and Mason, 1992).

Western Thule – This was defined at Point Hope on two burial lots and two house collections. The Western Thule inception has been dated at around 950 years BP on the basis of direct dating of artifacts apparently transitional from Birnirk to Western Thule (Morrison, 2001).

Tigara (late prehistoric/historic Iñupiat) – This was defined on the basis of more than 300 burial lots from a presumably long period, not directly dated but apparently postdating 750 years BP, possibly by centuries.

By way of a brief comparison with some Bering Strait results, and drawing on other ages from Gerlach and Mason (1992: Figures 2 and 3), dates for combined Okvik and Old Bering Sea cultural stages from the Siberian mainland provide central tendencies between 1,725 and 1,300 years BP, with the earliest reasonably possible at about 2,050 years BP; the same set from St. Lawrence Island are centrally 1,625 to 1,200 years BP, with the earliest reasonably indicated at around 1 CE. The Birnirk ages, for combined northwest Alaskan and Bering Strait, are all between about 1,450 and 900 years BP; unfortunately, many of these are almost certainly of doubtful significance, drawn from sites to which Birnirk was ascribed on the basis of very little artifactual evidence.

Whatever the final shortcomings, it is clear enough that the progression of cultures at Point Hope is very largely contemporary with that from the Bering Strait region. Clear also is that the presence of Norton – or Near Ipiutak – at Point Hope must be significantly earlier there than Ipiutak, just as it is earlier than the inception of the major Bering Strait sequence beginning with the Okvik aspect of Old Bering Sea.

East versus west

From a time before the dawn of the modern era (i.e. 1 CE), in the region including both sides of Bering Strait, there is evidence of a pervasive cultural distinction between possibly related peoples of Alaska on the one hand and Asia (including St. Lawrence Island) on the other. In terms of culture history, the Eastern or Alaskan sequence can be thought to have begun at least as early as the appearance of the Norton culture or tradition, sometime around or slightly after 2,450 years BP. Although clearly in evidence around the American shore of the Bering Sea and thence as far north as Point Barrow, this widespread cultural pattern is thus far not known to be represented along the Asian coast at all, despite the fact that the nature of the Norton pottery makes it clear that the ceramic modes were ultimately derived from an Asian model.

The simple presence of pottery, of course, is by no means the deciding characteristic, with ceramics found in the sequences of both Alaska and northeast Asia, presumably with a more ancient common origin, but with both stylistic and technological differences in evidence by the time of the known sequences. Unfortunately, however, before the time of the appearance of the Okvik or earliest Old Bering Sea culture on the Asian side of Bering Strait, the more ancient Asian cultural situation has not yet been elucidated sufficiently to allow specific comparisons to the very earliest Norton, some centuries BCE. With the appearance of Okvik/Old Bering Sea (possibly in the late centuries BCE, but equally possibly not until 1,750 years BP or so) the contrast becomes completely clear: to the west, in Asia, is Old Bering Sea and immediate descendants, to the east in Alaska is the Norton culture, which north of the Seward Peninsula was essentially replaced by (pottery-eschewing) Ipiutak at about the time of Old

Bering Sea appearance, or perhaps a century or two later. In any event, by the mid first millennium CE. it is clear that subsistence practices varied. The major pathways of migratory walrus and whales lay closer to the Asian side of Bering Strait, providing subsistence possibilities to people such as of Old Bering Sea and later Punuk culture that were unmatched in most of the near-coastal waters to the east, the exceptions there being the many fewer salient land projections such as at Wales, Point Hope, and – farther north – Point Barrow.

Implications

One very palpable result of the recognition of the existence of the parallel but contrasting sequences is that the researcher Owen Mason (see his contribution to this volume, Chapter 3) began to pose questions regarding relationships between Ipiutak and the Bering Strait cultures. Considerations of evidence of distinctions regarding wealth, of evidence for social complexity, of population size, and of possible hostilities, led him to wonder about the possibility of major inter-societal competition, and possibly warfare, between political units of Ipiutak at Point Hope and elsewhere in northwest Alaska, with Old Bering Sea units at East Cape, as represented in the Uelen and Ekven cemeteries with their many relatively wealthy burials (Mason, 1998). This led him into a series of papers exploring derivative possibilities (e.g., Mason, 2009), most of which are cited in his contribution to this volume – for which, indeed, the basis is similar. So, for instance, was the large number of arrowheads (made of caribou antler) in Ipiutak not evidence of an inland subsistence pursuit but evidence of warfare? Information bearing on this will come later.

Health and well-being at Point Hope

Three chapters in this volume attempt to assess the matter of general health at Point Hope, with consideration of the apparent active lifestyle of the people, of the general situation of humans in an arctic environment, and of the presumed variation in subsistence activities of the two temporally distinct Ipiutak and Tigara peoples of Point Hope. Of these, the chapters by both Cowgill and Shackelford find no statistically definitive distinction between peoples of the two separate occupations.

Cowgill (Chapter 9, this volume) studies growth evidence in skeletons of individuals aged from birth to about eighteen years. She finds their body mass to be generally intermediate in a sample of seven more-or-less worldwide populations. A general robustness appears in early age and persists. Skeletons attributed to the Ipiutak people and those to the Tigara people she considers separately in the context of her broader sample, but finds them not distinguishable from one another in a statistical sense. She finally concludes both to have been generally comparable to other populations from far northern locations.

Shackelford (Chapter 8, this volume), dealing especially with bone strength in the two Point Hope populations of adults, finds them to be not significantly distinct at the outset, and so merges the two when considering them with her comparative sample of five additional groups from Africa and Asia. In a way parallel to the conclusion of Cowgill, she finds the combined Point Hope populations to be robust in upper and lower limbs, with arm and leg proportions typical of peoples of arctic or near-arctic environments. Both of these contributors, then, find the skeletal samples from Point Hope to represent generally robust and apparently healthy populations representative of Arctic peoples.

Hilton *et al.* (Chapter 7, this volume), on the other hand, rather than assessing general health status given the foraging lifestyles, look at evidence for sickness and injury in the Ipiutak and Tigara skeletal samples. In this they approach the interest of Mason (Chapter 3, this volume), who has wondered whether the inventory of arrowheads in the Ipiutak deposits may be simply indicative of a chronic state of warfare with people to the west. Most obviously Hilton and his co-authors note their examination of what Larsen and Rainey (1948: 228, 243) recorded (respectively) as an adult male Tigara skeleton with embedded arrow point (of Tigara type) in the pelvis (Burial 12) and an adult male Ipiutak skeleton with two arrow points embedded in the sternum and many others (of uncertain type) in the chest cavity (Burial 89b). These numbers may tend to suggest that the Ipiutak people were at least somewhat more besieged (by foreign enemies or local rivals) than were Tigara people; of likely significance, the skeletal sample of the latter is more than three times the size of the former (i.e., eighty-two as against twenty-five). But with the cases limited to a single adult in each sample such a result is less than absolutely clear. Further, the consideration of so-called "parrying" injuries to the lower arms (a common indication of levels of strife and fighting) suggests that the two were essentially equivalent among both males and females grouped together (i.e., five in total in Tigara compared to one in Ipiutak, again with the sample sizes more than three to one in favor of Tigara). In terms of physical fighting, therefore, Hilton *et al.* provide no decisive information to indicate that the Ipiutak people were really any more beset by hostilities than were their Tigara successors, although the possibility is certainly not ruled out. With this aside, Hilton *et al.* do demonstrate an apparent relative increase in evidence for postcranial lesions in the Tigara sample, suggesting to them that the whaling way of life was physically more stressful than the presumably less maritime pursuits of the people of the Ipiutak sample.

Beyond this, however, they present some evidence from the skeletal sample of the appearance of tuberculosis among the Tigara population. Recognizing this apparent increase in the level of disease in the later Point Hope occupation, they mention the possibility of the responsible organisms being carried by certain sea mammals, an implied risk for the presumably more sea-mammal focused Tigara people. What they do not point out, however, is the possibility that the total contact population during the Tigara period had increased substantially over that of the Ipiutak period. On the one hand, as will be indicated presently, the development of the Thule culture with roots in the Bering Strait region, and its spread to north Alaska and thence to Arctic Canada and points east, implies a probable overall growth in population and a concurrent increase in extra-local contacts and hence an increase in the scale of the human population within which contageous organisms may reproduce. So, as Jensen (Chapter 2, this volume) indicates, by the beginning of the post-contact period in north Alaska, the occurrence of periodic and heavily attended trading fairs in a number of specific locations was well known. One must suppose that these predated the beginning of the contact period by a significant number of years – which, of course, would include the late precontact Tigara period.

The question of subsistence

The subsistence regimes that Larsen and Rainey (1948) adduced for the Ipiutak and Tigara peoples were in sharp contrast: caribou plus some sea mammals for the former, and sea mammals including (especially) whales for the latter. Their conclusions also included the seasons of occupation at Point Hope: for Ipiutak people, spring and summer (a seaside vacation?) with fall and winter occupied on the chase farther inland; for Tigara people, year-round residence at Point Hope as representative of the Arctic Whale Hunting culture. At Cape Krusenstern, however, Douglas Anderson (who was responsible for the major analysis of the Ipiutak cultural material) concluded that residence there by Ipiutak people had been largely a winter affair, and – while not flatly contradicting the conclusion of Larsen and Rainey – he wondered if the same might have been true also at Point Hope (Giddings and Anderson, 1986: 158–60), a supposition supported by the relatively substantial nature of driftwood-framed Ipiutak houses at both locations. Faunal remains at the Cape Krusenstern Ipiutak settlements were especially small seal, plus larger bearded seal and caribou. Caribou bones were of mature animals only, suggesting winter hunts (Giddings and Anderson, 1986: 155–6). In this connection one may note that Point Hope and Cape Krusenstern are both within the coastal edge of the annual range of the modern Arctic herd of Alaskan caribou (Hemming, 1971: Figure 1).

In the contributions to this volume the matter of subsistence regimes is addressed fairly directly by El Zaatari and by Krueger, both of whom provide analyses of aspects of Ipiutak and Tigara dentition directed toward comparing Ipiutak and Tigara diets. El Zaatari (Chapter 6, this volume), focusing on microwear texture of the molar occlusal surfaces, concludes that the diets of the two peoples varied, especially in that the Tigara folk were accustomed to tougher foodstuffs, with much included grit, probably a result of the open-rack drying of meat on the sandy Point Hope spit. This conclusion is indicated despite cited work by a previous researcher (Costa, 1982) who concluded tooth wear to have been actually faster among the people of Ipiutak; specifically, maximal wear in the Ipiutak sample was found to occur by age 26–30, while the Tigara exhibited maximal wear only after age 40. Whether or not this conflicts with El Zaatari's conclusion, this indication of itself would seem to be that the diets of the two were indeed somehow different.

Krueger (Chapter 5, this volume), looking at microwear texture on the incisor teeth, provides evidence of a distinction that she attributes both to a difference between Ipiutak and Tigara in customary diet as well as to different subsidiary uses of the front teeth – as a third hand, for instance, as in holding animal hides during processing. Compared to Ipiutak, she found in the Tigara sample less evidence of such non-dietary tooth use, and also somewhat less evidence of abrasiveness in foods. From her comparative samples she was able to compare the wear in Tigara teeth to that among precontact Aleut. Similarly, she could compare the tooth wear of the Ipiutak sample to that of Native people of Nunavuk Territory of the Canadian interior – specifically as exhibited by Thule-period remains from three archaeological sites near northwestern Hudson Bay. She is also able to cite results of previous stable isotope analyses of remains from the same Nunavuk sites that indicated a heavy reliance on terrestrial mammals (presumably caribou) as well as on some smaller marine animals and seabirds (e.g., Coltrain, 2009). This last element is especially significant, in that these Nunavut people had evidently enjoyed the diet attributed by Larsen and Rainey (1948) to the people of the Ipiutak site at Point Hope. One must also point out that this would appear as at least a partial reservation concerning the suggestion that the Point Hope stock of Ipiutak arrowheads was chiefly for use in warfare rather than in hunting.

Genetic relationships

Finally, Maley (Chapter 4, this volume) focuses on the question of a possibly direct genetic – that is, ancestor–descendant – relationship between the populations represented by the Ipiutak and Tigara skeletal samples from Point Hope, with hope pinned on revealing relatively specific connections between the two peoples, rather than the more generalized relationship of the two simply as precontact northwestern Alaskan Natives and presumed Eskimoan ancestors. He bases his consideration on the results of a battery of quantitative measures applied to a set of twenty-nine specific cranial measurements, in which the relatively early Point Hope Ipiutak sample and Barrow area Birnirk sample (the latter from presumably four different sites in the Point Barrow area; cf. Hollinger *et al.* [2004]) are examined in relation to more recent populations (including Point Hope's Tigara) to which were ascribed Thule-like material cultures; these "Thule" populations are from locations scattered over northwestern and western Alaska (see Figure 4.1). The result he concludes to be evidence of an attenuated relationship between Ipiutak and Tigara, with greater relationships of Ipiutak to samples from sites in western Alaska, and relationships of Tigara to Birnirk and so to samples from other recent sites in northwestern Alaska. This leads him then to propose routes of migration by Ipiutak people southward into western Alaska upon their abandonment of Point Hope in the later part of the first millenium, and to another by Birnirk people to the west and southwest (including Tigara and others) following the Birnirk departure from Point Barrow in the early second millennium (see Figure 4.5).

With his focus on possible relationships between Ipiutak and Tigara, in keeping with the aim of the present volume, he perforce ignores a question more often posed, which concerns the source and motivation for the apparent expansion of north Alaskan people eastward as the so-called Thule migration, an event or series of events that occurred close to the time of the departure of Birnirk people from the Point Barrow vicinity. That is, in Maley's present analysis there is no population representing the American Arctic east of the Mackenzie Delta region of far northwestern Canada, and hence no possibility of examining evidence of any major eastward expansion.

This question, however, the same author has raised earlier (Maley, 2011), in a work in which a similar quantitative analysis involving the same set of twenty-nine craniofacial measurements was applied to a total of twenty-seven populations, including representatives from eastern Canada and Greenland. The result there places Birnirk in close association with the samples from the eastern Arctic, and is embellished with a suggested migration track for Birnirk-derived people that, in addition to their spread around the north Alaskan coast to Wales (as in his chapter in this volume), includes treks past Hudson Bay to Labrador, across the Canadian Arctic, and over an incomplete ellipse that demarcates two-thirds of the entire coastline of Greenland (Maley, 2011: Map 7.2).

This earlier work of Maley's closely followed the question of eastern connections of Birnirk that was posed by Hollinger *et al.* (2009; see also Hollinger *et al.*, 2004), whose quantitative study used fourteen cranial measurements applied to eight populations from sites east of Point Barrow, as well as essentially the same set Maley used of Ipiutak remains from Point Hope and Birnirk-culture remains from sites near Point Barrow; the Hollinger set also included a historic-period Point Barrow population, remains of which at the time were held by the Smithsonian Institution, but were evidently repatriated before Maley made his study.[1] Hollinger *et al.* (2004, 2009) find the Birnirk and Ipiutak samples to provide quantitative results relatively far from one another, with Birnirk's closest relationships to populations in the far eastern Arctic, including some from Greenland; Ipiutak, on the other hand, finds in their analysis its closest relations with a combined Tigara and historic Point Hope sample and the historic population from (post-Birnirk) Point Barrow, suggesting significant post-Thule influence of Ipiutak people in late precontact- and contact-period north Alaska. Interestingly, however, the Ipiutak relationships are found almost as close to certain samples from the eastern Arctic (e.g., Hollinger *et al.*, 2009: Table 6.3), suggesting some Ipiutak relationships also in that direction and possibly related somehow to the Thule migration. This is mentioned not to criticize the careful analysis in Maley's present chapter, with its set of craniofacial measurements larger than that of Holloinger *et al.*, but simply to point out that so much depends on the context in which these quantitative analyses are rendered. This introduces, then, another aspect of the question: that is, the matter of context.

West versus east

The contrast in developments of different sequences of material cultures on opposing sides of Bering Strait, referred to above, was accompanied by a contrast in major subsistence economics. On the west, or Asian side, was a continuing development of maritime techniques, leading at least by Punuk-culture times (1,350 years BP or so; see Gerlach and Mason, 1992) to what appears to have been professional whaling. Birnirk culture, developed somewhere around this same time and apparently first concentrated on the Asian coast of Bering Strait and along the north shore of the Chukotsk Peninsula, was a benefactor (although its practice of whaling is still a matter of some disagreement).

In Alaska on the east, well before 2,050 years BP, people of Norton culture were scattered in settlements throughout much of western Alaska, from the vicinity of Point Barrow in the north, to the Alaska Peninsula in the south. North of Bering Strait, subsistence was apparently in hunting land mammals, some lake fishing, and seasonal pursuit of coastal sea mammals (with some uncertainty concerning possible whaling, an idea based on a pair of presumed whaling harpoon heads assigned by Larsen and Rainey to Near Ipiutak). In this region Norton was succeeded by Ipiutak before the middle of the first millennium CE. South of Bering Strait, Norton subsistence was based on hunting and especially on fishing. There, this focus continued, with larger and larger settlements of substantial houses appearing at coastal locations, especially in the vicinity of salmon-rich streams, and also in the interior at suitable locations along the same fish-rich watercourses. There were also, however, smaller southern sites of apparently seasonal and short-term residence presumably positioned for hunting, for caribou was obviously another important dietary resource. South of Seward Peninsula this situation endured until the end of the first millennium CE, some centuries after the Ipiutak people to the north had been replaced by others (Dumond, 2000).

Was it from this Norton base that Ipiutak developed? It is entirely likely. The material inventory of Norton and Ipiutak are similar enough that when small sites with limited physical remains are encountered, discrimination is difficult, as was mentioned above. So, for instance, when Helge Larsen, one of the excavators of the Ipiutak site at Point Hope, encountered a site in the Platinum vicinity on the east coast of the Bering Sea, he concluded it to be Ipiutak despite its partial departure from the tool industry from the type site in the presence of stone lamps (Larsen, 1950). This was before Norton culture had been described and long before its distribution was known, although had the collection included pottery, it is possible Larsen's conclusion would have been different. Much later, after the presence of the Norton culture was much more completely internalized by archaeologists, when Donald W. Clark reported on the "Ipiutak-related" occupation at Hahanudan Lake in the Koyukuk River drainage not far east of the base of Seward Peninsula, he was more circumspect. Although no pottery was found in any of the several houses he tested, some was turned up in an excavated cache pit in the general vicinity. While stressing the Ipiutak similarities, among his conclusions was his comment: "to a certain and perhaps large degree Ipiutak appears to be a localized development on a Norton base" (Clark, 1977: 97).

And so the cultural manifestations of the eastern side of that geographic division provided by the Bering Sea and Bering Strait, and contemporary with the Old Bering Sea progression to the west, were the somehow related Norton and Ipiutak cultures. One recurrent marker – seemingly minor, but pervasive, in Alaska – was the Norton and Ipiutak practice of wearing labrets or lip ornaments, with the absence of the practice in contemporary Asia. To provide a human model, those to the right, in America, wore lip plugs. Those to the left, in Asia, did not (see Dumond, 2009).

This relatively clear contrast between west and east endured in Alaska for something like a millennium, to be broken in the north when the (labret-wearing) Ipiutak people completely or largely disappeared, and (labret-less) people of Birnirk culture expanded eastward across the Chukchi Sea to Arctic Alaska, for this was an Asian or western transplant eastward into America. The question that relates to the present project, then, and to Maley's careful research, must be: Is the relationship he adduces between Ipiutak and later "Thule"-period occupations in western Alaska the result of migratory moves by Ipiutak, or does it simply expose a relationship derived from a common eastern (American) background? Similarly, is the apparent relationship between Birnirk- and "Thule"-period sites in northwestern Alaska the result of a migratory movement from Birnirk sites near Point Barrow or, again, simply that Birnirk and these more northerly Alaskan sites (including Tigara) share to a large extent a western or Asian background? Clearly, these questions cannot be resolved until an analysis similar to that performed by Maley includes collections from a period equivalent to that of the Birnirk sites. At the present time, according to my understanding, the necessary samples are lacking.

Interestingly, however, the evident increase of similarity between Ipiutak and very late precontact Barrow, reported by Hollinger *et al.* (2004, 2009), indicates a late resurgence of Eastern or Alaskan influence. This is a situation also in keeping with the contact-period practice of labret wearing in north Alaskan sites such as Barrow and, for that matter, Point Hope.

Accomplishments of the present collection

Although questions posed by the various contributors to this volume consistently include both Ipiutak and Tigara collections in the focus, it is clear from the history of research in north Alaska that the subject of Ipiutak has been of far the greater interest. This of course has to do with the still exotic-seeming nature of the Ipiutak decorative and ceremonial canon. This means questions about the day-to-day nature of Ipiutak (hailed for long as inhabitants of an ancient Alaskan metropolis, a notion now abandoned), its subsistence base, its seasons of occupation, and so on. And it also includes questions that are an outgrowth of researches on Bering Strait, north Alaskan and Northeast Asian prehistories, questions such as those posed by contributor Owen Mason both here and elsewhere, that deal more broadly with the nature of inter-social relationships throughout the entire region.

First of all, it seems clear that the Ipiutak population itself was healthy enough and relatively thriving in an environment that does offer something in the way of subsistence riches, but which most of us would regard as a most difficult place in which to live. Thus, there certainly is no indication that there would be any cessation of Ipiutak occupation at Point Hope without some external cause. This might have been based on conflicts such as those apparently envisioned by Mason, or perhaps equally likely by a period of uncomfortable climatic events in the late first millennium CE, such as several of the contributors have stressed.

Second, it appears to this commentator that the case for a partial-interior subsistence emphasis on the part of the Ipiutak people has been substantially advanced through contributions here. Especially desirable, of course, would be further efforts in the way of stable isotope analyses, given permission from the museums and descendent Native populations potentially involved.

Third, the questions of relatedness of the Ipiutak population to others of northwestern and western Alaska have been advanced, although much remains to be done, both in terms of comparisons with samples (if and when available) relating both to additional sites in all directions and to additional time periods. Further, comparative examinations of relatedness such as are made in this collection, when expanded to include additional comparative samples, can throw light on such questions as the sources of what now are suspected to be various and differing segments in the expansion of north Alaskan populations into the more easterly Arctic.

Notes

1. Editors' note: these remains were repatriated before Maley could measure them.

References

Arutiunov, S. A. and Sergeev, D. A. (2006a). *Ancient Cultures of the Asiatic Eskimos: The Uelen Cemetery*. Anchorage: National Park Service, Shared Beringian Heritage Program. [Translation by R. L. Bland of *Drevnie kul'tury aziatskikh eskimosov (Uelenskii mogil'nik)* (1969). Moscow: Akademiia Nauk SSSR.]

Arutiunov, S. A. and Sergeev, D. A. (2006b). *Problems of Ethnic History in the Bering Sea: The Ekven Cemetery*. Anchorage: National Park Service Shared Beringian Heritage Program. [Translation by R. L. Bland of *Problemy etnishiskoi istorii Beringomoria: Ekven mogil'nik* (1975). Akademiia Nauk SSSR, Institut Etografi Imeni, N.N. Moscow: Milkukho-Maklaia.]

Clark, D. W. (1977). *Hahanudan Lake: An Ipiutak-Related Occupation of Western Interior Alaska*. Mercury Series Archaeological Survey of Canada Paper 71. Ottawa: National Museum of Man.

Collins, H. B. (1937). Archaeology of Saint Lawrence Island, Alaska. *Smithsonian Miscellaneous Collections*, 96 (1).

Coltrain, J. B. (2009). Sealing, whaling and caribou revisited: Additional insights from the skeletal isotope chemistry of eastern Arctic foragers. *Journal of Archaeological Science*, 36, 764–75.

Costa, R. L. (1982). Periodontal disease in the prehistoric Ipiutak and Tigara skeletal remains from Point Hope, Alaska. *American Journal of Physical Anthropology*, 59, 97–110.

Dumond, D. E. (2000). The Norton tradition. *Arctic Anthropology*, 37 (2), 1–22.

Dumond, D. E. (2008). The Story of Okvik. In *Aspects of Okvik: Four Essays*. University of Oregon Anthropological Papers 68. Eugene, OR: University of Oregon, pp. 261–308.

Dumond, D. E. (2009). A Note on labret use around the Bering and Chukchi Seas. *Alaska Journal of Anthropology*, 7 (2), 121–34.

Dumond, D. E. (2011). *Archaeology on the Alaska Peninsula: The Northern Section, Fifty Years Onward*. University of Oregon Anthropological Papers 70. Eugene, OR: University of Oregon.

Ford, J. A. (1959). *Eskimo Prehistory in the Vicinity of Point Barrow, Alaska*. Anthropological Papers of the American Museum of Natural History 47:1. New York, NY: American Museum of Natural History.

Gerlach, C. and Mason, O. K. (1992). Calibrated radiocarbon dates and cultural interaction in the western Arctic. *Arctic Anthropology*, 29 (1), 54–81.

Giddings, J. L. (1949). Early flint horizons on the North Bering Sea coast. *Journal of the Washington Academy of Sciences*, 39 (3), 85–90.

Giddings, J. L. (1960). The archeology of Bering Strait. *Current Anthropology*, 1 (2), 121–38.

Giddings, J. L. (1964). *The Archeology of Cape Denbigh*. Providence, RI: Brown University Press.

Giddings, J. L. and Anderson, D. D. (1986). *Beach Ridge Archeology of Cape Krusenstern: Eskimo and Pre-Eskimo Settlements around Kotzebue Sound, Alaska*. National Park Service Publications in Archeology 30. Washington, D.C.: National Park Service.

Hemming, J. E. (1971). The distribution movement patterns of caribou in Alaska. *Alaska Department of Fish and Game, Wildlife Technical Bulletin*, 1.

Hollinger, R. E., Eubanks, E. and Ousley, S. (2004). *Inventory and Assessment of Human Remains and Funerary Objects from the Point Barrow Region, Alaska, in the National Museum of Natural History, Smithsonian Institution*. Manuscript report, Repatriation Office, National Museum of Natural History. Washington, D.C.: Smithsonian Institution.

Hollinger, R. E., Ousley, S. and Utermohle, C. (2009). The Thule migration: A new look at the archaeology and biology of the Point Barrow region populations. In *The Northern World, AD 900–1400*. Salt Lake City, UT: University of Utah Press, pp. 131–54.

Jenness, D. (1928). Archaeological investigations in Bering Strait, 1926. *National Museum of Canada, Bulletin*, 50, 71–80.

Larsen, H. (1950). Archaeological investigations in southwestern Alaska. *American Antiquity*, 15 (3), 177–86.

Larsen, H. (1982). An artifactual comparison of finds of Norton and related cultures. *Arctic Anthropology*, 19 (2), 53–8.

Larsen, H. and Rainey, F. (1948). *Ipiutak and the Arctic Whale Hunting Culture*. Anthropological Papers of the American Museum of Natural History 42. New York, NY: American Museum of Natural History.

Libby, W. F., Anderson, E. C. and Arnold, J. (1949). Age determination by radiocarbon content: World wide assay of natural radiocarbons. *Science*, 109, 227–8.
Maley, B. C. (2011). *Population Structure and Demographic History of Human Arctic Populations Using Quantitative Traits*. Ph.D. Washington University at St. Louis.
Mason, O. K. (1998). The contest between Ipiutak, Old Bering Sea and Birnirk polities and the origin of whaling during the first millennium AD along Bering Strait. *Journal of Anthropological Archaeology*, 17 (3), 240–325.
Mason, O. K. (2009). Flight from the Bering Strait: Did Siberian Punuk/Thule military cadres conquer northwest Alaska? In *The Northern World AD 900–1400*. Salt Lake City, UT: University of Utah Press, pp. 76–128.
Mathiassen, T. (1930). *Archaeological Collections from the Western Eskimos*. Report of the 5th Thule Expedition, 1921–1924, Volume 10 (1). Copenhagen: Reitzels.
Morrison, D. (2001). Radiocarbon dating the Birnirk–Thule transition. *Anthropological Papers of the University of Alaska* (New Series), 1 (1), 74–85.
Rasmussen, K. (1999). *Across Arctic America*. Fairbanks, AK: University of Alaska Press. [Reissue of the 1927 edition, G. P Putnam, New York.]

Index